FORCE AND GEOMETRY IN
NEWTON'S *PRINCIPIA*

FORCE AND GEOMETRY IN NEWTON'S *PRINCIPIA*

FRANÇOIS DE GANDT

TRANSLATED BY
CURTIS WILSON

PRINCETON UNIVERSITY PRESS
PRINCETON, NEW JERSEY

Published by Princeton University Press, 41 William Street,
Princeton, New Jersey 08540
In the United Kingdom: Princeton University Press,
Chichester, West Sussex

Library of Congress Cataloging-in-Publication Data

De Gandt, François, 1947–
[Force et géometrie. English]
Force and geometry in Newton's Principia / François De Gandt ;
translated by Curtis Wilson.
p. cm.
Includes bibliographical references and index.
ISBN 0-691-03367-6
1. Newton, Isaac, Sir, 1642–1727. Principia. I. Title.
QA803.G3613 1995
531—dc20 94-39315

This book has been composed in Times Roman

Princeton University Press books are printed on
acid-free paper and meet the guidelines for permanence
and durability of the Committee on Production
Guidelines for Book Longevity of the
Council on Library Resources

Printed in the United States of America

2 4 6 8 10 9 7 5 3 1

CONTENTS

TRANSLATOR'S INTRODUCTION

It is with much pleasure and satisfaction that I here present to American and British readers an English translation of François De Gandt's *Force et Géométrie: les "Principia" de Newton dans le XVIIème siècle*. Professor De Gandt proposes this book as an introduction to the reading of Newton's *Principia*, but the journey of exploration in which he engages us takes unexpected turns and leads to unexpected riches. The method throughout is that of *explication de texte*, long a specialty of French scholars; it combines close reading of texts with historically informed commentary. In applying this method to selected writings both of Newton and of his contemporaries and predecessors, De Gandt has given us a deeply original meditation on the sources and meaning of Newton's *Principia*—its idiosyncratic limitations as well as its essential originality.

The reader, I predict, will repeatedly gain unexpected insights into the relations of ideas that now seem to have become all too familiar. Newtonian scholars as well as newcomers to Newtonian studies should find De Gandt's work a source of illumination and a stimulus to ongoing investigation.

The Newtonian scholarship of the last half century—most of it British and American—is known territory for De Gandt, and he has taken it into account in planning the itinerary of his exploration. In the past, the reading of the *Principia* has been hampered by diehard myths, like that of Newton's "twenty years' delay" in the composition of his masterwork, or the notion, to which Newton himself sought to lend credence, that the book was first written in analytico-algebraic form then translated into "good geometry." The new scholarship has dismantled these myths. It has shown that until 1679, Newton's thought about orbital motion was framed in terms of Cartesian vortices and that it was Robert Hooke, corresponding with Newton in 1679, who first confronted him with the idea of constructing planetary orbits out of inertial motion along the tangent and an attraction to the center.

The exchange with Hooke triggered Newton's discovery of a crucial implication: that centripetal force implies equable description of areas about the center. This concept, coupled (as De Gandt shows us) with a generalization of Galileo's law of fall so as to make it applicable to nonconstant forces, enabled Newton to express geometrically the quantity of a centripetal force at a point. With centripetal force thus quantified, Newton could then determine the force laws in particular orbits.

Initially this insight led only to—silence; Newton, wounded by past controversies, had sworn in 1679 that his papers should see the light of day only after his death. But in 1684, a jovial Halley persuaded the saturnine Newton to present his findings on orbital motion to the Royal Society. The result, a small treatise of eleven propositions, was called *De motu*. The very act of inditing this little treatise revealed to Newton the multitudinous consequences of his discovery and the need to refine the formulation of its foundational principles. Thus the vast enterprise of the *Principia* began.

De Gandt has perceived that the seminal idea from which the *Principia* evolved was precisely the geometrization of force as it is found in *De motu*. This little treatise is presented in chapter I of De Gandt's book, along with a commentary highlighting the salient features of the concepts deployed and the modes of demonstration utilized. The advantage of this starting point is that it is unencumbered by the many elaborations of principle and consequence that the *Principia* comes to entail and that tend to obscure the essential outline and connection of its basic ideas. In chapters II and III De Gandt undertakes to trace the ideas and methods of *De motu* to their historical roots; chapter II deals with the notion or rather notions of "force" current before the *Principia* was written, and chapter III with mathematical methods—methods of "indivisibles," kinematics in geometry, fluxions, ultimate ratios—developed before and in the *Principia*.

I shall leave it now to the author to carry his readers forward and to the readers to assess the special riches of De Gandt's analysis: the demonstration of the crucial role for Newton of the Galilean tradition, both in dynamics and in mathematical methods; the examination of Newton's treatment of the "inverse problem" of central force in Propositions 39–41 of Book I, which is more detailed and discriminating than any earlier account; the contrast with Cartesian principles and linear logic; the discussion of Newton's careful articulation of the relations between mathematical methods on the one hand and the principles of natural philosophy on the other; and much, much more.

I hazard to say that, for those who would explore the *Principia* at first hand, De Gandt supplies the most careful, patient, insightful introduction yet available. But in accordance with Newton's "experimental philosophy," such a claim should be tested. *Probet lector!*

St. John's College
Annapolis, Maryland
December 1993

PREFACE

My aim in this book is to provide an introduction to the reading of Newton's *Principia.* The geometrical treatment of force is my principal thread through the works of Newton and those of some of his predecessors: How is "centripetal" force to be evaluated? What is the status of this notion? What mathematical tools are required for this theory? In order to explain the demonstrative structure of the fundamental reasonings, the physical ideas here put into play, and the mathematical methods employed, I have had recourse to a detailed study of the original texts, for which I provide new translations and a commentary. These translations and discussions can serve as materials, and (it is my hope) furnish solid and substantial grist for the history of scientific ideas as well as for a philosophical reflection on the constitution of classical science—a reflection that can only be sketched in the present work.

My first objective is to outline as clearly as possible the ruling ideas of the Newtonian edifice by following Newton's texts. The purpose of this study is truly introductory: having learned from experience how difficult the *Principia* is to read and having discovered over time, with astonishment and a certain inquietude, how few people have actually read a work so highly praised, I thought it indispensable to propose a path yielding access to it. Thus, chapter I presents a translation, accompanied with detailed commentary, of a first, simplified version of the *Principia* drawn from a manuscript of 1684. In some few pages, Newton there set out the foundations of what became in 1687 the first book of the *Principia.*

After this initial, perhaps elementary, exposition, I have attempted to place the *Principia* in the intellectual context of the seventeenth century and to elucidate the essential features of the Newtonian theory against the background of the thoughts and experimentation of other seventeenth-century scientists, in particular Kepler, Galileo, and Huygens. The originality and boldness of the *Principia,* and also Newton's debt to his predecessors, will appear by comparison with other important texts of the time: the *Epitome* of Kepler; the *Dialogo,* and above all, the *Discorsi* of Galileo; the *Horologium* and the *De vi centrifuga* of Huygens. I have also cited texts of other scientists and philosophers of the seventeenth century (such as Torricelli, Descartes, Boulliau, Wallis, Barrow, Hooke, etc.) but have sought clarity of characterization more than abundance of historical information. In general, it is as a function of concern with Newton's achievement that other authors of the seventeenth century are interesting, either because of the extent to which Newton may have been directly or indirectly indebted to them, or because a comparison between these authors and Newton reveals something about the originality of the *Principia.* Often historians, in tracing influences and placing works in context, become iconoclasts because they dislodge the genius from an exalted position. Here the result is more nearly the reverse: compared with his predecessors and contemporaries, Newton seems greater yet, and the *Principia* appears as an exceptional creation.

The concern for clarity and the attention focused on intellectual constructions can lead to excessive simplifications; sometimes the history of scientific ideas risks being reduced to a dialogue on the Elysian Fields, between two or three great geniuses, neglecting the meanders and happenstance of the actual transmission of ideas and discounting personages of lesser importance and minor circumstances of the creative work. In any case, I hope that the reader will always be able to rely on the texts that I provide, in order to supply nuances to the commentary. The richness of the texts that I have cited is perhaps one of the chief values of this work.

An important part of these texts is mathematical in nature, and I have attempted to comment on them while respecting their terms and content and, if the expression may be allowed, their mathematical literalness, in order to avoid anachronisms in commenting on the mathematical texts of the past. The procedures actually adopted by Newton, his language and his notations (or the lack of them), his appeals to geometric or kinematic intuition when it is indubitably required—in a word, the mathematical *style* proper to these texts—must be given their value in the commentary.

Chapter III is exclusively devoted to the discussion of mathematical procedures—those which were employed in the mid-seventeenth century (indivisibles, geometry of motion) and those to which Newton recurs in the *Principia*. I have had to confront, in particular, the difficult problem of the distinction between a differential or fluxional calculus and the modes of reasoning of the *Principia*. Laplace and many others believed it impossible that Newton should have been able to write, or even conceive, the *Principia* without using his method of fluxions or the equivalent of an infinitesimal calculus; he must have afterwards retranscribed his reasonings into the terms of "synthetic" geometry. The case is otherwise, as will be shown, and it must be admitted that Newton, from his first sketches, reasoned as in the *Principia*—in geometrical terms; nevertheless it is a very special geometry whose nature must be described.

Beside this question of the mathematical mode of expression, certain themes have oriented the discussion of the texts presented: the heritage of Galileo in the construction of the Newtonian theory, the measure of force by deflection or deviation, the place of time in the definition of force, and the possibility of determining motion from the force (the "inverse problem"). The conclusion of the book attempts to define the status of the new Natural Philosophy that had its origin in the *Principia*. Newton claimed to treat forces in a purely mathematical mode; by deferral, which in a sense turned out to be final, he left in suspense the properly philosophical or physical questions concerning the causes of gravitation and the ontological reality of force. This neutrality (or "secularism") of "centripetal" force in face of the controversies on the cause of gravitation is the essential characteristic of the new science.

Why present a new book on Newton after the classical works of A. Koyré, J. W. Herivel, A. R. and M. B. Hall, I. B. Cohen, R. Westfall, D. T. Whiteside? It seemed to me that there did not yet exist a detailed technical study on the theme of the mathematical treatment of force in which Newton's contribution was placed in the context of the seventeenth century. Midway between the more synthetic studies of Westfall in his *Force in Newton's Physics* and the learned notes of Whiteside in volume 6 of *The Mathematical Papers of Isaac Newton*, there was a place for a work of conceptual history based on the most recent publications of manuscripts and on the critical editions now available.

Among the books on the new science of the seventeenth century, the work of Alexandre Koyré stands out. He has commented on the texts of Galileo and Kepler with a remarkable penetration and with the finesse that breadth of learning has made possible. His analyses of the new conception of motion and of the role of mathematics have opened new paths and nourished the reflection of historians and philosophers, even if he can be reproached, rightly, with a certain prejudice in characterizing Galileo as a platonist. I also owe to him, as to I. B. Cohen, the critical edition of the *Principia*, an indispensable tool for a work such as mine. Yet his *Études newtoniennes* remains a little inferior to his *Études galiléennes*; his study of the opposition between Newton and Descartes constitutes the most valuable part of the former work, in which he did not enter into the Newtonian machinery.

Over some twenty years, a vast collection of the Newtonian texts has been published, especially by D. T. Whiteside with regard to the mathematical texts. The eight volumes of *The Mathematical Papers of Isaac Newton* are a veritable monument raised by a single man. Among the treasures thus put at the disposal of amateurs, I have made only a very restricted choice, except perhaps from volume 6 of these *Mathematical Papers*, in which Whiteside has assembled the manuscripts relative to the *Principia*. Among the successive layers in the writing of the great work, it is the initial one that will interest us most, the family of manuscripts entitled *De motu* dating from 1684 and 1685.

A number of other publications have been essential for this inquiry: the *Unpublished Scientific Papers of Isaac Newton*, edited by A. R. and M. B. Hall; the *Introduction to Newton's "Principia"* by I. B. Cohen, which retraces with the greatest precision the circumstances of the publication of the work in its successive editions; the *Correspondence of Isaac Newton*, edited in seven volumes by H. W. Turnbull and several successors; *The Background to Newton's "Principia"* by J. W. Herivel; the two books by Westfall—*Force in Newton's Physics* and *Never at Rest*; J. E. Hofmann's *Leibniz in Paris*; and finally the *Patterns of Mathematical Thought in the Later Seventeenth Century* by Whiteside.

I cite the *Principia* from the third edition. The seventeenth-century authors are cited from the standard editions: the Edizione Nazionale for Galileo, the *Gesammelte Werke* for Kepler, the *Oeuvres Complètes* for Huygens, the *Opera* of Torricelli, and the Adam-Tannery edition for Descartes. All the translations have been made afresh.

This work owes much to conferences, shared study, and discussions. Philippe de Rouilhan and the late Dominique Dubarle were incomparably stimulating partners in an attentive and systematic reading of the *Principia*. François Russo first suggested to me the writing of this book, in which I have been guided in various aspects by the encouragement and criticisms of Yvon Belaval, Pierre Costabel, Jean-Toussaint Desanti, Gilles-Gaston Granger, Roger Martin, Yves Michaux, Pierre Souffrin, and Jean-Marie Souriau. I have profited from the counsel of I. B. Cohen and William Shea in America. The discussions with Emily Grosholz, Bernard Goldstein, Charles Larmore, Eric Aiton, Eberhard Knobloch, Jean Pierre Verdet, Alain Segonds, Michel Tixier, Joel Roman, Michel Blay, and Georges Barthélémy, have been extremely valuable. The group of mathematical friends who have become historians of their discipline, like Jeanne Peiffer, Amy Dahan, Jean Luc Verley, and the members of the Institut de Recherche pour l'Enseignement des Mathématiques (IREM), have constituted a medium for attentive and cordial critique. I owe much to the remarks of Curtis Wilson, who translated the book, and of Gail Schmitt, who was the copy editor. Finally it is to the French Centre National de la Recherche Scientifique (CNRS) that I am indebted for the leisure to engage in this long undertaking.

CONVENTIONS AND ABBREVIATIONS

"AB ∝ CD" signifies that AB is proportional to CD.

The most frequent references are abbreviated:

Princ.: Isaac Newton, *Philosophiae Naturalis Principia Mathematica*, 3rd ed. (reproduced with the variations in the other editions), 2 vols., ed. A. Koyré and I. B. Cohen (Cambridge: Harvard University Press, 1972).

AT: R. Descartes, *Oeuvres*, ed. C. Adam and P. Tannery (Paris, 1897–1913).

Cajori: *Sir Isaac Newton's Mathematical Principles of Natural Philosophy and his System of the World*, trans. Andrew Motte, ed. and rev. Florian Cajori (Berkeley: University of California Press, 1966).

Cohen: I. B. Cohen, *Introduction to Newton's "Principia"* (Cambridge: Harvard University Press, 1971).

Corresp.: *The Correspondence of Isaac Newton*, ed. H. W. Turnbull (Cambridge: Cambridge University Press, 1959).

EN: G. Galilei, *Opere*, Edizione Nazionale (Florence: Barbèra, 1890–1909).
The following volumes include the listed works:
EN 2—*Mechanics*
EN 4—*Discorso intorno*
EN 7—*Dialogo*
EN 8—*Discorsi* and related fragments

GW: J. Kepler, *Gesammelte Werke* (Munich: 1937).
The following volumes include the listed works:
GW 3—*Astronomia nova*
GW 7—*Epitome*

Hall: *Unpublished Scientific Papers of Isaac Newton*, eds. A. R. and M. B. Hall (Cambridge: Cambridge University Press, 1962).

Herivel: J. Herivel, *The Background to Newton's "Principia"* (Oxford: Oxford University Press, 1965).

HO: C. Huygens, *Oeuvres*, 22 vols. (La Haye: Société hollandaise des sciences, 1888–1950).
The following volume includes the listed work:
HO 16:255–301—*De vi centrifuga*

Hofmann: J. E. Hofmann, *Leibniz in Paris* (Cambridge: Cambridge University Press, 1974).

Koyré *EG*: A. Koyré, *Études galiléennes* (Paris: Hermann, 1966).

Koyré *EN*: A. Koyré, *Études newtoniennes* (Paris: Gallimard, 1968).

Koyré *RA*: A. Koyré, *La révolution astronomique* (Paris: Hermann, 1961).

LMS: G. W. Leibniz, *Mathematische Schriften*, 7 vols., ed. C. I. Gerhardt (Hildesheim: Olms, 1971).

NMP: *The Mathematical Papers of Isaac Newton*, 8 vols., ed. D. T. Whiteside (Cambridge: Cambridge University Press, 1967).

TO: E. Torricelli, *Opere*, 4 vols., eds. G. Loria and G. Vassura (Faenza, 1919–44).
ULC: University Library of Cambridge.
Westfall *FNP*: R. Westfall, *Force in Newton's Physics* (London: Macdonald, 1971).
Westfall *NR*: R. Westfall, *Never at Rest: A Biography of Isaac Newton* (Cambridge: Cambridge University Press, 1980).
Whiteside "Patterns": D. T. Whiteside, "Patterns of Mathematical Thought in the Later Seventeenth Century," *Archive for History of Exact Sciences* 1, no. 3 (1961): 179–388.

Extracts from the *De motu* that are cited in this book are transcribed from the manuscript. A photographic reproduction can be found in Isaac Newton, "The fundamental 'De motu corporum in gyrum' (Autumn 1684) (Add. 3965.7, fols. 55r–62*r)," in *The preliminary manuscripts for Isaac Newton's 1687 "Principia," 1684–1685: facsimiles of the original autographs now in Cambridge University Library, with an introduction by D. T. Whiteside.* Cambridge: Cambridge University Press, 1989. 3–11. These extracts are designated by "Add. 3965.7, fols. 55r–62*r."

FORCE AND GEOMETRY IN NEWTON'S *PRINCIPIA*

PREAMBLE

The Forty-Shilling Book of Sir Christopher Wren

This account could begin in the manner of a typical British detective story: an architect, an inventor, and an astronomer had become fascinated with a certain enigma whose complete solution seemed to hover ever within their grasp, but always escaped them.

The architect, Christopher Wren, and the inventor, Robert Hooke, saw one another frequently. They had collaborated in the vast enterprise of reconstructing London after the great fire of 1666, and they also met in various clubs with colleagues they had known in their Oxford days. In 1662 one of these clubs had assumed a more official form—it became the Royal Society of London, under the patronage of the king, and as such brought together weekly a group of learned scientists and fortunate amateurs. Their aim was to encourage the advancement of philosophy through the examination of "all systems, theories, principles, hypotheses, elements, histories, and experiments of things naturall, mathematicall, and mechanicall."[1] Hooke was the "curator" of this society, charged with presenting at each meeting "three or four considerable experiments." This great task gave him the opportunity of deploying his talents as mechanic and of presenting new and bold ideas on respiration, colors, clocks, fossils, and so forth.

The third member of the trio, Edmond Halley, had been educated at Oxford a generation later. Upon returning from a voyage to the island of St. Helena, he had dedicated to the king a catalogue of the stars of the southern hemisphere and since then had participated in the meetings of the Royal Society.

What was this enigma that fascinated Halley, Hooke, and Wren? Nothing less than "the system of the world," as it was soon to be called (such is the title of the third book of Newton's *Principia*). The initial idea was that the planets traverse their paths under the action of a certain force directed toward the sun. What was the nature of this force? No one quite knew. It might "attract" the planets or it might "push" them. In the first case, the sun itself would be the source and cause of the force. In the second case, the sun would occupy the geometric center of the configuration only by coincidence, without being the cause of the motion.

Some theorists sought to explain this sun-directed force by the action of a space-filling fluid or "aether." There might, for example, be an invisible substance swirling with great velocity about the sun and driving bodies toward the center of the vortex, much as the fragments of tea leaves are driven to congregate at the bottom of the cup when the tea is stirred. Or again, differences in pressure in this interplanetary medium might push bodies toward the center. It was on the Continent that hypotheses featuring such invisible fluids were especially in favor, whereas in England, the intellectual milieu was more open to "attractionist" theories:[2] the sun might act at a distance on the planets and, like a great magnet, draw them toward itself. This concept had come to be called the Magnetic Philosophy. More than a half-century earlier, it had inspired important works: the *De magnete* of William Gilbert (1600), and Kepler's *Astronomia nova* (1609) and *Epitome astronomiae Copernicanae* (1618–21).

It was thought that this attractive force might vary from point to point in space and might diminish with distance from the sun according to some determinate law; for instance, it might vary inversely as the square of that distance.[3] Diminution inverse to the square of the distance was plausible because light itself diminishes according to this law. If, from a point source of light, the rays spread out in straight lines in all directions without loss or modification, as in figure P.1, a spherical surface surrounding the source will receive the same amount of light whether it is very close to the source or far distant from it. Under what conditions will surface areas belonging to two concentric spheres receive the same quantity of light? It suffices that the lines bounding the rays fall on the same cone, whose apex is at the luminous source (fig. P.2). In the part of the cone nearest the

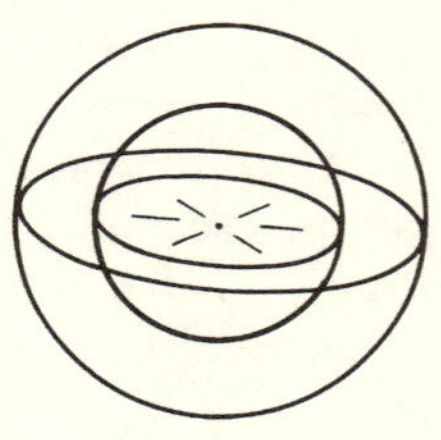

Figure P.1

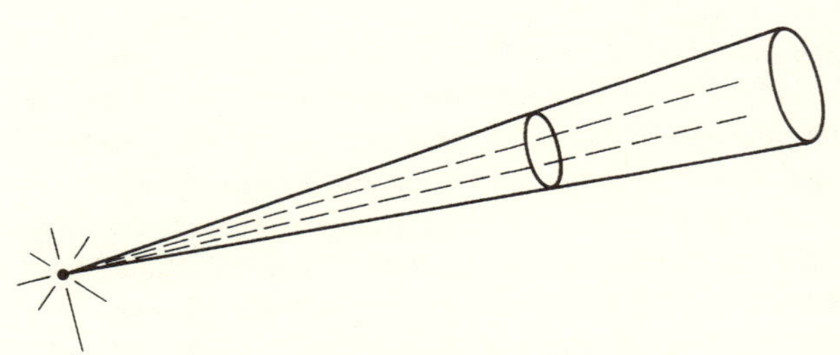

Figure P.2

source, the rays are more closely packed, while in the more distant portion they are more widely spaced, but in each there are always "as many." The same amount of light is thus spread out over larger and larger surfaces that increase as the squares of their distances from the source (for the areas of concentric spheres are as the squares of their radii). A unit area of surface therefore receives a quantity of light inversely proportional to the square of its distance from the source. If the force that pushes or attracts the planets toward the sun is analogous to light, it must diminish according to this same law. Kepler himself had envisaged this hypothesis only to reject it.[4]

The decisive step that the three Londoners desired to take was this: to relate the precise trajectory of the planets to this law of diffusion. Would a weakening of the force proportional to the square of the distance suffice to account for the characteristics of the celestial motions, in particular the elliptical form of the orbits and the relations of the planetary periods? It had been observed that the planets move more slowly as they are more distant from the sun,[5] and Kepler had given a precise law for this dependency: the velocities of the different planets vary with distance from the sun in such a way that the cubes of the radii (r) of the orbits are proportional to the squares of the periods. Might these characteristics of the celestial motions be derivable from the action of a force varying as $1/r^2$?

Ultimately, as is well known, it was a fourth personage who resolved the enigma: Isaac Newton was the Sherlock Holmes of this affair. He lived in Cambridge, in solitude, having succeeded Isaac Barrow in the Lucasian chair of mathematics. Newton had become recognized for his ideas about light, which had occasioned a polemic from Hooke (already!), and had also circulated among acquaintances a number of letters and mathematical manuscripts of a very innovative character, particularly with regard to the development in infinite series of certain algebraic quantities.[6]

When Halley and the Royal Society learned that Newton had found the solution, they encouraged him to publish it in a book. But Robert Hooke, when informed of Newton's claims, considered that he was being deprived of his part in the discovery. Newton, outraged in turn and threatening to withhold the third book of the *Principia*, demanded Halley's advice. Halley's response is an invaluable recital of the fruitless attempts that preceded the *Principia*:

> I waited upon Sr Christopher Wren, to inquire of him, if he had the first notion of the reciprocall duplicate proportion from Mr Hook, his answer was, that he himself very many years since had had his thoughts upon making out the Planets motions by a composition of a Descent towards the sun, & an imprest motion; but that at length he gave over, not finding the means of doing it. Since which time Mr Hook had frequently told him that he had done it, and attempted to make it out to him, but that he never satisfied him, that his demonstrations were cogent.
>
> and this I know to be true, that in January 83/4, I, having from the consideration of the sesquialter proportion of Kepler [Kepler's proportionality between the cubes of the radii and squares of the periods], concluded that the centripetall force decreased in the proportion of the squares of the distances reciprocally, came one Wednesday to town, where I met with Sr Christ. Wrenn and Mr Hook, and falling in discourse about it, Mr Hook affirmed that upon that principle all the Laws of the celestiall motions were to be demonstrated, and that he himself had done it; I

> declared the ill success of my attempts; and Sr Christopher to encourage the Inquiry sd, that he would give Mr Hook or me 2 months time to bring him a convincing demonstration thereof, and besides the honour, he of us that did it, should have from him a present of a book of 40s.
>
> Mr Hook then sd that he had it, but that he would conceale it for some time that others triing and failing, might know how to value it, when he should make it publick; however I remember Sr Christopher was little satisfied that he could do it, and tho Mr Hook then promised to show it him, I do not yet find that in that particular he has been as good as his word. The August following when I did my self the honour to visit you, I then learnt the good news that you had brought this demonstration to perfection, and you were pleased, to promise me a copy thereof, which the November following I received with a great deal of satisfaction from Mr Paget. (Halley to Newton, 29 June 1686, *Corresp.* 2:441–42)

What became of the 40-shilling book? History does not say. No one of the three colleagues deserved it; it was Newton who should have received it.

The failures and the promises not kept bear witness to how difficult it was to draw the detailed implications from the first intuition and to transform the general idea of a force varying as $1/r^2$ into a theory with determinate consequences.

Wren required truly deductive demonstrations, and what Hooke proposed to him did not convince him. But what would a demonstration in these matters be like? What examples could be used? (What Kepler considered to be an a priori deduction of the elliptical motions of the planets is presented later [pp. 82–84].) In geometry, the criteria of proof were well established by virtue of a culture nourished by the books of the ancients and cultivated in discussions, courses, challenge contests, and discoveries. But what would constitute proof when it was a question of forces and motions? What indubitable principles could be adopted as a foundation? What mathematical tools should be used?

The initial intuition in Halley's account is not yet that of "universal gravitation." The envisaged force acts on the planets, entraining them toward the sun. There is no reference in the question either to the moon or to terrestrial weight, unless these were included by Hooke in his mention of "all the celestial motions."

What cases did Hooke embrace within his hypothesis? It is difficult to know. In Halley's letter, Hooke is a braggart, fearing that he is not appreciated at his true value, promising more than he can deliver, and proposing reasonings that do not convince. (Later on, Newton will make him feel it cruelly, minimizing his role in the series of events leading to the discovery.)

As for Wren, he appears to have initially posed a broader question: could the planetary motions be explained by combining inertial motion along the tangent with a descent toward the sun? The question may have had its origin for him in Galileo's idea that at the creation, the planets had been launched in a straight line with an accelerated motion of fall, then turned aside into circular orbits.[7]

HALLEY'S VISIT

The close of Halley's letter sketches the dénouement of the intrigue: Halley visited Newton in his Cambridge retreat and was surprised to learn that he, Newton, possessed the solution. This visit was the initial stimulus that gave birth to the *Principia*. An account of it was transmitted forty years later by the mathematician Abraham de Moivre to John Conduitt, husband of Newton's niece:

> In 1684 Dr Halley came to visit him at Cambridge, after they had been some time together, the Dr asked him what he thought the Curve would be that would be described by the Planets supposing the force of attraction towards the Sun to be reciprocal to the square of their distance from it.
> Sr Isaac replied immediately that it would be an Ellipsis,
> the Doctor struck with joy & amazement asked him how he knew it,
> Why saith he I have calculated it,
> Whereupon Dr Halley asked him for his calculation without any further delay, Sr Isaac looked among his papers but could not find it, but he promised him to renew it, & then to send it him. (Cohen, 50, 297–98)

The date of the visit is uncertain. Most students of the question put it in the middle of the summer of 1684, in agreement with Halley's "August 1684," but some indications would suggest a date in May.[8] It matters little; the earlier date would mean only that Newton took longer to respond.

Newton in fact did wait several months before keeping his promise. And in the end it was not a demonstration that Halley received, but a small treatise. The exemplar actually sent in November 1684 has not been recovered, but there exist several more or less augmented copies or versions in the Cambridge University Library and in London. Their common title is *De motu* (On motion).[9] These are as yet very unrefined and simple versions of what will become the *Principia*. Newton corrected them, re-wrote them, amplified them, and inserted into them passages from others of his writings. His course of lectures at Cambridge, given in two consecutive years, was made up of these materials.[10]

The little treatise *De motu* is therefore the kernel from which sprang an enormous work. The first manuscripts in the series that culminated in the *Principia* undoubtedly resemble very closely the text sent to Halley in November 1684 and contain only four theorems and four problems in ten or so pages, while the *Principia* contains nearly two hundred propositions in more than five hundred pages. The text had swollen to unbelievable proportions between November 1684 and July 1687, the publication date of the *Principia*. Halley's visit had unleashed a creative storm.

The Inverse Problem

Consider the question that Halley posed to Newton: assuming that the force varies as $1/r^2$, what will be the resulting trajectory? The *Principia* responds, in fact, to a different question: if the trajectory is an ellipse, what must be the law of the variation of the force? It is important not to confuse the two:

A. Given the trajectory, how to find the law of force?
B. Given the law of force, how to determine the trajectory?

The second question is called "the inverse problem." Newton himself, in the *Principia*, does not always distinguish the two questions clearly. He sometimes evokes a response to question (A) as if it were a response to question (B).

The inverse problem, that is to say, the passage from the law of the force to the trajectory, is mathematically a much more difficult enterprise than the passage from the trajectory to the law of force. A known trajectory gives a geometrical representation of the situation from which reasoning is easier. If, on the contrary, only the law of force is known, the geometrical object must be constructed (whether in one step, or gradually in several).

In the *De motu* and in the *Principia*, Newton gives a detailed response to question (A). But it was question (B) that Halley had posed. Did Newton believe he had resolved the inverse problem? He was undoubtedly cognizant of the difference between a theorem and its converse (as, for example, in Propositions 1 and 2 of the *Principia*).

Johann Bernoulli reproached him twenty-three years later with having supposed without demonstration that the trajectory must be a conic section (see below, p. 264). However, in the *Principia*, Newton had advanced very far along the path of deducing trajectories from forces. Is it right to suppose that he gives, if not in the *De motu* (Problem 4), then at least in the *Principia* (sec. 8, especially Prop. 41), a truly general and complete solution of the inverse problem? Opinions are as divided today as they were when the rivalry was so inveterate between the Continental scientists[11] and Newton's defenders.[12]

Is the delay in the response to Halley an index of this difficulty? If the initial demonstration swelled to become, first, a small treatise, then an enormous work, might this be because Newton sought to give a bit more substance to a reply that had escaped his lips too quickly? Conscious of the lacunae in his proof, might he have been attempting, in his expansion of his first exposition, to give a true response to question (B)?[13]

But other motives, as well, impelled Newton to expand his initial outline. He saw here the occasion to get to the bottom of certain questions.[14] Difficulties in various applications had forced him to return to the fundamental enunciations.[15] Above all, the new instruments that he had forged were proving rich in applications. In particular, the law of areas (Theorem 1) opened up possibilities that were completely new for the study of curved trajectories, rotating orbits, vertical trajectories of fall, the two-body problem, and other problems.[16] The prodigious work incorporated in the *Principia* originated not so much in the pursuit of an impossible proof as in the pioneering of virgin territory.

CHAPTER I

THE *DE MOTU* OF 1684

THE ELEMENTS OF NEWTON'S SOLUTION

I turn now to the demonstration actually given by Newton toward the end of 1684 in the little treatise that so delighted Dr. Halley.

How to pass beyond the vague idea of an attractive force to its geometrical expression—to its evaluation along a trajectory? Newton saw in the mathematical elaboration of this problem the difference that separated him from Hooke. It is one thing to propose a conjecture concerning the variation of force, and quite another to enter into the detail of the geometrical determination, the observations, and the calculations. Hooke had spoken and written as if all of those—the working out of the mathematical details—were only a subsidiary chore, a piece of drudgery that he did not himself have the time to carry out.[1] To carry out the task, it would in fact have been necessary to construct the entire edifice of a theory of central forces, starting from materials scattered throughout the scientific writings of the seventeenth century.

Inertia and Deviation

In a schematic view of the text of the *De motu*, the elements of Newton's solution are the following:

> Left to themselves, the planets would move uniformly in straight lines (this is called the principle of inertia, which will be Law I of the *Principia*).
>
> The incurvation of their trajectories is due to an exterior force directed toward the Sun (Newton speaks of a deviation or deflection).
>
> To evaluate the force, it is necessary to measure the incurvation, that is, the difference between the virtual rectilinear trajectory and the real, incurved trajectory.[2]

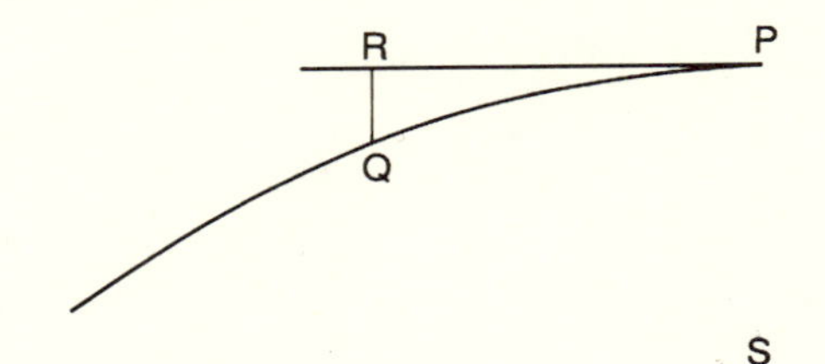

Figure 1.1

The supposition that the deviation enables evaluation of the force, because force and deviation are proportional to one another, remains implicit in the first manuscripts of 1684 but will become Law II of the *Principia*:

> The change of motion is proportional to the motive force impressed.

The segment QR is therefore the index and geometrical measure of the force that draws P toward S.

The Generalization of Galileo's Law of Fall

How then to measure the deviation QR? Besides the intensity of the force, it is necessary to take account of other factors to determine how much the moving body departs from a rectilinear trajectory. The separation between the tangent PR and the curve PQ is greater if, for example, the moving body is farther from P along the arc. Newton chose time as the basic variable: the deviation depends on the time elapsed.

The *De motu* poses the hypothesis that the deviation QR is proportional to the square of the elapsed time. Whence comes this relation? It is a generalization of Galileo's law of fall: the space traversed by a body in free fall starting from rest is proportional to the square of the time.[3] To be able to apply this law in the present case, it is necessary to accept several assumptions:

> The force that attracts the planets toward the sun is analogous to terrestrial weight.
>
> The length QR represents a kind of path of descent.

In other words, the curved trajectory PQ must be considered as a combination of two motions: one that is uniform and rectilinear from P to R, and another that is accelerated from P towards S. The length QR represents this second component, viewed by itself and in abstraction from the first component.

It was Galileo, again, who had made such a decomposition possible and legitimate. In the Fourth Day of the *Discorsi*, he had shown how the trajectories of projectiles could be analyzed into a uniform, rectilinear motion (either horizontal or oblique) and a vertical, accelerated motion. This operation of

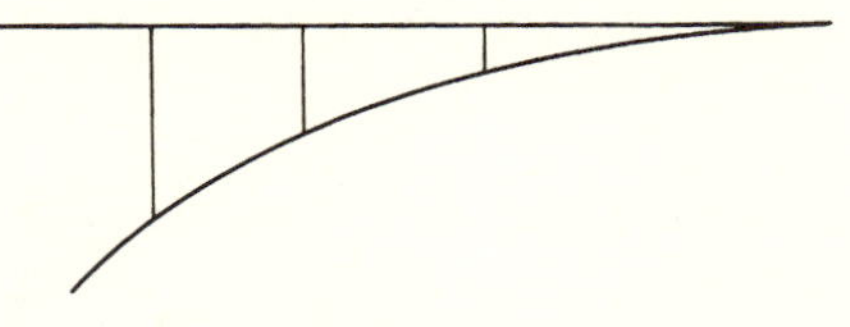

Figure 1.2

abstraction in the decomposition of motions had become—notably in the work of Huygens[4]—a very precious tool.

Between the case of the planets and that of terrestrial weight, however, there are several important differences. The fall or quasi-fall of the planet is not vertical with constant direction but is instead directed toward a point, the fixed center S. Moreover, the intensity of the force varies from point to point in space. With the planet's change in position, its "weight" is no longer the same; the moving body is subject to a force that varies, though perhaps only slightly, along its path. In the planetary case, then, Galileo's law is applicable only in the infinitely small, very close to point P. As Newton put it, the law of fall is here true only "at the commencement of the motion." His reasoning is thus valid only if arc PQ is very small or "nascent."

In sum, in figure 1.3—a figure inspired by those in Newton's text—the length QR, which is the index and measure of the force directed toward S, is proportional to this force and proportional also to the square of the time, provided that Q is very close to P.[5]

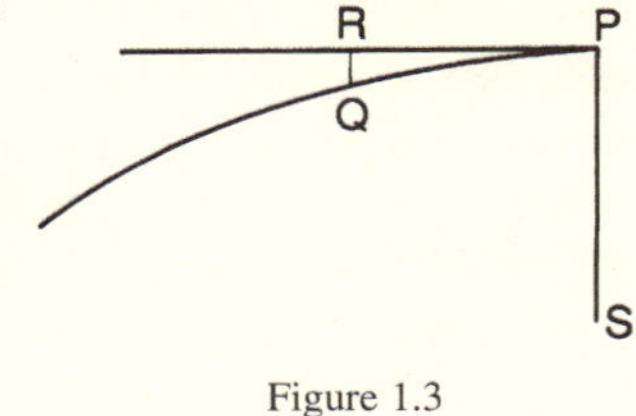

Figure 1.3

The Measurement of Time by Area

But what to say of the time itself? How to introduce it into the diagrams and the calculations? The geometrical diagram shows the path traversed but not the time elapsed. If length of path were exactly proportional to time, the latter could be replaced by the length of path traversed; the planet, however, does not move constantly at the same speed. To evaluate its velocity it would be necessary to know its initial velocity and the force at the different points of its path.

Kepler's law of areas provides an escape from this circular trap. It asserts that the time required for a planet to traverse an arc can be evaluated by measuring the area of the sector swept out by a radius connecting the planet to the sun. Whatever the intensity of the force acting at P, the surface area swept out

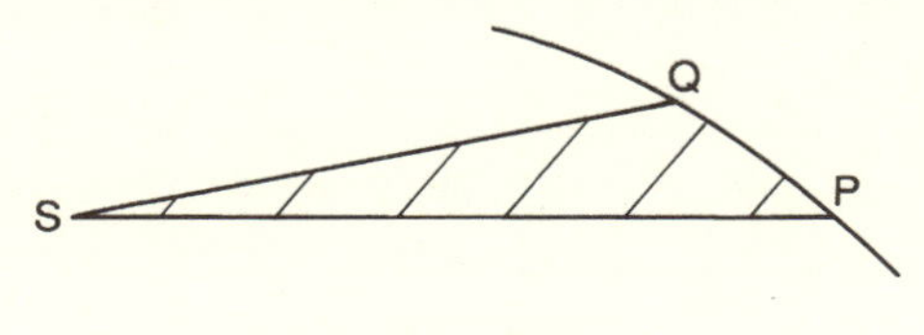

Figure 1.4

by the radius SP (traditionally called the *radius vector*) is always proportional to the time that elapses. The infinitely small triangle SPQ is therefore a measure of the time required to traverse the arc PQ. In place of the time it is thus possible to substitute the area of the triangle SPQ, and if this area is expressed in terms of the base SP of the triangle and its height QT, the length QR is proportional to the square of the product of SP and QT.

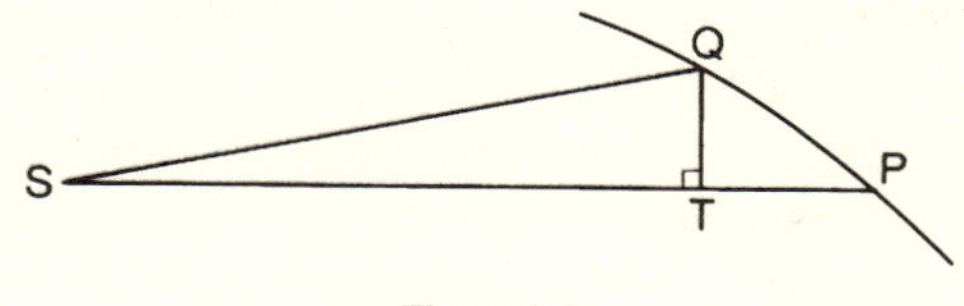

Figure 1.5

The General Formula

Thus a relation is obtained that connects the force, the deflection QR, and the time as represented by the area of the triangle:

QR is as the force and as $(SP \times QT)^2$,

or equivalently,

the force is as the deflection QR and inversely as the square of the area $SP \times QT$.

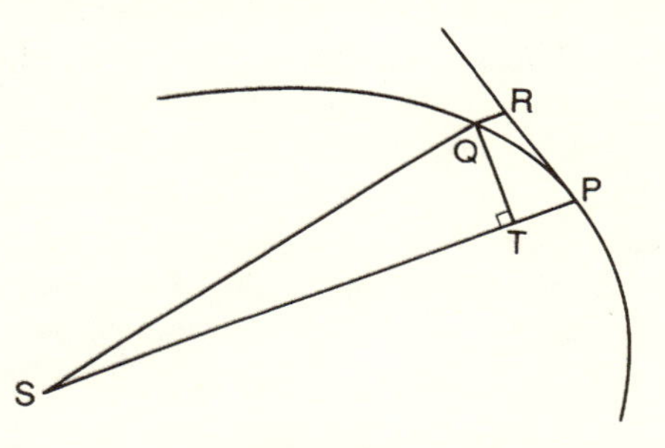

Figure 1.6

Here is a completely geometrical expression of the force—at least if, into geometry, are admitted "nascent" arcs and traversed paths that are "very small." In order to evaluate the force to which a body is subject at a point P of its trajectory, it suffices to determine the value of $QR/(SP^2 \times QT^2)$. And to prove that the force varies from point to point according to a certain law, for example, as a function of distance from the center of force, it suffices to calculate how $QR/(SP^2 \times QT^2)$ depends on the distance SP when P varies.[6]

It is this that Newton did—in contrast to the three other scientists—in the writing that he sent to Halley. The variation of $QR/(SP^2 \times QT^2)$ when P traverses an ellipse and the force is directed toward a focus S is inversely proportional to the square of the distance SP:

$$\frac{QR}{SP^2 \times QT^2} \propto \frac{1}{SP^2}.$$

Thus he demonstrated that if the planets move in ellipses, with the sun at a focus of the ellipse in each case, then the force attracting them to the sun diminishes as the square of the distance.

WHY READ THE *DE MOTU*?

It is this fundamental schema of reasoning in the *De motu* that Newton follows in the *Principia*, but in the latter work it is less prominent because of all the complements, amplifications, refinements in demonstration, and philosophical scholia that Newton added to his pristine text. Between 1684 and its publication in 1687 he expanded the text inordinately—from eleven propositions to more than two hundred.

Newton was already ahead of his time in 1684 when he arrived at a mathematical formulation of the idea of attraction—the idea that his three colleagues had been discussing. He was yet more advanced in 1687 when he at last submitted Book III of the *Principia* to the printer—so far had his speculations led him beyond his contemporaries. In three years he had produced an immense work, which the eighteenth century would have the task of understanding, developing, and verifying. What Proust has said of literary works is true here also: a veritably original work must create its own public. Newton's true contemporaries, Euler and Laplace, belonged to the next century.

Few seventeenth-century readers found the *Principia* accessible. The main results became known indirectly through reviews, popularizations, and simplified expositions. Even the most determined and enthusiastic readers found their patience tried by Newton's book. John Locke, an admirer, wanted to comprehend it, and at length Newton composed for him in four pages a demonstration that the planets by their gravity move in ellipses.[7] For the benefit of discouraged readers, Newton declared in the preamble to Book III:

> However, since the propositions there [in Books I and II] are many, and could cause too much delay even to readers well trained in mathematics, I do not suggest that anyone read them all. It is enough if one carefully reads the definitions, the laws of motion, and the first three sections of Book I, then passes to this book concerning the system of the world, and consults at pleasure such of the remaining propositions of the first two books as are cited here. (*Princ.*, 386)

The *De motu* is much more modest in its proportions and much more accessible as well. There is great advantage in reading one of its versions before plunging into the meanders and refinements of the *Principia*. An initial orientation is indispensable—a pinpointing of principal results, connections, and methods—to follow the *Principia*.

The first pages of the *Principia* have been much read and discussed, especially by philosophers: the preamble of the book, setting forth the laws of motion and giving the scholia on absolute time and space, has nourished debates on the foundations of mechanics and on the nature of space and time. From the point of view of this discussion, these texts no longer occupy so essential a place. The *De motu*, in all its early versions, contains neither the three famous laws of motion nor any mention of absolute time and space but presents, above all, the first evidence of a geometrical translation of the concept of central force.[8] This is its primary interest: how did force come to be expressed geometrically, and how did mathematics become capable of translating dynamics?

TRANSLATION AND COMMENTARY

The Definitions of Force in the *De motu*

In the version discussed in this book (the *De motu corporum in gyrum*[9] as given in MS B of Hall and text I, para. 1 of *NMP*, vol. 6), which is probably the earliest, the *De motu* is divided into four parts:

> Three fundamental theorems (the law of areas, a formula for the force in uniform circular motion and then a formula for the force in any curvilinear motion whatever)
>
> Three problems illustrating the preceding theorems
>
> Various complements (Kepler's third law, the determination of orbits, the inverse problem, Kepler's problem, rectilinear fall)
>
> Study of motion in a resisting medium[10]

I leave aside the fourth part, because, being the germ of Book II of the *Principia*, it has no implications for Book I and because the analysis of the theorems on resisting media would stray too far afield.

The work is presented in a deductive form, starting from three definitions and three hypotheses. The definitions are as follows:

> *Definition 1. Centripetal force* I call that by which a body is attracted or impelled towards some point viewed as a center.
>
> *Definition 2.* And the *force of a body* or the *force innate in a body*, that by which the body endeavors to persevere in its motion along a straight line.
>
> *Definition 3.* And *resistance*, a regular impeding due to a medium. (Add. 3965.7, fol. 55r)

The reader who seeks here a definition of force will be disappointed. Newton defines, not *force*, but the terms *centripetal*, *innate*, and *resistance*. The two first definitions concern the adjectives: "*Centripetal force* I call that by which . . ." and "the *force innate in a body*, that by which . . ." The third proceeds likewise on the basis of a tacit understanding of the concept of force. These definitions, therefore, bear not on force per se, taken generally, but only on the specifications that Newton has chosen to give to it.

The word *force* at the time Newton was writing was a vague term, unspecialized and poorly defined. John Wallis, one of the few authors at this time to specify a meaning for the word, defined *force* (*vis*) as a "power productive of motion" (*potentia efficiendi motum*).[11] It will be assumed for the moment that this definition covers the usage that Newton made of the word.

Most people know more or less what a force is. Newton's definitions specify what constitutes a centripetal force, an innate force, and a resistance. Innate force (*vis insita*), or *the force of a body*, was not an absolute novelty in 1684; the concept had been accepted and the term received. Descartes,[12] Galileo's disciples, Wallis, and Huygens spoke commonly of the force with which—or by virtue of which—a body continues its motion in a straight line with constant speed. The term *force* was thus not reserved for an external cause. Uniform rectilinear motion was associated with a force; like all motion it required a power that produced it and maintained it. Today it would be said: of itself the body perseveres in its uniform rectilinear motion. Newton and his contemporaries preferred to say: by the force that is inherent in it, which is properly its own, the body perseveres.

The adjective *inherent* (*insita*) had already been employed by Kepler and Gassendi to designate the force naturally implanted in, or belonging properly to, bodies: *insita* is the participle of the verb that means "to graft," "to implant." If this force is inherent, it is because the other force is not: centripetal force is not essential to bodies—it is not inherent or implanted. The opposition of the two forces underlies the use of the word *insita*. In later editions and in *Opticks*, Newton makes clear that weight does not belong essentially to matter and that the only truly inherent, or truly natural or essential force, is the "passive" force by which a body continues its uniform rectilinear motion and resists changes of motion.[13]

The notion of centripetal force is the great innovation of this text, and as Definition 1, it is given pride of place. Newton invented the word in an act of conscious imitation: Huygens had conceived of "centrifugal force," and Newton honored him and corrected him in inverting the point of view.[14] The force that accounts for curvilinear motions is directed toward the center, "tending toward the center" (*centripeta*).

The mode of action of this force is left indeterminate; it could be a push or a pull (i.e., an attraction): *impellitur vel attrahitur*. Various explanations remain possible. But a dynamical analysis of incurved motions does not require determination of the "physical" cause that curves the moving body in toward the central point.

Very little is known about the central point itself. It is not a center in a strict geometrical sense but is any point that can be considered as a center because the trajectory incurves around it; the path need not be perfectly circular. Is there a virtue that emanates from this privileged point, a flux of magnetism that gushes forth from it? This question cannot be answered. However, it is known that the centripetal force resembles weight. Hypothesis 4 of the *De motu* stipulates that the effects of centripetal force are similar to those of weight, at least locally. Centripetal force varies from point to point, but at each point its action, like that of weight, causes the body to traverse a distance proportional to the square of the time. The relationship between weight or gravity and centripetal force was so essential in Newton's eyes that he substituted the one for the other in different versions of the *De motu*.[15]

Centripetal force is no longer so prominent in the *Principia*. Its conceptual scope is more encompassing and more richly articulated: centripetal force is ranged in the larger class of "impressed forces" (*vis impressa*, *Princ*., Definition 4, 2), which also includes both impact and pressure. The pair *innate force–centripetal force* is thus replaced by the pair *innate force–impressed force*.[16]

Centripetal force—which might seem to be a verbal and ad hoc creation, almost one of the occult scholastic qualities that Newton rejected so vigorously—thus comes in the *Principia* to be fitted into a more elaborate conceptual scheme. The theoretical apparatus of 1687 corresponds to a wider ambition: with subject matter no longer confined to bodies moving in orbits, the work aims to embrace the whole of natural philosophy—albeit that theories of electricity, cohesion, or chemistry remain at the stage of program or promise.

The three types of force defined in the *De motu* make possible the reconstruction of various kinds of motion. The third, resistance, is first mentioned in an addition to the original manuscript and is employed only in the last part of the text: in Problem 6, which treats of a body subject only to innate force and resistance, and in Problem 7, which adds a uniform centripetal force. It is the first two forces that are fundamental. Orbital motions are analyzed as a resultant of these two forces combined: the innate force engenders, or tends to engender, a uniform rectilinear motion, and the centripetal force incurves the trajectory, continuously or in a series of punctiform actions, in the direction of the point that plays the role of center. The curvilinear motion of a body presupposes these two elements and only these: an innate force and a centripetal force, the second serving to deflect the inertial motion due to the first. As Newton writes further on:

> By their innate force alone [the revolving bodies] would describe the tangents. . . . Centripetal forces are those which continually draw bodies back from the tangent to the circumference.[17] (Add. 3965. 7, fol. 55r)

Here is the fundamental contribution of this manuscript: for the first time curvilinear motion is given a detailed analysis in terms of its two components, the inertial one and the deflection. Galileo had outlined a closely related idea (EN, 7:242), and others had envisaged this conception but as though in passing and without drawing from it all its fruitful consequences. Five years earlier, Newton himself was still attached to a different representation: according to his second response to Hooke, in 1679, the circulation of a body round a center of attraction would be due to the combined action of a centrifugal force and gravity, with each, by turns, exceeding the other (*Corresp.*, 2:307; see below, p. 151).

In the text of 1684, centrifugal force is no longer present. The revolving motion requires only centripetal force and innate force. It was probably Hooke who had made Newton understand this by proposing his hypothesis of "compounding the celestiall motions of the planetts of a direct motion by the tangent & an attractive motion towards the centrall body" (*Corresp.*, 2:297, 306). To the two elements by which he proposed to account for a circular trajectory, Hooke gave the names *direct motion* and *attractive motion*. (As to the "power" of attraction of the central point, Hooke's references are vague.) These motions are components that can be added and combined in various ways, the one counterbalancing the other. Newton preferred to apply the name *force* to both these elements. Thus doing, he established them as of the same rank and so capable of being combined. He was aware, to be sure, that these two actualities were very different, that they were not forces in the same sense. The *Principia* makes clear that innate force is a "power," while centripetal force (or any other impressed force) is an "action" which exists only at the instant of impact or pressure and does not endure beyond (*Princ.*, Defs. 3 and 4, 2). This external and instantaneous action is henceforward called *force* and ranked with the permanent and internal power or force associated with inertia. Thus it becomes possible to combine them and so to reconstitute curvilinear motions.

The Hypotheses of the *De motu*

Hypothesis 1. In the first nine of the following propositions the resistance is nil; in the remaining propositions it is proportional jointly to the speed of the body and to the density of the medium.

Hypothesis 2. Every body by its innate force alone proceeds uniformly to infinity in a straight line, unless it be impeded by something extrinsic.

Hypothesis 3. A body in a given time is carried by a combination of forces to the same place to which it is borne by the separate forces acting successively in equal times.

Hypothesis 4. The space that a body, urged by any centripetal force, describes at the very beginning of its motion is in the doubled ratio of the time. (Add. 3965.7, fol. 55r)

What an odd assemblage! Under the rubric *hypotheses*, Newton brought together four quite different enunciations. The first is merely a convention assumed for the remainder of the work: resistance is to play no role before the last two theorems, and then it will be taken to be proportional to speed. The last two enunciations are lemmas that are demonstrated in the *Principia*:

The rule of the composition of forces
A generalization of the theorem of Galileo

Isolated amidst this series of hypotheses is an axiom of the greatest importance that will become Law I of the *Principia*: the principle of inertia.

The admixture of these hypotheses is reassuring to others: Newton the genius did not attain in one stroke the clear formulation of his three laws as they appear in the *Principia*:

> I. Every body perseveres in its state of rest or uniform rectilinear motion except insofar as it is compelled to change its state by forces impressed upon it.
>
> II. The change of motion is proportional to the motive force impressed, and is made in the direction of the straight line in which that force is impressed.
>
> III. To an action there is always a contrary and equal reaction; that is, the mutual actions of two bodies are always equal and oppositely directed. (*Princ.*, 13–14)

The third law is completely absent from the *De motu* in its early versions; it appears only much later in the series of redactions that culminate in the *Principia*. Moreover, it has no use here: there is no study of impact, no action of the planets on one another, no truly universal gravitation.[18]

Nor is the second law present, at least explicitly, in the early version of the *De motu* being considered here. It is at work, however, in the reasonings of this version, as will be shown later. The change of motion depends on the centripetal force: in the first theorem, the deflection is directed toward the center of force; in the two following theorems, the intensity of the deflection is proportional to the centripetal force exercised at the point where the body is. Only in much later versions does Newton come to articulate this concept, so essential to his reasoning, as a primary axiom in its own right.[19]

The first law is the principle of inertia in one of the clearest and most concise formulations of the seventeenth century. From this law as formulated in the *Principia*, Hypothesis 2 of the *De motu* differs in a number of respects. The hypothesis states that the body "proceeds uniformly to infinity" if nothing impedes it; the restrictions with which Galileo had hedged his earlier enunciation of the principle have here disappeared,[20] and Newton accepts the principle as an abstract and absolute axiom in all its bold unimaginableness. But the hypothesis is more restricted than Law I—it does not mention *rest*, and it relates inertial motion to the existence of innate force. The *Principia*, in contrast, states that the body persists "in its state" of rest or uniform rectilinear motion, and so it introduces a note of greater generality and perhaps the germ of a justification in the manner of Descartes: motion and rest are states of the same rank. In Descartes' *Principia philosophiae*, which the young Newton had read attentively, the immutability of God guarantees that each thing remains in the same *state*; consequently—motion being a state—any body that moves will move forever insofar as it depends on itself (*quantum in se est*); hence each part of matter, considered in isolation, tends to continue its motion straight onward, rectilinearly, unless another part of matter comes to *deflect* it.[21]

Hypothesis 3 enunciates a rule of composition of forces that does not correspond exactly to what is called "the parallelogram of forces," a designation that agrees better with Corollary 1 of the Laws in the *Principia*, to wit:

> By conjoined forces a body describes the diagonal of a parallelogram in the same time as it would describe the sides by those forces separately. (*Princ.*, 14)

In the *De motu* there is no mention of a parallelogram, and the two forces are supposed to act in succession: the body is "carried . . . to the same place to which it is borne by the separate forces acting successively." A body goes from A to B in a certain interval of time under the action of a force F, then in another interval of time of equal length, from B to C under the action of another force G. If the forces F and G are applied to this same body simultaneously, the body will traverse, during an interval of time of the same length as the preceding intervals, a path leading from A to C.

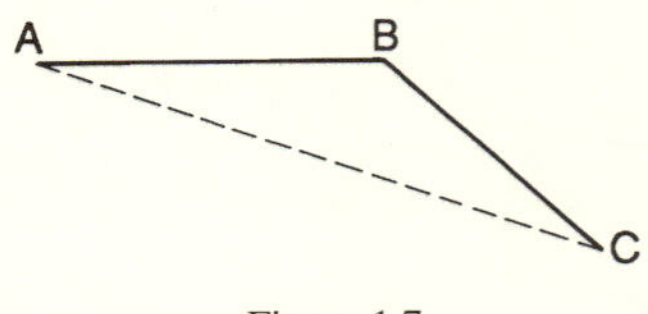

Figure 1.7

Newton says nothing about the form of the resulting path AC. The motions to be compounded are not specified as rectilinear and uniform. As formulated, Hypothesis 3 makes possible the compounding of motions in any case whatever and thus is much more abstract and vague than Corollary 1 of the Laws of the *Principia*. Nor does Newton describe the nature of the forces to be considered: are they innate forces or centripetal forces? For instance, a portion of a parabola compounded of uniform rectilinear motion and the motion caused by gravity could be taken as an application of Hypothesis 3.[22] This indeterminacy as to the form of the path and the nature of the forces persists in the succession of redactions up to the published demonstration of Corollary 1. Only in the second edition of the *Principia* does Newton stipulate that the motions are rectilinear and uniform and that the forces are forces impressed in a single instant at point A.

Hypothesis 4 returns to Galileo's theorem on falling bodies and generalizes it so that it applies to all variable forces. Galileo had shown in the *Dialogo* and the *Discorsi* that a falling body traverses distances proportional to the squares of the times. In the vocabulary of the seventeenth century, the distances traversed are "in the doubled ratio" of the times. Newton asserts that this is true of every centripetal force: centripetal forces are analogous to weight. But Galileo's law supposes that the weight remains constantly the same. Near the surface of the earth, weight *is* approximately invariable, but for a body falling from a very great height, the weight varies from point to point of the path, being greater the nearer the body comes to the earth. And for the solar system, the wager of Wren, Hooke, Halley, and Newton was precisely that the variation of the attractive force as a function of distance makes it possible to account for the paths of the planets.

If the force is variable from point to point of space, it is necessary to consider it in its very first effects, at the beginning of the motion (*in ipso motus initio*). At each point of a planet's path, one can apply a law identical to the law of falling bodies but valid only for the very beginning of the motion.

What is true of Galileo's falling body will also be true, at the limit or infinitesimally, for the action of any centripetal force. A variable force can locally be assimilated to a constant force, the word *locally* signifying a double restriction:

1. To a single point of space, since the force varies from point to point
2. To the very beginning of the motion

The properties of the fall of bodies are thus verified in every point of space for a motion of infinitesimal duration.

Newton does not speak explicitly of infinitesimals, at least not here. He considers the path traversed "at the very beginning of . . . the motion." This concept is a remarkable novelty rich with consequences. With or without infinitesimals, here is the first indication of a new mathematics that will clothe itself in various vestments: the method of fluxions or differential calculus.

No justification, no trace of a proof is put forward in this version of the *De motu* either to support the extension of Galileo's law or to provide a foundation for similar reasonings with infinitesimals. What is meant, in all rigor, by "the beginning of the motion"? How to manipulate entities so fugitive? What sort of demonstration or calculus may be hoped for in such a matter?

In the later versions of the *De motu* directly before the *Principia* (MS D in Hall), Newton gives a quasi-geometrical demonstration modeled on that of Galileo (see below, pp. 163–67). The corresponding geometrical figure for a constant force is deformed up to the disappearance of certain elements. The relations between evanescent (or nascent) magnitudes are "read" on a sort of finite equivalent of the vanishing figure, each infinitely small line or triangle being reproduced in the finite by virtue of a corresponding finite element that remains always similar to it. This demonstration is progressively augmented to become the germ of the entire section 1 of the *Principia*, which concerns the "ultimate ratios of nascent or evanescent quantities" (*Princ.*, 28–38). The generalization of Galileo's law was thus the motive and kernel of the little mathematical excursus that Newton placed at the beginning of Book I of the *Principia*; it was in extending Galileo's law of force that Newton developed his account of ultimate ratios.

The Law of Areas

Theorem 1. All orbiting bodies describe, by radii drawn to the center, areas proportional to the times.

Let the time be divided into equal parts, and in the first part of the time let the body by its innate force describe the straight line AB. Again, in the second part of the time, if nothing were to impede it, it would proceed straight on to c, describing (by Hypothesis 1) the straight line Bc equal to AB, so that, when the radii AS, BS, cS are drawn to the center, the completed areas ASB, BSc will be equal.

But when the body arrives at B, let the centripetal force act with a single but large impulse and cause the body to deflect from the straight line Bc and proceed along the straight line BC.

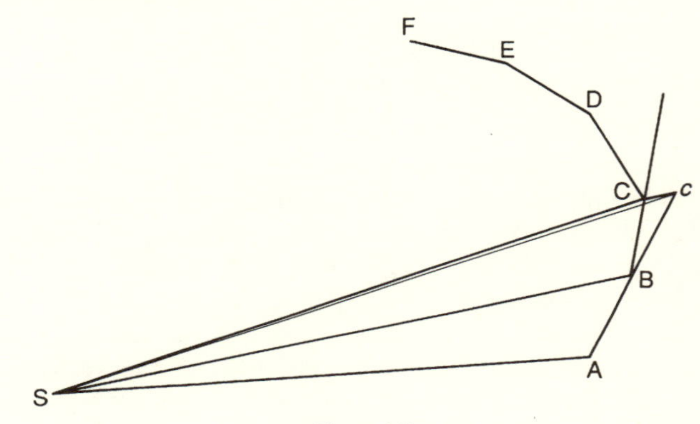

Figure 1.8

Let cC be drawn parallel to BS meeting BC in C, and when the second interval of time is ended, the body will be found at C (Hypothesis 3). Join SC, and the triangle SBC, because of the parallels SB, Cc, will be equal to the triangle SBc, and therefore equal to the triangle SAB.

By a similar argument, if the centripetal force acts successively in C, D, E, etc., causing the body in separate moments of time to describe the separate straight lines CD, DE, EF, etc., the triangle SCD will be equal to the triangle SBC, SDE to SCD, and SEF to SDE.

In equal times, therefore, equal areas are described.

Now let these triangles be infinite in number and infinitely small, so that each triangle corresponds to a single moment of time, and with the centripetal force then acting unremittingly, the proposition will be established. (Add. 3965.7, fol. 55r)

To open the demonstrative sequence of the *De motu*, Newton chose the "law of areas," often referred to nowadays as the "conservation of orbital angular momentum."

In order to measure the time of passage of a planet along a portion of its orbit, it is necessary to evaluate the surface area swept out by an imaginary radius connecting the planet to the center of force. The triangle or sector SAB (fig. 1.9) thus swept out is proportional to the time that elapses as the planet goes from A to B. One consequence is that, in an elliptical orbit with the sun at one focus of the ellipse, a planet will move more rapidly when near the focus—in going from C to D—and more slowly when far from the focus—in going from E to F (fig. 1.10). The heights of the triangles or sectors SCD and

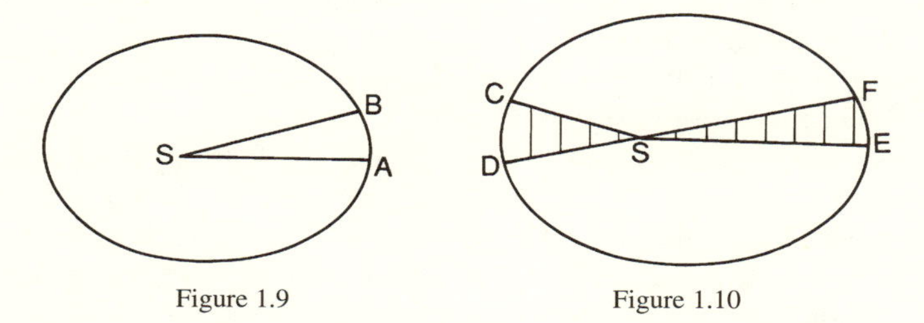

Figure 1.9 Figure 1.10

SEF are different, but the area must be the same if the times to traverse the two arcs are the same, the inequality of the heights being compensated for by an inverse inequality of the bases CD and EF. Given equal times, CD will be longer than EF, and the planet will move more swiftly from C to D than from E to F. In the final text of the *Principia*, Newton makes explicit this consequence of the law of areas, showing how by means of it the ratio of the speeds of a body at different points of its orbit can be determined (*Princ.*, Prop. 1, Cor. 1, 40).

Kepler had formulated this law in his *Astronomia nova* of 1609 and had discussed and used it further in his later works. In appearance the principle was not new: Newton was redemonstrating a well-known principle. But to put the matter in its true light, Newton's demonstration was incredibly new. Kepler had proposed his method of areas as an approximative procedure for calculating the times required by planets to traverse different arcs of their orbits. He had demonstrated it only very imperfectly (and he had even noticed that one of the consequences of the enunciation contradicted one of the premises!). Moreover, no one before Newton had accepted this "law" as an indubitable principle; for the astronomers between Kepler and Newton, it was at best a convenient calculating trick, but preference was given to other methods of evaluation that were more rapid and just as plausible.[23] Yet Newton took this principle as the point of departure for his theory, enunciating it in complete generality and demonstrating it with an unprecedented economy of means—at least if one accepts the passage to the limit, by which the demonstration is concluded.

To demonstrate a known theorem by means that are altogether new and different is already remarkable. When Newton succeeded in proving the law of areas, he undoubtedly saw in this accomplishment a confirmation that he was on the right track. In his known manuscripts prior to 1684, there is never any mention of the law of areas. Whenever Newton discovered it, he did not admit it as true before he was able to prove it.[24]

In its generality, Newton's proposition goes far beyond the enunciation of Kepler. Here the principle is valid for every centripetal force, and not only for the case of planets orbiting about the sun. Moreover his demonstration presupposes no particular law of force, while Kepler's attempt at a demonstration was based on the linear attenuation of force with distance. In Newton's text, by contrast, the force can vary in any arbitrary way from point to point of space; it is only supposed that it is always directed toward point S. Newton's genius was in this restraint: from the beginning he established the law of areas on the level of generality that properly belongs to it.

The dynamical premises are simple: the body, left to itself—that is, left solely to its innate force—will in equal times traverse equal distances along the same straight line. It will therefore go from A to B, and then from B to c.[25] If the centripetal force acts at the end of the first interval of time, it will divert the body (Newton like Descartes used the verb *deflect*) and make it take a new direction, say Bz, which need not be specified. Where will the body be on Bz at the end of a second interval of time equal to the first? It is Hypothesis 3 that

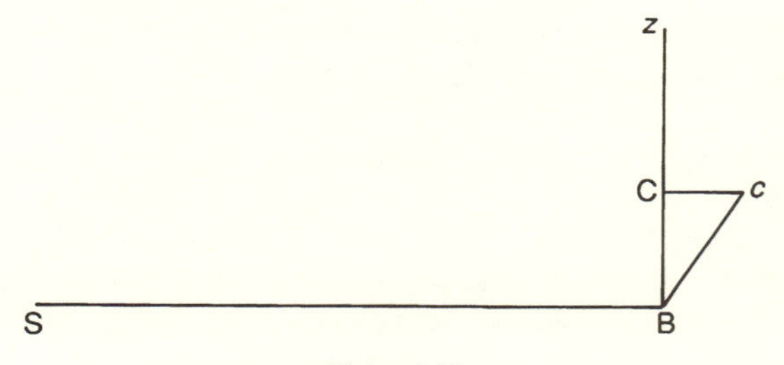

Figure 1.11

answers this question: the innate force and the centripetal force acting together will carry the body to the same place as if they were to act separately and in succession. The innate force by itself carries the body to c, and the effect of the centripetal force is represented by the line cC. It is not useful here to ask about its size; it suffices that its direction, which by hypothesis is the direction of BS, is known. The segment cC is therefore parallel to BS, and point C is determined as the intersection of cC with Bz.

The action of the central force is absolutely punctiform or instantaneous; it is produced in the manner of an impact at point B (*impulsu unico sed magno*), then at point C, at point D, and so on. Immediately after this action, the innate force takes over and the motion is again uniform. There is therefore no accelerated motion, for which one would have to use Hypothesis 4 (the reasoning presupposes nothing as to the length of cC). The orbit is made up of a series of rectilinear segments, and at each vertex of the polygonal path the body is "deflected" by an impulsion.

The geometrical part of the proof is also very simple and very elegant. A single theorem from Euclid's first book is twice applied: the areas of triangles are equal if the bases are equal and also the heights are equal. For the triangles constructed on the rectilinear path ABc—SAB and SBc (figure 1.12)—the equality of the bases, AB and Bc, results from the uniformity of the motion,

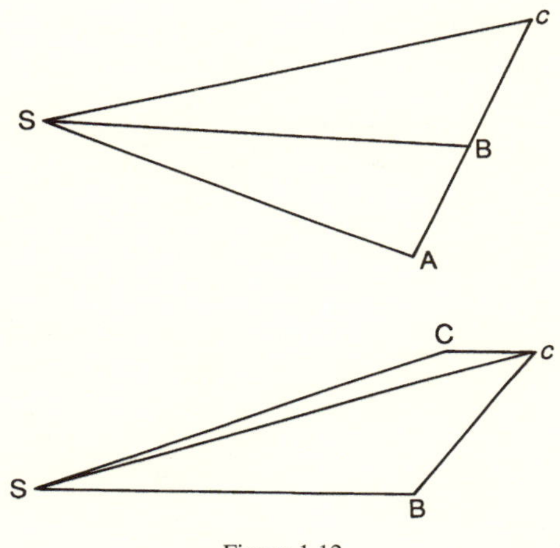

Figure 1.12

and the height of both triangles is the same (it is the perpendicular let fall from S onto Ac). As for the two triangles with horizontal bases—SBC and SBc, their bases are identical (SB), and the height is the same because these triangles are between the parallels SB and Cc. This demonstration can be reproduced an indefinite number of times for successive segments of the polygonal path. The area of the consecutive triangles will be the same as long as the intervals of time are the same and the deflection is always directed toward S.

The law of areas is thus demonstrated for the case of force acting in discrete, instantaneous impulsions at regular intervals; the trajectory that results is made up of successive rectilinear segments. One might imagine here an underlying causal or physical hypothesis, say of Cartesian inspiration (see below, p. 120ff.): the weight would result from a series of impacts. But it appears instead that the true motive for this approach is mathematical: it is first necessary to demonstrate the proposition in a discrete form before varying the conditions so as to obtain the real case as a limit.

How is the demonstration to be extended to cover the less simple case of a curvilinear orbit with a centripetal force acting continuously? Newton did not show himself very hard to please on this point: it sufficed to take very small intervals of time! "Let the triangles be infinite in number . . . and the proposition will be established." Thus, a sector of area corresponds to each moment of time, the polygonal contour becomes a curve, and the discrete impacts become a continuously acting force.

Uniform Circular Motion

After the opening presentation of the law of areas, Newton proposes a first case, both simple and fundamental, of centripetal force: uniform circular motion.

> *Theorem 2.* For bodies orbiting uniformly on the circumferences of circles, the centripetal forces are as the squares of the arcs simultaneously described, divided by the radii of the circles.
>
> Let the bodies B and b, orbiting on the circumferences of the circles BD and bd, describe in the same time the arcs BD and bd.

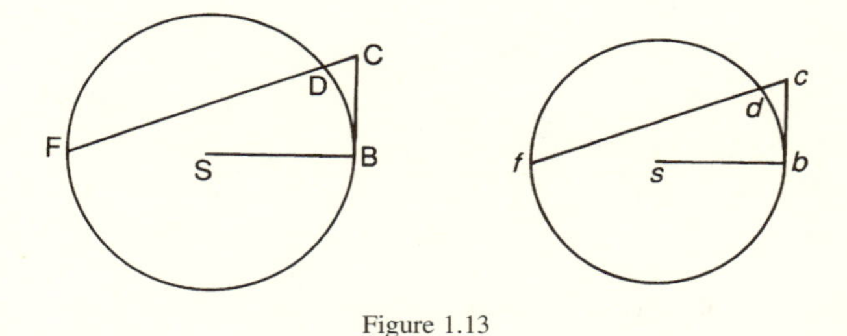

Figure 1.13

> By their innate forces alone they would describe the tangent lines BC and bc equal to these arcs. It is the centripetal forces that perpetually draw the bodies back from the tangents to the circumferences, and hence they are to each other as the distances CD and cd gained by the bodies; that is, if we prolong CD and cd to F and f, they are as BC^2/CF to bc^2/cf, or as $BD^2/\frac{1}{2}CF$ to $bd^2/\frac{1}{2}cf$.
>
> I speak of the distances BD and bd as very small and diminishing to infinity, so that for $\frac{1}{2}CF$ and $\frac{1}{2}cf$ may be written the radii of the circles SB and sb. Which being done, the proposition is established.
>
> *Corollary 1.* Consequently the centripetal forces are as the squares of the velocities divided by the radii of the circles.
>
> *Corollary 2.* And inversely as the squares of the periodic times divided by the radii. (Add. 3965.7, fols. 55r–56)

This result is not entirely new. Huygens had already analyzed the same situation and obtained a formula for the evaluation of the force using mathematical procedures that were more refined and rigorous (see below, p. 124ff.). But he had published (in 1673) only his results without the demonstration. A crucial difference from Newton was that Huygens supposed a force acting in the opposite sense: his force was "centrifugal" rather than "centripetal."

The fundamental idea of the present demonstration is the same as that in Theorem 1: the bodies left to themselves—left to their innate forces—would move along the tangents; if they are drawn back from the tangents and constrained to deviate from rectilinear paths, it is because centripetal forces act on them: "By their innate forces alone they would describe the tangent lines. . . . It is the centripetal forces that perpetually draw the bodies back from the tangents to the circumferences. . . ." In the demonstration of Theorem 1 the deflective influence of the force made itself felt in punctiform impulses; here the situation is more difficult to apprehend intuitively because the centripetal force acts "perpetually." But if the deviating action in the preceding proposition and the ultimate transition from the polygonal to the curved trajectory are comprehended and admitted, then so also can be the continuous deflective action in the case of the circle.

Newton assumed without discussion that the deflective force is directed toward the center of the circle. In the definitive text of the *Principia*, he proves that such is indeed the case: the force that causes the body to deviate tends toward the circle's center because the equal areas are described around this point (*Princ.*, Prop. 4, 43). The converse of the law of areas (*Princ.*, Prop. 2, 41) states that if areas are uniformly swept out around a point of space, this point is the center of force—the point towards which the centripetal force is constantly directed.

How to evaluate or measure the centripetal force? (It was not necessary to do so in Theorem 1, where all that mattered was the direction of the force and of the resulting deflection.) Since the centripetal force is the cause of the deviation, the one must be proportional to the other: "It is the centripetal forces that perpetually draw the bodies back . . . and hence they are to each other as the distances . . . gained by the bodies."[26]

The decisive element in the reasoning thus appears only in passing: centripetal forces can be evaluated by comparing the deflections, the departures from the inertial path. The idea first appeared in the *Dialogo* of Galileo and was taken up again by Huygens in his study of "centrifugal" force. Newton makes it Law II of his *Principia*: the impressed force (impact or centripetal force) finds its expression in a change of motion, or in other words, the change of motion is the index and measure of the force impressed.

To study the size of the deviation, Newton compares two different circles on each of which a body circulates; the velocities of the bodies are different but uniform (fig. 1.13). The radii SB and sb are of different lengths, and the arcs traversed, BD and bd, are of different lengths for the same interval of time.

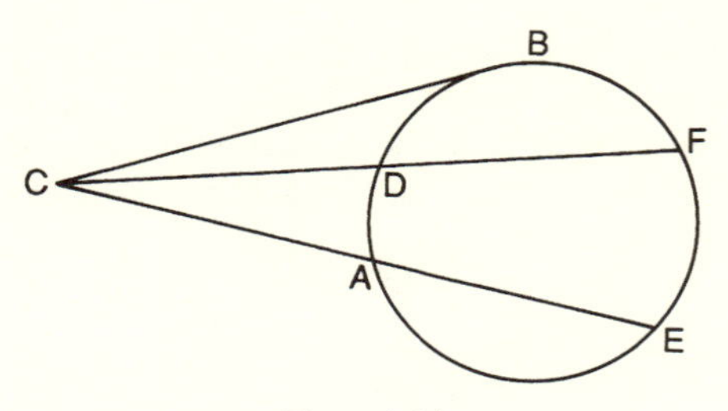

Figure 1.14

Elementary geometry permits putting the departures CD and cd, which represent the centripetal forces, into relation with other segments of the figure. According to Propositions 36 and 37 of Euclid's Book III,

$$CA \cdot CE = CD \cdot CF = CB \cdot CB.$$

If, in addition, these relations are considered when the arcs traversed are very small, a formula is obtained that relates the deviation to the arc traversed and the radius of the circle in a very simple way: the deviation, and therefore the centripetal force, is proportional to the square of the arc and inversely proportional to the radius. The arc BD and the tangent BC are equal by hypothesis; also, $\frac{1}{2}$CF is nearly equal to SB, and the two can the more justifiably be set equal as D is closer to B (that is, as the arc traversed is smaller). CF can therefore be replaced by the diameter or by the radius (the coefficient $\frac{1}{2}$ is of no importance, since we are reasoning in the language of proportions). Algebraically (and using the letters of fig. 1.13),

$$\text{Force} \propto CD = CB^2/CF \cong DB^2/2BS \propto DB^2/BS.$$

The force is therefore proportional to the square of the arc traversed and inversely proportional to the radius.

The reasoning implies a reference to time. In reality, time is the basic variable. Newton writes; "I speak of the distances BD and bd as very small and diminishing to infinity." It thus appears that he is allowing the arc to vary. But he has previously stipulated that BD and bd are arcs traversed in the same time. To say that they are very small implies that the interval of time common to the two displacements is also very small. The time is effaced and no longer appears from the second line of the demonstration onwards because it is represented and measured by the length of arc traversed in uniform motion.

Though some might wish to discuss the direction of the segments CD and cd,[27] Newton does not do it. He is content with saying that if BD is very small, CF is nearly identical with 2BS. Later, in the *Principia*, he will demonstrate that the deflection (or more precisely the *sagitta* or arrow) in its nascent or ultimate state is constantly directed toward the center of forces;[28] he will also show that the diversity of inclination of the several vanishing subtenses can be

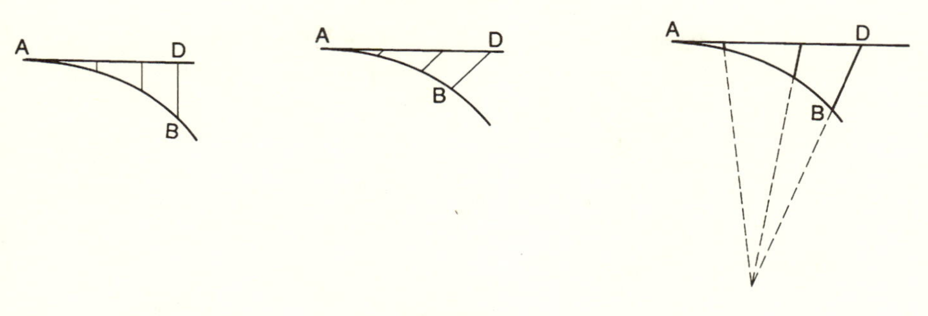

Figure 1.15

neglected (*Princ.*, Lemma 11, 34–36): the vanishing subtense BD is proportional to the square of the arc AB whether BD remains constantly perpendicular to AD or inclined at a constant angle or constantly directed toward a fixed point (the latter will be precisely the case of the arrow in the second and third editions).

Corollaries concerning Various Laws of Force

Newton has thus demonstrated a relation between the centripetal force, the radius, and the angular velocity (or the period). If the velocity itself depends on the radius, the centripetal force will be given as a function of the radius. Corollaries 3–5 of Theorem 2 deal with three conceivable cases:

> *Corollary 3.* Therefore if the squares of the periodic times are as the radii of the circles, the centripetal forces are equal; and conversely.
>
> *Corollary 4.* If the squares of the periodic times are as the squares of the radii, the centripetal forces are inversely as the radii.
>
> *Corollary 5.* If the squares of the periodic times are as the cubes of the radii, the centripetal forces are inversely as the squares of the radii; and conversely.
>
> *Scholium.* The case of Corollary 5 holds for the celestial bodies. The squares of the periodic times are as the cubes of the distances from the common center about which they revolve. That this obtains in the major planets circling round the Sun and in the minor planets orbiting round Jupiter, astronomers are now agreed. (Add. 3965.7, fol. 56)

Imagine several bodies in uniform circular revolution around the same center of force. Each body has its own constant distance from the center and its own period. If it can be determined how the period of revolution varies with radius, the variation in the intensity of the force with distance from the center can be evaluated.

Newton states a couple of propositions as examples. If the square of the period is proportional to the distance, the force does not vary with distance (there are only constants in the expression of the force as a function of the period and distance). If the square of the period is as the square of the radius, the centripetal force diminishes proportionally to the increase of the radius. These can be inferred very quickly from an algebraic expression of the formula determining the centripetal force, in which v represents the velocity, r the radius, and T the period or time of revolution (the three are related by the identity $v = 2\pi r/T$):

$$\text{Force} \propto v^2/r = (2\pi r/T)^2/r = 4\pi^2 r/T^2.$$

If r is proportional to T^2, the expression contains only constants; if r^3 is proportional to T^2 we find that

$$\text{Force} \propto (4\pi^2 r)/r^3 = \text{constant}/r^2.$$

Of these instances, the latter is far from being merely hypothetical. To suppose that the cube of the radius is proportional to the square of the period is to admit Kepler's third law. In 1618, Kepler, having observed that the planets move more slowly the farther they are from the sun, discovered in the context of his "harmonic" speculations that the cubes of the major axes of the orbits are proportional to the squares of their periods of revolution.

Newton then introduces a factual claim into the reasoning: "the case of Corollary 5 holds for the celestial bodies." He adds to the case of the major planets that of the satellites of Jupiter, which is in agreement with information he has received from Flamsteed.[29] Among the different instances that the mathematician can study, Nature has chosen one.

The affirmation of a universal gravitation according to the inverse-square law ($1/r^2$) thus results from the formula for the centripetal force restricted to uniform circular motion combined with Kepler's third law. Further on, Kepler's third law will appear in a different context: it will be demonstrated in the case of elliptical orbits by supposing that the force varies as $1/r^2$ (see Theorem 4 below). The "harmonic" law of Kepler then becomes a consequence, and the variation of the force as $1/r^2$ has to have a different demonstration (Problem 3).

It is a miniature system of the world that is proposed in this version of Theorem 2, which along with its five corollaries and scholium, provides a shortened version of the *Principia* for a simplified case—the celestial motions are supposed circular and uniform. This passage can be considered in isolation because it does not, in all rigor, depend on Proposition 1 and could have been set forth independently.

If Newton had published only this theorem and its corollaries, he would have thereby outstripped all his contemporaries. To be sure, Huygens had made known his formula for "centrifugal force" in 1673, but he did so without demonstration, and above all, he had not envisaged a gravity extending to the planets (see *HO*, 21:472).

The General Formula for Measuring Forces

Newton now presents the general formula for evaluating a centripetal force, which is in fact the heart of the whole theory:

> *Theorem 3.* If a body P orbiting about a center S describes any curved line APQ; and if the straight line PR is tangent to that curve at any point P, and to the tangent from any point Q is drawn QR parallel to SP, and the perpendicular QT is dropped to the line SP; I say that the centripetal force will be inversely as the solid $(SP^2 \times QT^2)/QR$, provided that one always takes the ultimate quantity of that solid when the points P and Q come to coincide.

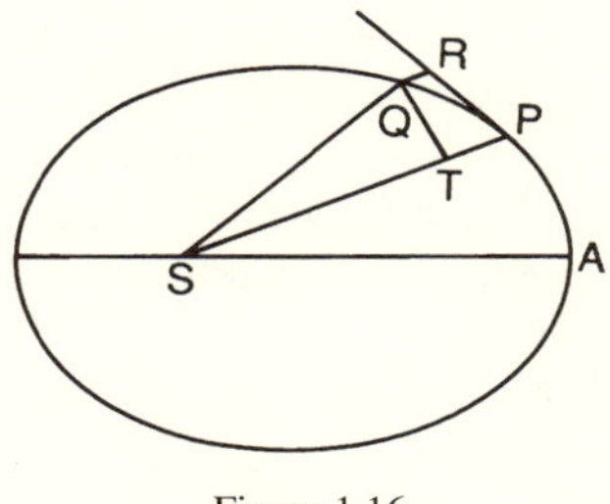

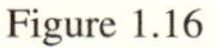
Figure 1.16

> For in the indefinitely small figure QRPT, the little line QR is, in a given time, as the centripetal force, and if the force is given, as the square of the time (Hypothesis 4), and hence if neither is given, it is as the centripetal force and the square of the time conjointly, that is, as the centripetal force and the square of the area SPQ, proportional to the time, or of its double $SP \times QT$.
>
> Let both sides of this proportionality be divided by the line QR, and unity will be as the centripetal force and $(SP^2 \times QT^2)/QR$ taken together, that is, the centripetal force is inversely as $SP^2 \times QT^2/QR$. (Add. 3965.7, fol. 56)

The enunciation is rather awkward and makes no sense independently of the figure. Given any curve AP traversed by a body subject to the action of a centripetal force, and also given the center S toward which the centripetal force constantly tends, the centripetal force acting at the point P can be evaluated. To do that, another point, Q, a little farther along the trajectory must be considered. It is not too far—in fact it is infinitely near, since everything here will have validity only "ultimately," that is, "when the points P and Q come to coincide."

The principle is always the same: if the centripetal force were not to act, the body would continue its motion along the tangent (here PR). The force deflects it from the tangent in the direction of the point S. To represent and evaluate the centripetal force, the segment QR is used, which is the displacement between the tangent and the actual trajectory and is taken in the direction PS (and therefore parallel to PS).

Attention must be focused on this small line, QR. Concerning it, Newton sets forth the following two properties:

1. QR is proportional to the force;
2. QR is also proportional to the square of the time.

The first property was already made use of in the preceding proposition in comparing the different deviations between circles and their tangents. It is the assumption, still tacit but which later becomes Law II of the *Principia*, that the change of motion is proportional to the centripetal force.

The second property is a consequence of Hypothesis 4, which generalizes Galileo's law of fall: the distances traversed under the action of a centripetal force are proportional to the square of the time, at least if the analysis is confined to the very commencement of the motion. The segment QR is thus a very small—nascent or infinitesimal—trajectory of fall; it is the line that the body would traverse during the time considered under the sole action of the centripetal force if it started from rest. The force that acts at the point is analogous to weight.

The next step consists in replacing the time by its geometrical representative: the area of the nascent sector SPQ is proportional to the time, by virtue of Theorem 1. It is therefore possible to express the square of the time in the form of the square of the area $\frac{1}{2}SP \times QT$ (one-half the base times the height of the triangle, considered as rectilinear). The coefficient $\frac{1}{2}$ can be omitted in the relation of proportionality.

After a manipulation of ratios that is a bit archaic in flavor (why does Newton give a detailed explication of this step, which holds no mystery, while passing in silence over other steps that are both decisive and enigmatic?), the final result is announced: the centripetal force is inversely as the solid $SP^2 \times QT^2/QR$ in its ultimate state.

Why inversely as $SP^2 \times QT^2/QR$ and not directly as $QR/(SP^2 \times QT^2)$? It is probably because of a concern (already a little anachronistic) for dimensional realism, which makes the first formulation more acceptable in Newton's eyes: $SP^2 \times QT^2/QR$ is of dimension 3, and hence a "solid," while $QR/(SP^2 \times QT^2)$ is of dimension −3, and without intuitive signification.

The three first theorems of the *De motu* have a common subject: the deviation caused by the action of a centripetal force. The first (the law of areas) says nothing as to the size or extent of this deviation; it depends only on the deviation's being directed toward the center of force. In the second theorem (the case of uniform circular motion), the size of the deviation enters into consideration, but the geometry of the circle suffices to determine it very simply: the deviation is proportional to the square of the arc traversed, the latter itself being proportional to the time. Finally, in the third theorem (the general formula), the dependence between the deviation and the time cannot be determined without recourse to the generalization of Galileo's law of fall (Hypothesis 4).

The First Applications: Problem 1

The text following Theorem 3 is devoted to three applications of the general formula, the first of which follows.

Corollary. Consequently, if any figure is given, and in it a point toward which the centripetal force is directed, the law of the centripetal force can be found that will cause a body to orbit on the perimeter of that figure. It will be sufficient to calculate the solid $SP^2 \times QT^2/QR$ which is inversely proportional to this force. Of this we give examples in the following problems.

Problem 1. A body revolves in the circumference of a circle. It is required to find the law of the centripetal force that tends toward a point of the circumference.

Let SQPA be the circumference of the circle, S the center of the centripetal force, P the body borne along in the circumference, Q a closely proximate position into which it shall move.

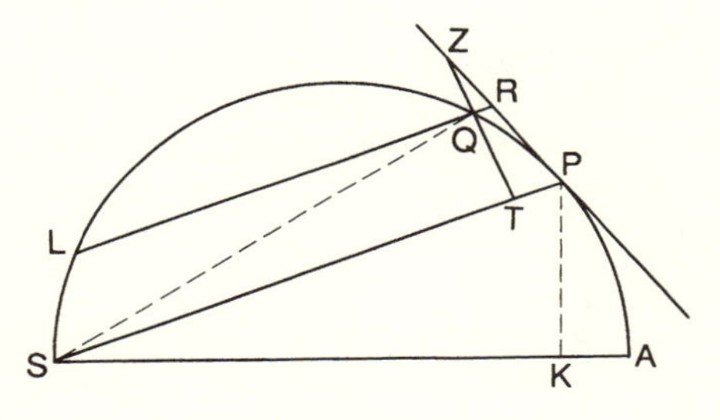

Figure 1.17

On the diameter SA and on SP drop the perpendiculars PK, QT; then through Q draw LR parallel to SP, meeting the circle in L and the tangent PR in R.[30] We shall then have RP^2 (that is, QRL,[31]) to QT^2 as SA^2 to SP^2. Therefore $QRL \times SP^2/SA^2 = QT^2$. Multiply these two equal quantities by SP^2/QR and, with the points P and Q coinciding, write SP for RL. Thus we obtain

$$SP^5/SA^2 = QT^2 \times SP^2/QR.$$

Therefore the centripetal force is inversely as SP^5/SA^2, that is (since SA^2 is given), as the fifth power of the distance SP. Which was to be found.

Scholium. In this and similar cases, however, you must conceive that after the body reaches the center S it will no more return in its orbit but will depart along the tangent. In a spiral that cuts all its radii at a given angle, the centripetal force tending toward the spiral's starting point is reciprocally as the cube of the distance; but in this starting point there is no straight line fixed in position that touches the spiral. (Add. 3965.7, fols. 56–57)

It is now a matter of finding, in different cases, "the law of the centripetal force." What does that entail? The centripetal force varies with the position of the body that is subject to it. This situation is different from that of the group of corollaries following Theorem 2 (uniform circular motion); in that case it was necessary to compare the centripetal force on several distinct, concentric orbits. Here it is the force at different points of the same orbit that must be examined. On an ellipse, parabola, or spiral, for instance, the distance to the center or focus is not constant, and the intensity of the centripetal force must vary from point to point.

The wager of the Newtonian theory is that this variation of force obeys a simple law when the body follows a particular determinate trajectory. (There is no reason that this should be the case in general.) With each point of space is associated a force analogous to weight; it is conjectured that this generalized weight varies solely as a function of the distance to a central point; it is further conjectured that the determinate form of the trajectory permits the deduction of a particular and very simple law regulating the variation of intensity of the centripetal force.

How to evaluate the variation of the centripetal force as a function of distance? The scholium after Theorem 3 proposes the method that is then applied in the three following problems: application of the general formula given in Theorem 3. Since the centripetal force at the point P is inversely proportional to $SP^2 \times QT^2/QR$, an expression equal to this product can be sought, and then, in virtue of the relations following from the situation (elliptical or circular orbit, fixed position of S in a particular point, etc.), reduced to the simplest possible form. It is necessary that the magnitudes QT and QR disappear from the final expression, since they are nascent magnitudes, and further, that this final expression contain, aside from constants, only the variable SP. It will then be clear how the centripetal force depends on the variable distance SP between the center of force and the body.

Problem 1 is quite unreal; it is rather an exercise in applying the formula. In this example, Newton examines the case of a body revolving on a circle under the action of a force that tends toward a point on the circumference itself. What happens to the body when it arrives at this point? The center of force is on the trajectory and the attraction near this point is considerable, growing as $1/r^5$. Newton briefly discusses this question in the following scholium and claims, without supporting argument, that the body will depart along the tangent after having passed the center of forces.

The demonstration assumes, first of all, the similarity of the three rectilinear triangles ZQR, ZTP, SPA (fig. 1.18). To verify this similarity, it suffices to prove the equality of the angles PZT and PSA. Newton gives no indication of a route for doing this, but several are possible. For instance, there is recourse to auxiliary angles, in showing, say, the equality of the angles PZT, YPA, OPS, and OSP (fig. 1.19). Then the homologous sides of the similar triangles are proportional:

$$ZP : ZT :: ZR : ZQ :: SA : SP.$$

Subtraction of corresponding terms in the ratios preserves the proportionality:

$$(ZP-ZR) : (ZT-ZQ) :: RP : QT :: SA : SP.$$

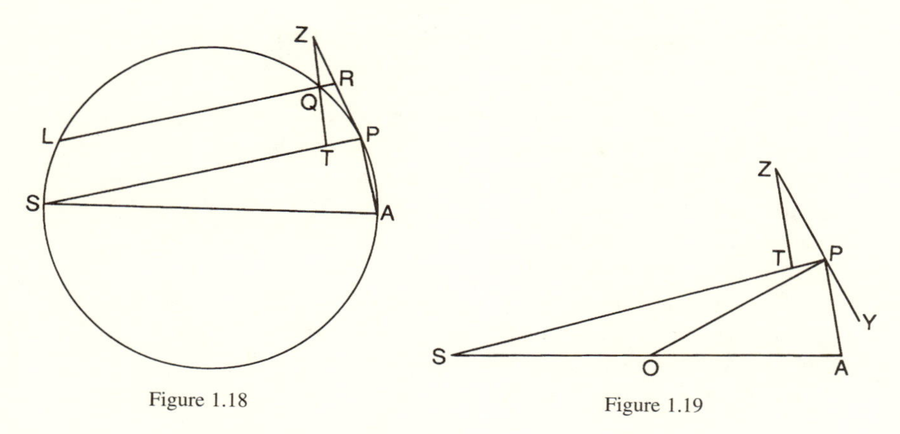

Figure 1.18

Figure 1.19

On the other hand, by Euclid III.36:

$$RP^2 = RQ \times RL,$$

and therefore, by virtue of the preceding result,

$$RQ \times RL : QT^2 :: SA^2 : SP^2$$

or

$$QT^2 = RQ \times RL \times SP^2/SA^2.$$

An expression equal to $SP^2 \times QT^2/QR$ can be obtained by multiplying both sides by SP^2/QR.

The last step consists in replacing, in the expression just obtained, the length RL by the length SP, since these two finite magnitudes differ only by a vanishing quantity when Q and P coincide. Then the expression equal to $SP^2 \times QT^2/QR$ contains only a constant of the trajectory (the diameter SA of the circle) and the variable distance SP raised to the fifth power. The centripetal force is therefore inversely proportional to r^5.

This result and this demonstration will pass almost without change into Proposition 7 of the first edition of the *Principia* then become Corollary 1 of Proposition 7 in the second edition. The case of the equiangular or "logarithmic" spiral, which is evoked in passing in the following scholium, will become Proposition 9 of the *Principia*.

Elliptical Motion: First Approach (Problem 2)

The second application of the general formula is devoted to motion on an ellipse. It is a first approach to the crucial problem; here the center of force is not yet at the focus of the ellipse but is rather in its center.

Problem 2. A body orbits in the ellipse of the Ancients: it is required to find the law of the centripetal force tending to the center of the ellipse.

Let CA and CB be the semi-axes of the ellipse, GP and DK conjugate diameters, PF and QT perpendiculars on the diameters, QV an ordinate on the diameter GP, and QVPR a parallelogram. With this construction it follows (from the *Conics*) that[32] $PVG : QV^2 :: PC^2 : CD^2$ and $QV^2 : QT^2 :: PC^2 : PF^2$, and by the compounding of ratios, $PVG : QT^2 :: (PC^2 : CD^2) \times (PC^2 : PF^2)$; that is, $VG : QT^2/PV :: PC^2 : CD^2 \times PF^2/PC^2$. Write QR for PV and $BC \times CA$ for $CD \times PF$, and in addition (as the points Q and P coincide) 2PC for VG, and when the extremes and means are multiplied into each other we obtain: $QT^2 \times PC^2/QR = 2BC^2 \times CA^2/PC$; the centripetal force is therefore inversely as $2BC^2 \times CA^2/PC$. That is to say, since $2BC^2 \times CA^2$ is given, as 1/PC. It is therefore directly as the distance PC. Which it was necessary to find. (Add. 3965.7, fol. 57)

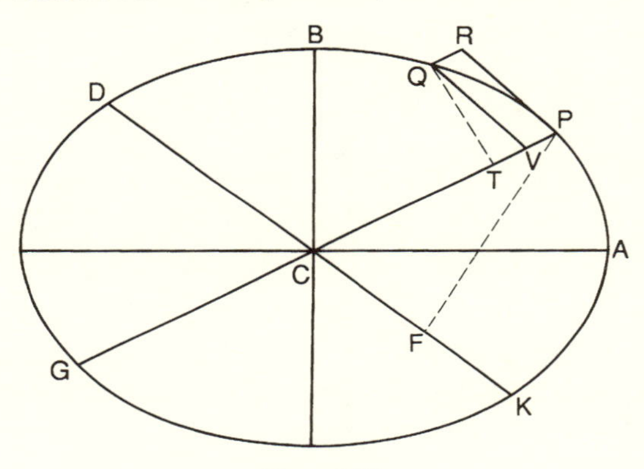

Figure 1.20

The law of the force resulting from this situation corresponds to a very important and fundamental class of forces: the forces directly proportional to the distance (nowadays called "harmonic oscillators"). Newton will treat them in the *Principia* as one of two major and important cases, the other being the case of forces varying as $1/r^2$.

In his demonstration, Newton utilizes some ordinary properties of conics. To comprehend the whole it is necessary to have an idea of what conjugate diameters are: given the tangent at a point of an ellipse, the parallel to this tangent passing through the center of the ellipse is "conjugate" to the straight line passing through the point of tangency and the center. (According to Apollonius (*Conics*, bk. 1, Def. 4, in *Opera*, 1:6), a diameter can also be defined as the locus of the midpoints of the "ordinates" parallel to a given direction.)

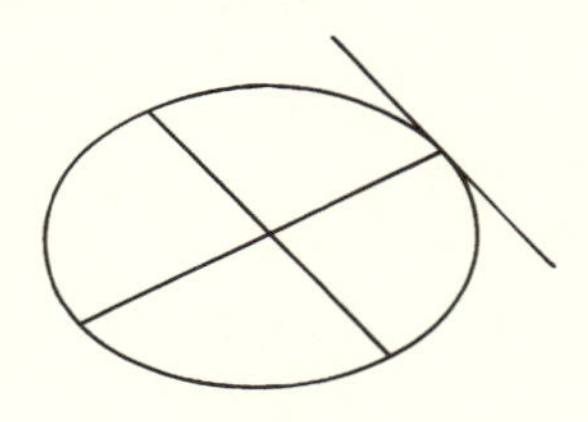

Figure 1.21

Apollonius's result (*Conics*, bk. 1, Prop. 21, in *Opera*, 1:72–74) relative to a point V taken on one of these diameters must be accepted (fig. 1.22),

$$PV \times VG : QV^2 :: PC^2 : CD^2,$$

as well as a proposition that certain versions of the *De motu* give explicitly in the form of a lemma placed after the hypotheses (Hall, Lemma 4, 245): the rectangles constructed on diameters are all equal, for example $CD \times PF = BC \times CA$ (see fig. 1.23).

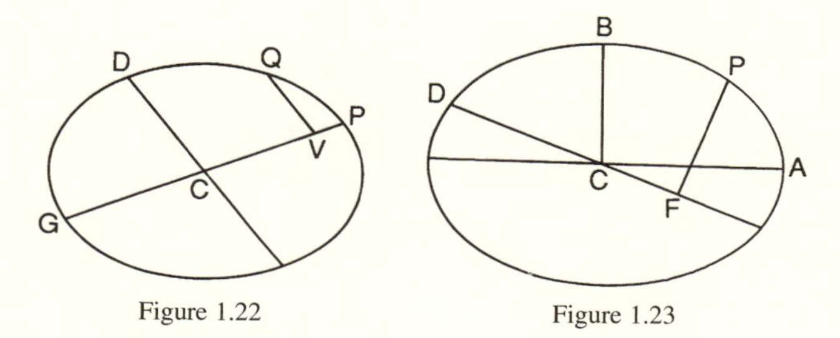

Figure 1.22 Figure 1.23

Because QRPV (fig. 1.20) is a parallelogram (by construction), and RP is a tangent corresponding to DK (therefore parallel to this diameter), QV is parallel to CF; hence the angles QVC and VCF are equal and the right-angled triangles QTV and PFC are similar. Whence follows the proportion QV : QT :: PC : PF.

Finally, as Q tends to P ("as the points Q and P coincide," *punctis Q et P coeuntibus*), VG can be replaced by PG or 2PC.

Thus is obtained the inverse expression of the force, that is, of $PC^2 \times QT^2/QR$ (in which C plays the role of S in the general formula, the letter S being reserved for the focus which will become the center of force in the next proposition). This expression is equal to constants divided by PC. The force is therefore directly proportional to the distance.

The demonstration of this result combines the very classical geometry of conics with the procedures of "ultimate" substitution. When the later position Q of the body approaches its initial position P, then in the relations demonstrated by Apollonius it is possible to replace certain segments by others. In the ratio $VG : QT^2/PV$, Newton retains the vanishing magnitudes QT and PV (which correspond to QT and QR in the general formula of Theorem 3), and he replaces VG by PG, because these two quantities differ only by the vanishing magnitude PV.

Newton employs the language of proportions in its traditional form. In particular, he avoids confusing ratios with quotients (I have sought to maintain the distinction by writing a:b for the former and a/b for the latter). Thus the proportionality Newton arrives at makes use of both relations while keeping them distinct; in Newton's Latin[33] it appears as:

$$\text{VG ad QT}^{q}/\text{PV ut PC}^{q} \text{ ad CD}^{q} \times \text{PF}^{q}/\text{PC}^{q}$$

and has been transcribed as:

$$\text{VG} : \text{QT}^2/\text{PV} :: \text{PC}^2 : \text{CD}^2 \times \text{PF}^2/\text{PC}^2.$$

Elliptical Motion with the Force Tending to the Focus (Problem 3)

Here finally is the famous proposition that was the foremost aim of the entire work: if the planets orbit on ellipses with the sun at one focus, the force varies as r^{-2}.

> *Problem 3.* A body orbits in an ellipse. It is required to find the law of the centripetal force tending toward the focus of the ellipse.[34]
>
> Let S be the superior focus of the ellipse, and let SP be drawn cutting the diameter DK of the ellipse in E and the line QV in X, and let the parallelogram QXPR be completed.

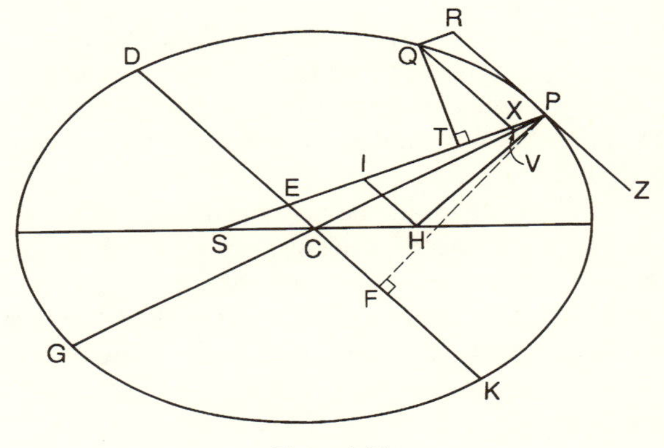

Figure 1.24

> It is evident that EP is equal to the semimajor axis AC, for if, starting from the other focus H of the ellipse, we draw the line HI parallel to EC, then because of the equality of CS and CH, the lines ES and EI will be equal; so that EP is the half-sum of PS and PI, that is (because of the parallels HI, PR and the equal angles IPR and HPZ) the half-sum of PS and PH which taken together are equal to the whole axis 2AC.
>
> On SP let fall the perpendicular QT. Then, calling L the principal latus rectum of the ellipse (that is to say, $2\text{BC}^2/\text{AC}$), we shall have
>
> $$\text{L} \times \text{QR} : \text{L} \times \text{PV} :: \text{QR} : \text{PV} :: \text{PE (or AC)} : \text{PC}$$
> $$\text{and L} \times \text{PV} : \text{GVP} :: \text{L} : \text{GV}$$
> $$\text{and GVP} : \text{QV}^2 :: \text{CP}^2 : \text{CD}^2;$$

and, say, $QV^2 : QX^2 :: M : N$,

and $QX^2 : QT^2 :: EP^2 : PF^2$, that is, $CA^2 : PF^2$ or $CD^2 : CB^2$.

When all these ratios are compounded, we shall have

$$L \times QR : QT^2 :: (AC : PC) \times (L : GV) \times (CP^2 : CD^2) \times (M : N) \times (CD^2 : CB^2)$$
$$:: (AC \times L \text{ or } 2BC^2 : PC \times GV) \times (CP^2 : CB^2) \times (M : N)$$
$$\text{or} \quad :: (2PC : GV) \times (M : N).$$

But as the points Q and P coalesce, the ratios $2PC : GV$ and $M : N$ become ratios of equality, so that $L \times QR$ and QT^2 are equal. Multiply both sides by SP^2/QR and we shall have $L \times SP^2 = SP^2 \times QT^2/QR$.

Therefore the force is inversely as $L \times SP^2$, that is, reciprocally in the double ratio of the distance. Which was to be found.

Scholium: Therefore the major planets orbit in ellipses having a focus at the center of the Sun; and by radii drawn to the Sun describe areas proportional to the times, just as Kepler supposed. And the latera recta of these ellipses are QT^2/QR, the points P and Q being distant from one another by the least possible and as it were an infinitely small distance. (Add. 3965.7, fol. 57)

The same figure (1.24) is used as for the preceding proposition. This time the centripetal force tends towards S, the focus of the ellipse. The task is the same as in the preceding problems: to find an expression for the centripetal force at any point P; that is, an expression equal to $SP^2 \times QT^2/QR$, containing, besides various constants, only the variable SP. To do this, it is necessary to transform the expression of the general formula in terms of the characteristic magnitudes of the ellipse, such as the length of an axis or the latus rectum or parameter which depends on the quotient of the axes (the latus rectum, which Newton designates L, is equal to $2BC^2/AC$).

Newton first proves an auxiliary result: EP is equal to the semimajor axis (E is on the diameter KD conjugate to PC. See fig. 1.24.). He does this by applying the theorem of Thales to the triangles SEC and SIH and by using the "optical" property of tangents to an ellipse: the tangent reflects toward the focus H the rays issuing from the focus S, so that the angle SPR equals the angle HPZ (see Apollonius, *Conics*, bk. 3, Prop. 48, in *Opera*, where the property is stated without optical connotation). Hence,

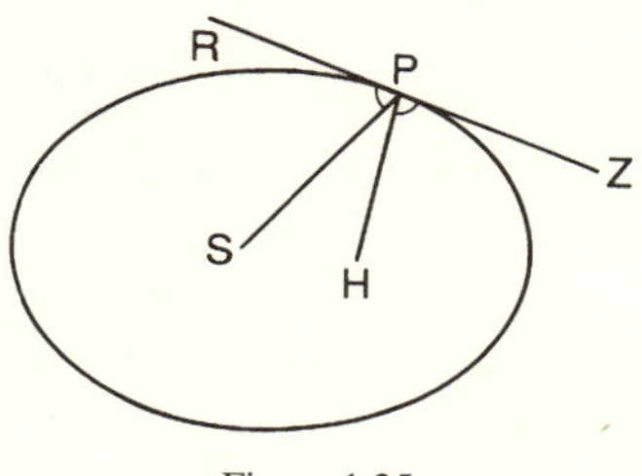

Figure 1.25

$$PE = \tfrac{1}{2}(PS + PI) = \tfrac{1}{2}(PS + PH) = \text{the semimajor axis.}$$

What follows is a disconcerting sequence of ratios. Newton announces, one after another, five different proportionalities, which he finally combines. The sequence is incomprehensible, until it is seen that the ratios on the left are so placed in succession that in the subsequent compounding of ratios the intermediate terms are eliminated, leaving only the first and last terms. Thus if the ratio a : b is compounded with the ratio b : c, then with c : d, d : e, and finally e : f, the whole combination reduces to the ratio a : f.[35]

Newton gives in succession (with L denoting the latus rectum) the following proportions:

1. $L \times QR : L \times PV :: (QR : PV = PE : PC =)\ AC : PC$.
2. $L \times PV : GV \times PV :: L : GV$.
3. $GV \times PV : QV^2 :: CP^2 : CD^2$.
4. $QV^2 : QX^2 :: M : N$.
5. $QX^2 : QT^2 :: (EP^2 : PF^2 = CA^2 : PF^2 =)\ CD^2 : CB^2$.

In line 1, the first equality is justified by the similarity of the triangles PVX and PCE, and the second equality results from the auxiliary demonstration: $PE = AC$. Line 2 is justified because the value of a ratio is not changed when the antecedent and the consequent are multiplied by the same quantity. Line 3 gives us the property of conics already employed in the preceding proposition (Apollonius, *Conics*, bk. 1, Prop. 21, in *Opera*). Line 4 displays a simple notational convention, designating by M : N a ratio that becomes equal to 1 as Q tends toward P (Newton will eliminate this notation in the *Principia*; see *NMP*, 6:47 n. 46–47). Finally, line 5 is justified by the similarity of the triangles QTX and PFE and by the equality of the "rectangles" $CD \times PF$ and $CA \times CB$.

The compounding of all these ratios results in:

$$L \times QR : QT^2 :: (2PC : GV) \times (M : N).$$

The two ratios compounded on the right both become ratios of equality when Q coincides with P. Therefore the ratio of $L \times QR$ to QT^2 is also a ratio of equality. Multiplying the two sides of the equality by SP^2/QR yields the expression of the inverse of the force:

$$SP^2 \times QT^2/QR = L \times SP^2.$$

L is a constant of the ellipse. Therefore the centripetal force at the point P is inversely as the square of the distance.

The miracle of this proof is somewhat obscured in two respects. In the first place, the machinery of the demonstration, for the first time, is ponderous; but the importance of the result can overcome the cumbrous character and the detours of the proof. In the second place, the scholium is unquestionably improper: how, starting from what has just been demonstrated, can Newton conclude that the planets revolve in ellipses with the sun at one focus? This is to confound the direct problem with the inverse problem.

This scholium sounds like a conclusion. Newton seems to have wanted to assemble in a few lines the "three laws" of Kepler: the elliptic trajectory, the equality of the areas swept out in equal times, and—in Theorem 4, two lines after the scholium—the proportionality between the squares of the periods and the cubes of the major axes. The list of the three laws is thus complete within a few lines.[36]

The end of the scholium mentions a result that will be used immediately. Since $L \times SP^2 = SP^2 \times QT^2/QR$, there is a relation between the parameter of the ellipse and the nascent, or "so to speak infinitely small" elements of the trajectory:

$$L = QT^2/QR.$$

From the form of the nascent arc, the form of the complete ellipse can be inferred.

The Demonstration of Kepler's Third Law (Theorem 4)

Here now are the complements that Newton adds to his principal result. After having proved that the elliptical orbit is traversed under the action of a centripetal force varying as $1/r^2$, he shows that the third law of Kepler is a consequence of this force and then adds some further results concerning the determination of orbits.

> *Theorem 4.* Supposing that the centripetal force is inversely proportional to the square of the distance from the center, the squares of the periodic times are as the cubes of the transverse axes.
>
> Let AB be the transverse axis of the ellipse, PD the other axis, L the latus rectum, S one of the two foci. From the center S, with the radius SP, describe the circle PMD. And in the same time, let the two orbiting bodies describe the arc PQ of the ellipse and the arc PM of the circle, by virtue of the centripetal force that tends towards the focus S. Let the ellipse and the circle have PR and PN for tangents at the point P. Draw QR and MN parallel to PS, meeting the tangents in R and N.
>
> Let the figures PQR and PMN be indefinitely small, so that (by the Scholium to Problem 3) there comes to be $L \times QR = QT^2$ and $2SP \times MN = MV^2$. Because of their common distance SP from the center, and the resulting equality in the centripetal forces, MN and QR are equal. Therefore QT^2 is to MV^2 as L is to 2SP, and so QT is to MV as the mean proportional between L and 2SP (that is, PD) is to 2SP.
>
> Hence the area SPQ is to the area SPM as the total area of the ellipse is to the total area of the circle. But the parts of these areas generated in individual moments are as the areas SPQ and SPM, and therefore as the total areas, and consequently, when they are multiplied by the number of the moments, they will become simultaneously equal to the total areas.
>
> Therefore the revolutions are accomplished in the same time on ellipses and on

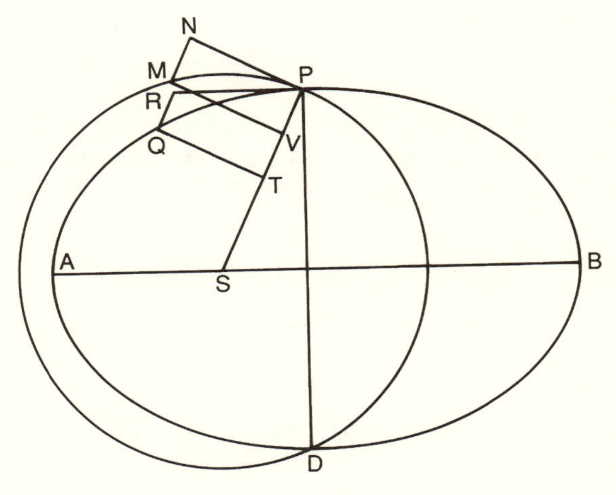

Figure 1.26

circles whose diameters are equal to the transverse axes of the ellipses. But (by Corollary 5 of Theorem 2) the squares of the periodic times on circles are as the cubes of the diameters. Hence also on ellipses. Which was to be proved. (Add. 3965.7, fol. 58)

Kepler's third law has already been introduced—in the Scholium of Theorem 2 it was admitted as a fact of observation and resulted in the conclusion that the force varied as r^{-2}. Here, on the contrary, it is deduced from the law of force.

Another important difference comes from the form of the orbit: previously the trajectories were concentric circles, now they are ellipses. But the case concerning circles serves as the point of departure in the reasoning. What was said previously as to the variation of the force in the case of concentric circular orbits is here extended to ellipses. The objective in this theorem thus is to transfer to ellipses what was previously proved for circles.

A method of this kind will frequently be employed in the *Principia*: to evaluate a force, Newton takes as reference a fictive body subject to this force (it could be called a test particle placed in the field) and evaluates the real orbit starting from knowledge of the simpler fictive orbit.

In the present case, the period of an imaginary body moving on a circle of radius SP is compared with that of the actual body moving on the ellipse with focus S. Point P is common to the two orbits (it is not just any point whatever of the ellipse but is the endpoint of the minor axis: the circle therefore has a diameter equal to the major axis of the ellipse). Newton proves that the period of the circular motion around S is the same as that of the elliptical motion.

To demonstrate it, he employs a rather remarkable procedure: the comparison of increments. The reasoning turns on the parts of area generated in the same very small interval of time (a "moment"). If the parts generated in one and the same instant are known, the total magnitudes can be compared—more precisely, the times required to generate the total magnitudes can be compared. The reasoning is rather ordinary: the elementary parts are multiplied by the number of moments (*partes . . . singulis momentis genitae . . . per numerae momentorum multiplicatae* [ibid.]). On the other hand, these parts are infinitely small, and the number of moments is infinite (*Sint autem figurae PQR, PMN indefinitae parvae* [ibid.]).

This manner of reasoning is not far removed from the method of fluxions. Here, as in his fluxional treatises, Newton compares the instantaneous increments or the velocities of generation of several magnitudes (see below, p. 209ff.) then makes his way back from the contemporaneously generated parts to the total magnitudes. (The details concerning these various procedures can be found later in this book (chap. 3).) There is, however, an essential difference: the fluxional reasonings are more rigorous and do not involve multiplying the instantaneous elements by the number of instants; in other words, the element or fluxion remains incomparable with the finite magnitude.

Here the ratio between the contemporary increments is the same as that between the total magnitudes: the triangles or sectors SPQ and SPM are to one another as the whole ellipse is to the whole circle. Since, on the other hand, these elements of area are described in the same interval of time, the ellipse and the circle are "finished off" in the same time, after the same number of intervals. The period is therefore the same for the circle as for the ellipse: "the parts of these areas generated in individual moments are as . . . the total areas, and consequently, when they are multiplied by the number of the moments, they will become simultaneously [*simul*] equal to the total areas."

How to demonstrate the proportion

$$\text{area SPQ} : \text{area SPM} :: \text{elliptic area} : \text{circular area?}$$

The two triangular sectors have the same base SP, hence they are to one another as their heights QT and MV. It therefore suffices to show that the ratio of these heights is the same as that of the area of the ellipse to the area of the circle:

$$QT : MV :: \text{ellipse} : \text{circle}.$$

In the case of the circle, since the latus rectum is simply the diameter 2SP, the equation between the latus rectum and the expression QT^2/QR (from the last lines of the preceding proposition; it is here that the hypothesis of an inverse-square force intervenes)

$$L = QT^2/QR,$$

becomes

$$2SP = MV^2/MN.$$

The next step is based on physics or dynamics. Because the force is the same—the distance SP being the same—the deviation will be the same in the same interval of time, and therefore MN = QR. Newton is here again implicitly making use of the proportionality between force and change of motion, which will be enunciated in Law II.

The ratio between the latera recta can then be expressed more simply as

$$\mathrm{L} : 2\mathrm{SP} :: \mathrm{QT}^2 : \mathrm{MV}^2.$$

Newton completes the reasoning by speaking of the mean proportional between L and 2SP, that is,

$$\sqrt{\mathrm{L} \cdot 2\mathrm{SP}}.$$

The modern reader will no doubt prefer a different mode of presentation. If a and b represent the axes of the ellipse, by definition, $\mathrm{L} = 2b^2/a$ and $2\mathrm{SP} = 2a$ (because the sum of the distances from a point on an ellipse to the two foci is equal to the major axis). Hence,

$$2b^2/a : 2a :: \mathrm{QT}^2 : \mathrm{MV}^2$$

or

$$\mathrm{QT} : \mathrm{MV} :: b : a.$$

In other words, the ratio of the areas of the infinitesimal sectors $\frac{1}{2}(\mathrm{SP} \times \mathrm{QT})$ and $\frac{1}{2}(\mathrm{SP} \times \mathrm{MV})$ is exactly the ratio of the ellipse to the circle constructed on the major axis. The period is therefore the same, and what is true of circles is also true of ellipses: the square of the periodic time is proportional to the cube of the major axis.

The Determination of Orbits

The relation between the period and the size of the orbit is put to use in the following scholium in order to determine the trajectories of planets. The question is of an exclusively geometric nature: how to trace an ellipse of which the major axis and certain points are known?

> *Scholium.* Hereby, in the celestial system, from the periodic times of the planets may be known the proportions of the transverse axes of the orbits. Let one axis be assumed; then the others are known.
>
> But from given axes the orbits may be determined in the following way. Let S be the place of the Sun, that is, one of the foci of the ellipse; let A, B, C, D be positions of the planet found by observation, and Q the transverse axis of the ellipse.
>
> From the center A, with the radius Q – AS, draw the circle FG, and the other focus of the ellipse will lie on this circumference. Similarly, from the centers B,

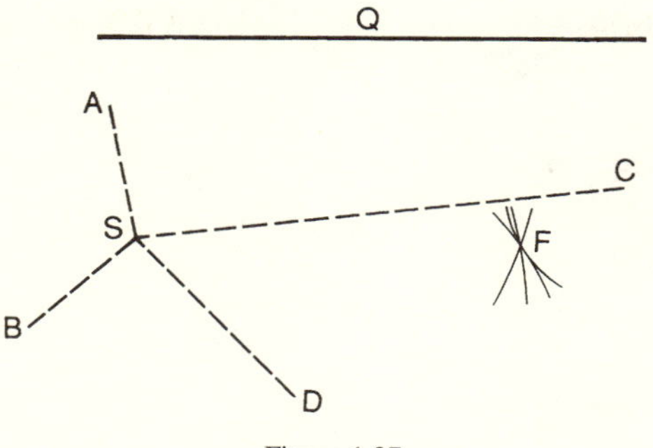

Figure 1.27

C, D, etc. with radii Q – BS, Q – CS, Q – DS, etc., let there be described in the same way any number of other circles, and the other focus will lie on all the circumferences, and therefore in the common intersection F. If all the intersections do not coincide, it will be necessary to take the mean point as focus.

The advantage of this procedure is that for eliciting one conclusion a large number of observations can be applied and easily compared together.

How single places A, B, C, D, etc. of a planet may be determined from two observations, if one knows the great orb [*magnus orbis*] of the Earth, has been shown by Halley. If this great orb has not yet been determined exactly enough, one can, starting from this orbit approximately known, determine more exactly the orbit of a planet, say Mars, then by the same method determine more exactly the orbit of the Earth starting from the orbit of the planet. Then from the orbit of the Earth will be determined the orbit of the planet much more exactly than before; and so on turn and turn about until the intersections of the circles meet in the foci of both orbits exactly enough.

By this method the orbits of the Earth, Mars, Jupiter and Saturn may be determined; but those of Venus and Mercury may be determined as follows.

From observations of the maximum digressions of the planets from the Sun, the tangents of the orbits are obtained. To such a tangent KL from the Sun is dropped a perpendicular SL, and with center L and radius equal to the semi-axis of the ellipse, the circle KM is described. The center of the ellipse will lie on this circle;

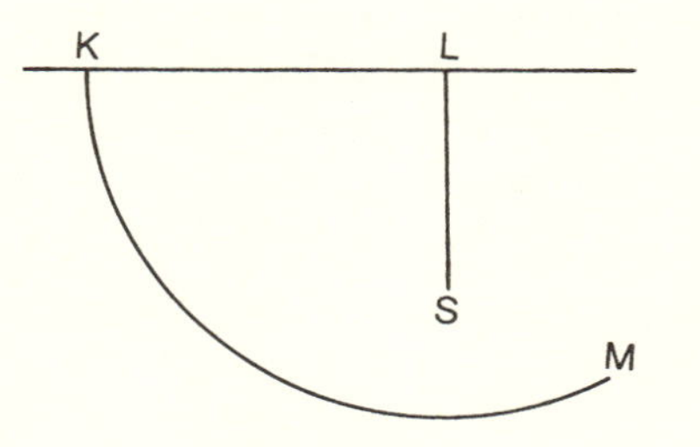

Figure 1.28

> and when several circles of this sort have been described, we shall find the center at their common intersection.
>
> Finally, when the dimensions of the orbits are known, the longitudes of these planets will then be determined very exactly by their transits across the disk of the Sun. (Add. 3965.7, fols. 58–59)

Newton does not explain the underlying astronomical procedures and confines himself to geometrical reasonings. One of the foci of the ellipse, that corresponding to the Sun, is known, as are the length of the major axis—represented by the length Q—and several successive positions of the planet—A, B, etc., determined observationally (for example, as Newton suggests, by the method proposed by Halley in 1676; see *NMP*, 6:52 n. 63). To find the other focus, it is sufficient, with these points A, B, etc., as centers, to draw circles having for radii the major axis minus the distance of the point from the focus S.

The significance of the passage concerning tangents is less clear: what is a tangent in terms of astronomy? Is it a matter solely of tangents at the aphelion, that is, at the highest point of the ellipse, as suggested by the mention of the maximal elongation from the sun? (If L is the aphelion of the ellipse, found by dropping the perpendicular from S onto the tangent KL, it is clear that the center of the ellipse will be in the direction LS at a distance from the aphelion equal to the semimajor axis; but this is not what Newton says. If KL is any tangent whatever, then L is not a point of the ellipse, and it is necessary to infer a much more complex reasoning, such as that proposed by Whiteside; see *NMP*, 6:53 n. 65.)

Newton will return to these procedures and complete them in sections 4 and 5 of the *Principia*, which form a small treatise on conics almost independent of the rest of the work.

The Inverse Problem, Restricted to Conics (Problem 4)

It is Problem 4 in the *De motu* that comes closest to responding to Halley's question: if the force varies inversely as the square of the distance, what will be the trajectory? The response, however, is restricted in import since the very enunciation of the problem presupposes that the orbit will be an ellipse ("there is required the ellipse"). Nevertheless, the ensuing discussion deals with the various possible forms of trajectory: by modification of certain of the quantities initially given, other conics are obtained.

> *Problem 4*. Supposing that the force is inversely proportional to the square of the distance from its center, and with the quantity of that force known, there is required the ellipse that a body will describe if it is launched from a given place with a given velocity along a given straight line.

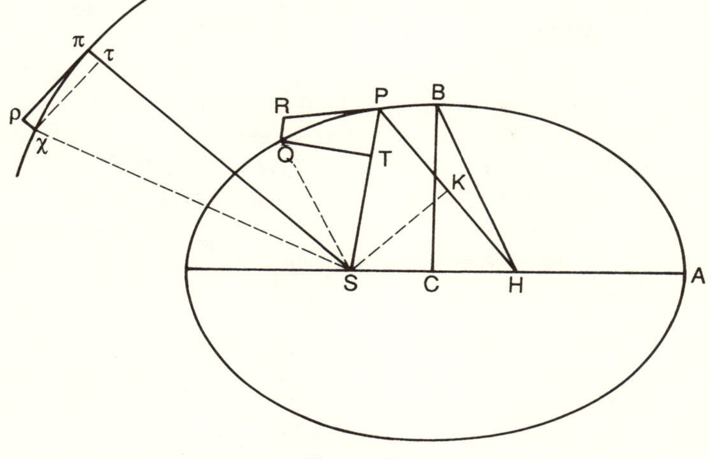

Figure 1.29

Let the centripetal force tending toward the point S be such that it would cause the body π to orbit in the circle $\pi\chi$ described with center S and any radius $S\pi$. Let the body P be launched from the point P along the line PR, and directly thereafter under the action of the centripetal force be deflected into the ellipse PQ. The straight line PR will therefore be tangent to the ellipse at P. Let the straight line $\pi\rho$ correspondingly touch the circle at π, and let PR be to $\pi\rho$ as the initial velocity of the body P when launched to the uniform velocity of the body π. Parallel to SP and $S\pi$ draw RQ and $\rho\chi$, meeting respectively the circle in χ and the ellipse in Q, and from Q and χ to SP and $S\pi$ let fall the perpendiculars QT and $\chi\tau$.

QR is to $\chi\rho$ as the centripetal force in P is to the centripetal force in π, that is, as $S\pi^2$ is to SP^2, and therefore their ratio is given. Also given is the ratio of QT to $\chi\tau$, because the ratios QT : RP and RP : $\chi\tau$ are given, and hence also given is the ratio compounded of them. Let this latter ratio squared be divided by the given ratio of QR to $\chi\rho$, and there will remain the ratio of QT^2/QR to $\chi\tau^2/\chi\rho$, that is (by the scholium of Problem 3), the ratio of the latus rectum of the ellipse to the diameter of the circle.

Therefore the latus rectum of the ellipse is given; let us call it L. There is given, furthermore, the focus S of the ellipse. Let the complement to two right angles of the angle RPS be RPH, and there will be given in position the line PH, in which is located the other focus. On letting fall the perpendicular SK to PH and erecting the semiminor axis BC, we shall have:[37]

$$\begin{aligned} SP^2 - 2KPH + PH^2 = SH^2 = 4CH^2 &= 4BH^2 - 4BC^2 \\ &= (SP + PH)^2 - L(SP + PH) \\ &= SP^2 + 2SPH + PH^2 - L(SP + PH). \end{aligned}$$

Add $2KPH + L(SP + PH) - SP^2 - PH^2$ to each side and we shall have: $L(SP + PH) = 2SPH + 2KPH$, that is, $SP + PH$ is to PH as $2SP + 2KP$ to L.

Whence the other focus H is given. But given the foci along with the transverse axis SP + PH, the ellipse is given. As was to be found.

Matters are thus when the figure is an ellipse. For it can happen that the body moves in a parabola or hyperbola. Indeed if the velocity of the body is such that the latus rectum L is equal to 2SP + 2KP, the figure will be a parabola having its focus at the point S, and all its diameters parallel to the line PH.

But if the body is projected with a yet greater velocity, it will move in a hyperbola having one focus at the point S and the other focus on the other side of the point P, with the transverse axis equal to the difference between the lines PS and PH. (Add. 3965.7 fols. 59–60)

The great interest of this text resides in the geometrical translation of all the magnitudes under discussion. The intensity of the force is symbolized by a circular orbit of arbitrarily fixed diameter (the masses never enter into these reasonings). The velocities are represented by segments tangent to the orbits, and the forces by deflections parallel to the *radii vectores*.

It is supposed that the body is projected from the point P with a velocity PR, so that PR is to $\pi\rho$ as the velocity in P—Newton calls it "the initial velocity" (*prima celeritas*)—is to the constant velocity on the circle, the latter being fixed since the circular motion serves to translate the determinate intensity of the force. Newton could have said that PQ and $\pi\chi$ are traversed in the same very small interval of time, as in the preceding propositions, but he preferred to speak here of velocities.

Since the force depends on the distance, and its intensity is supposedly known for the distance $S\pi$, it is also known in P. Therefore the deflections RQ and $\rho\chi$ are in a given ratio—the inverse ratio of SP^2 to $S\pi^2$. The ratio of QT to QR is also given since the angle SPR and the segments PR and QR are also determined. By a series of steps, the ratio of QT to $\rho\pi$ is therefore given. On the other hand, $\chi\tau$ is equal to $\rho\pi$ (because the orbit is a circle). Therefore the ratio QT : $\chi\tau$ is given. The ratio $QT^2/QR : \chi\tau^2/\chi\rho$ is also given, and it is equal to the ratio of the latera recta (scholium of Theorem 3). The diameter of the circle (which is equal to its latus rectum) was fixed arbitrarily at the beginning; hence the latus rectum of the ellipse is also known. One focus and one tangent of the ellipse are also known.

Thus, starting from a fictive circular trajectory that characterizes the quantity of the force, the elliptical orbit corresponding to a position P and an initial velocity PR is in principle determined.

The ellipse can be in fact described given the focus S, the latus rectum L, and the angle SPR between the tangent and the radius vector. To construct it, Newton employs what is called the optical property of conics (see p. 39 above), which gives the direction PH along which is found the second focus. Then he expresses the distance PH as a function of SP, L, and the perpendicular SK let fall from S onto the direction PH (the minor axis BC is only an intermediary in the reasoning and disappears from the final expression; it is not supposed known). The result is expressed by

$$SP + PH : PH :: 2SP + 2KP : L.$$

The geometric style of the reasoning does not preclude a discussion of other possible figures. The point of departure is the proportionality that has been enunciated, which must remain valid and retain a significance in the case of orbits other than the ellipse. D. T. Whiteside, however, remarks that Newton has not yet demonstrated that the quotient QT^2/QR remains equal to the latus rectum in the cases of the parabola and the hyperbola (*NMP*, 6:56 n. 73).

The discussion turns on what can happen to the fraction (SP + PH)/PH. If this quotient equals 1, then SP must be negligible in relation to PH, or PH is then infinitely great, and the second focus is pushed back to infinity, which transforms the ellipse into a parabola. If the quotient (SP + PH)/PH is less than 1, it is necessary that SP + PH be smaller than PH; then it is not the sum of the distances to the foci but rather their difference that is constant, and the body must traverse a hyperbola.

This quotient depends on the initial conditions, as may be seen from what happens to point K in different possible configurations. But it is not easy in this case—in contrast to an analytico-algebraic mode of presentation employing differential equations and constants of integration—to see how the initial position and velocity enter into the expression (2SP + 2KP)/L. Moreover, the reasoning depends on the presupposition that the trajectory must be a conic section.

The Orbits of Comets and Kepler's Problem

Newton is pursuing the exposition of what might be called a program: astronomy reformulated in dynamical terms. The study of centripetal forces should make it possible to determine more exactly the motion of planets, to decide whether comets may return, and finally to evaluate the orbital arcs traversed after a given time. The last problem ("Kepler's problem") reduces to cutting from an ellipse a sector of given area.

> *Scholium.* By virtue of the solution of this problem it is possible to define the orbits of comets and consequently their periods of revolution; and from a comparison of their orbital magnitudes, eccentricities, aphelia, inclinations on the plane of the ecliptic, and nodes, it is possible to ascertain whether the same comet returns with some frequency to us.
>
> Indeed, starting from four observations of a comet's positions, under the hypothesis that the comet is moving in a straight line, we must determine its rectilinear path. Let us call it APBD, with A, P, B, D the positions of the comet in that path at the times of observation and S the position of the Sun.

Imagine that the comet, with the same velocity with which it traverses uniformly the straight line AD, is projected from one of the positions P, and, subjected immediately to the centripetal force, is deflected from the rectilinear path, going off in the ellipse Pbda. This ellipse is to be determined as in the above problem. In it let a, P, b, d be the positions of the comet at the times of observation. The longitudes and latitudes of these places as seen from the Earth are known. To the extent that the observed longitudes and latitudes are greater or less than these, by so much take new longitudes and latitudes greater or less than the observed ones, and from these new ones let the comet's rectilinear path be found anew, and from this, as before, the elliptical path. And the four new positions on the elliptical trajectory, increased or diminished by the previous errors, will now agree exactly enough with the observations. Or if perchance the errors still remain sizable, the whole process can be repeated. And in case astronomers find the computations troublesome, it will be sufficient to determine all these things by a geometrical procedure.

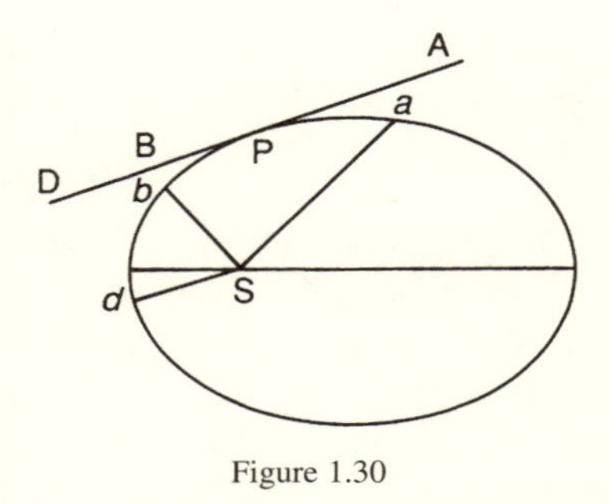

Figure 1.30

But it is difficult to assign areas aSP, PSb, bSd proportional to the times. On the major axis of the ellipse describe the semicircle EHG. Take the angle ECH proportional to the time. Draw SH and CK parallel to it and meeting the circle in K. Draw HK and take a triangle SKN equal to the segment HKM of the circle (by means of a table of segments or otherwise). On EG let fall the perpendicular NQ, and in it take PQ to NQ as the minor axis of the ellipse is to the major axis; the point P will then be on the ellipse, and the straight line PS once drawn will cut off from the ellipse an area EPS proportional to the time.

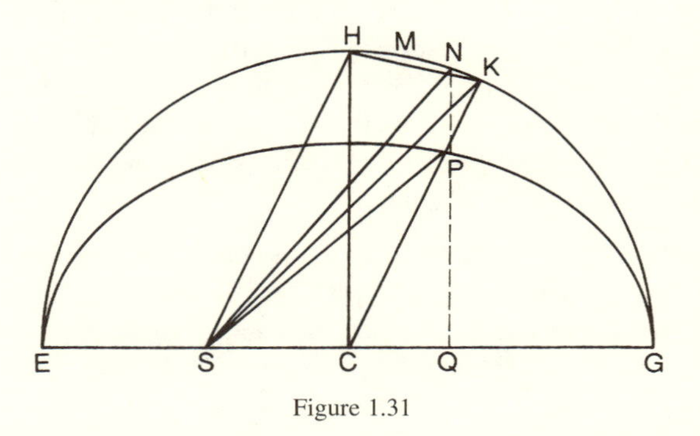

Figure 1.31

For the area HSNM, augmented by the triangle SNK and diminished by the segment HKM which is equal to it, becomes equal to the triangle HSK, that is, to the triangle HSC. By adding these equal magnitudes to the area ESH, we shall obtain the equal areas EHNS and EHC.

Consequently, since the sector EHC is proportional to the time, and the area EPS is proportional to the area EHNS, the area EPS will also be proportional to the time. (Add. 3965.7, fols. 60–61)

The first part of the scholium concerns the trajectories of comets. Newton supposes four observations made at different dates, from which a rectilinear trajectory can be deduced (for instance, by means of the procedures that Newton gives in his treatise on algebra, *Arithmetica universalis*, Problem 56). This straight line is a first approximation; it serves as a tangent to the trajectory at one of these points: if the comet leaves point P in the direction PBD, it is immediately "deflected" onto an incurved orbit, which can be determined by means of the preceding problem. It can be calculated where on this new trajectory the comet must be at the dates considered (this calculation presupposes knowing how to evaluate the arc of the ellipse or parabola traversed in a given time, that is, how to cut off a sector aSP of given area, a procedure to be explained in the second part of the scholium).

On this ellipse, for the four dates, are thus calculated new longitudes and latitudes (as seen from the earth) that do not agree with the values actually obtained from observation. Then corrections are made (the meaning of the original Latin phrase is not perfectly clear) in order to obtain new positions, which make possible the calculation of a new rectilinear trajectory, then an ellipse, and so on. The procedure supposes that the distance traversed along the ellipse as a function of time can be determined: where will the body be on the ellipse after such and such an interval? For this the law of areas is needed, again, but this time in a practical form that is concretely applicable.

It is with the approximate solution of "Kepler's problem" that the remainder of the Scholium is concerned: namely, how to cut from an ellipse a sector centered on an eccentric point S, the area thus obtained being to the whole area of the ellipse in a given ratio. Actually it is a circle rather than an ellipse that is to be considered: the result is the same whether it is the position P on the ellipse or the position N on the associated circle (which is constructed on the major axis as diameter) that is determined.[38] In formulating the problem all reference to the ellipse can be omitted: from a semicircle EHG is cut an eccentric sector ESN whose area is a certain part of the total area EHG.

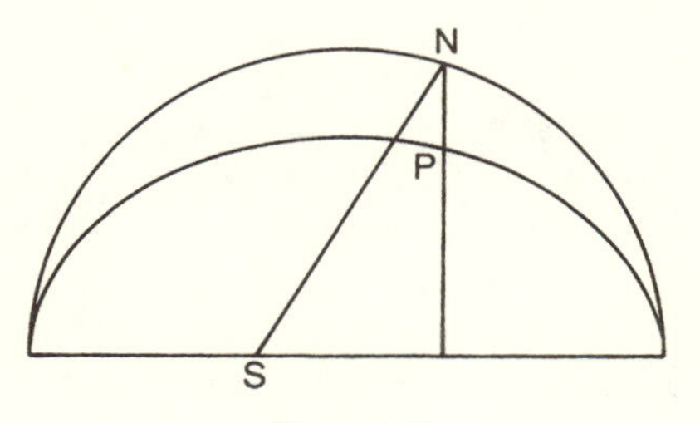

Figure 1.32

The *Principia* will give several approximate solutions of this problem. Here, in the first versions of *De motu*, Newton supposes an angle at the center described uniformly, with the uniform rotation representing time, like the "mean motion" of classical astronomy.

Suppose that the point C, at the middle of the axis EG, serves as center for the uniform motion (as "equant point"): the rotation of the ray CH, that is the variation of the area ECH, will serve as a measure of the time (ECH has to the entire circle the same ratio as the time required to traverse arc EH does to the entire period).

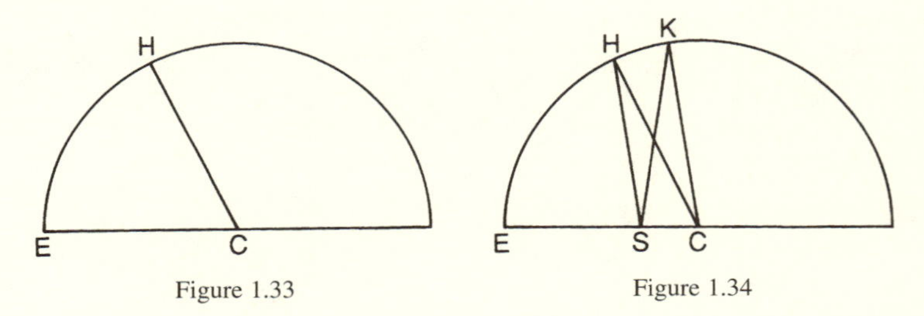

Figure 1.33 Figure 1.34

The area of the sector centered at C is proportional to the time. It is necessary now to find a sector centered at S with the same area as ECH, and thus replace the central sector with an eccentric sector. The area ESH is a little too small, it lacks the area SHC. It would therefore suffice to add to ESH a sector centered at S and equal to the difference SHC.

Drawing CK parallel to SH will give approximately what is needed: the area SHK is nearly equal to the area SHC. It lacks a little: the area SHC is entirely rectilineal, while the area SHK is curved between H and K. It would suffice to be able to subtract from SHK a very small sector, centered in S, and equal to the segment HMK. Newton supposes that the reader knows how to determine a sector SKN equal to this segment.

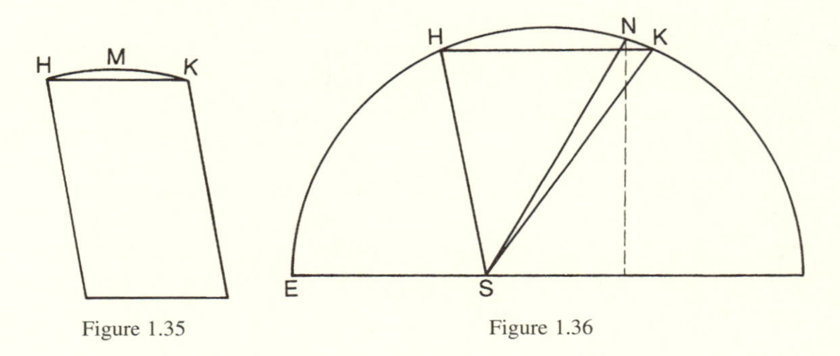

Figure 1.35 Figure 1.36

Adding to the sector ESH another sector HSK diminished by KSN yields a sector ESN equal in area to ECH and therefore proportional to the time. N is therefore the point sought (at least on the associated circle; one must then return to the ellipse at P).

The Ellipse Deformed into a Vertical Trajectory

Newton has shown how the law of areas is to be used practically for calculating arcs traversed on elliptical orbits. But if the orbit is rectilinear, how to evaluate the distance traversed in a given time? This is the case of a body falling directly toward the center of force or moving directly away from it in a straight line. The law of areas remains still applicable and useful by virtue of a mathematical artifice that is very characteristic of the Newtonian style: a procedure of deformation that enables consideration of the rectilinear trajectory as an infinitely flattened ellipse, the "ultimate" state of an ellipse that is progressively diminished in width.

> *Problem 5.* Supposing that the centripetal force is inversely proportional to the square of the distance from the center, to define the distances that a body falling in a straight line describes in given times.[39]
>
> If the body does not fall perpendicularly, it describes an ellipse, for instance APB, of which the inferior focus (say S) coincides with the center [of the earth]. That is evident by what has already been demonstrated.
>
> On the major axis AB of the ellipse describe the semicircle ADB, and let the straight line DPC perpendicular to the axis pass through the falling body. With DS and PS drawn, the area ASD will be proportional to the area ASP, and hence to the time as well.
>
> Keeping the axis AB the same, let the width of the ellipse be continually diminished, and the area ASD will always remain proportional to the time. Let the width be diminished infinitely, and the orbit APB then coinciding with the axis AB, and the focus with the end B of the axis, the body will descend on the straight line AC, and the area ABD becomes proportional to the time.
>
> Thus the distance AC that the body falling perpendicularly from A will describe in a given time may be defined, if one takes an area ABD proportional to the time and drops from the point D a perpendicular DC on the straight line AB. Which it was required to do. (Add. 3965.7, fol. 61)

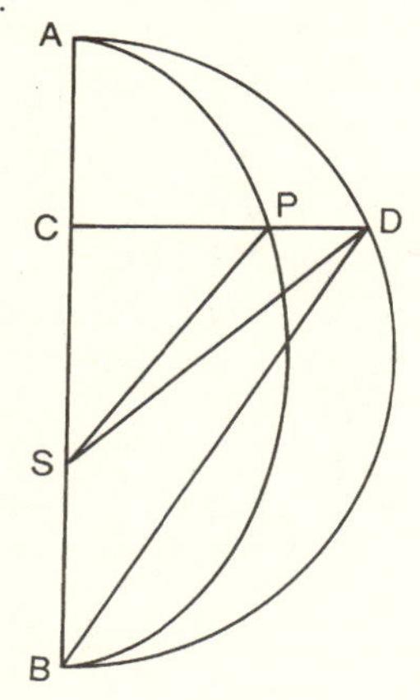

Figure 1.37

Imagine a fall during which the weight would be variable, as in the case of a body falling toward the earth from a very great height. Since the trajectory is rectilinear, the law of areas seems inapplicable: the time of travel can not be evaluated by the sector swept out. Newton nevertheless succeeds in extending the law of areas to this particular case.

First, suppose that the orbit is not a straight line passing through the center of force but rather an ellipse having this point as one of its foci. With the same major axis AB, there can be ellipses that are more or less narrow, like APB or AP′B; their foci S or S′ are more or less close to the end B of the axis.

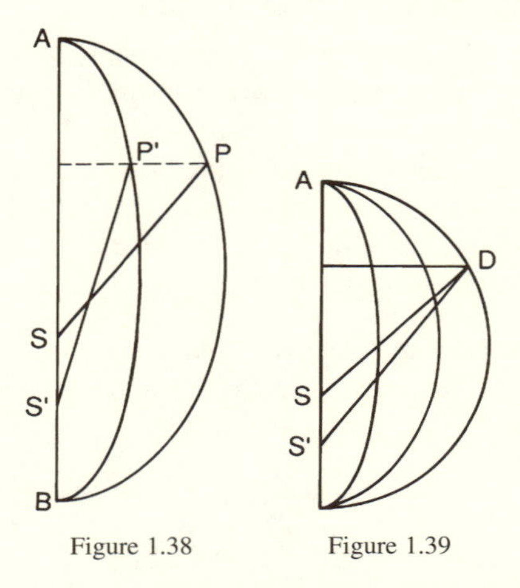

Figure 1.38 Figure 1.39

For all these ellipses, it is possible to calculate the time of traversal of an arc by measuring the area of the eccentric sector ASP, AS′P′. In each case there is the advantage, as in the preceding problem, of replacing the area in the ellipse by the area in the associated circle that is constructed on the major axis. The areas ASD, AS'D will be the measure of the time elapsed.

This circle is the same for all the ellipses that have AB as major axis. The only thing that varies is the center of the sectors, which is the focus in the ellipses: it approaches the endpoint B in the measure that the ellipse flattens. The permanence of this circle is decisive for the reasoning: whatever the thickness of the ellipse, the "witness" circle serves to represent and measure the time of travel.

Why should it not be the same at the last stage, when the ellipse, now infinitely flattened, becomes identical with the axis AB?[40] Then the point P, which represents the falling body, is on the vertical axis, and the focus S has been pushed back to the endpoint B. In this particular case the sectors must be cut off from B, and it can be seen, for example, that the times of traversal for the last portions of the path AB are shorter and shorter (the velocity becomes infinitely great as P approaches B).

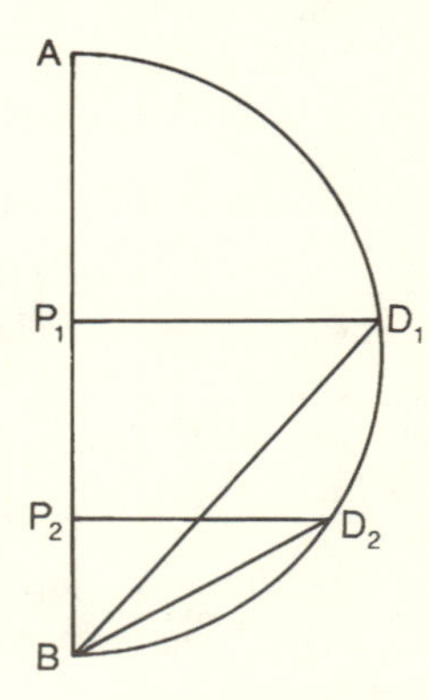

Figure 1.40

This manner of reasoning is at once simple and audacious: the ellipses are deformed to such an extreme that there is no longer a true ellipse or focus or area swept out, and yet it remains possible to follow the sweeping out of area on a "witness" circle that is not affected by the deformation.

THE PROCEDURES OF DEMONSTRATION IN THE *DE MOTU*

Such, then, is the small tract that so pleased Doctor Halley.[41] Not only does it reply to the question posed, in taking it, to be sure, the wrong way round—if a body orbits on an ellipse under the action of a centripetal force directed toward a focus, is it demonstratively true that this force diminishes as the square of the distance from the focus?—but it goes much further in that it sketches a reformulation of astronomy as a whole in terms of forces.

The *Principia* will pursue that approach in detail while extending it into other domains: simple machines, impact, pendular motions, optics, motions in resisting media, fluid dynamics. The interest of the *De motu* is that it permits a grasp of this theory, so new and so rich in consequences, at the first moment of its emergence and in condensed form—before Newton has developed it in the "downstream" (diverse applications) and "upstream" (axiomatic foundation) directions.

But has Newton *demonstrated* what was required, in the sense in which "geometers" understand the word "demonstration"? Should Sir Christopher Wren—the man who did not accept the reasonings of Hooke—have been satisfied? In the ways of demonstration followed by Newton, most of the sequences are perfectly acceptable in classical geometry—at least for those adept in Euclid and Apollonius. What is essential here is a series of proportions enunciated in relation to determinate figures.

Certain of the logical sequences are less in conformity with the traditional canons—not the mechanical or trial-and-error procedures of approximation, such as are used in the determination of the orbits of comets or in the solution of Kepler's problem (where it is necessary to assume that an angular sector equal to a curvilinear segment can be cut out), but rather in the new procedures that could be called infinitesimal or ultimate. Here is a list of these new procedures as they appear in the *De motu*.

1. The evaluation of the space traversed under the action of a centripetal force is valid only at the commencement of the time (*ipso motus initio*, Hypothesis 4; see above p. 18). Consequently the measure of the centripetal force by means of the formula $SP^2 \times QT^2/QR$ is acceptable only if the configuration QRPT is infinitely or indefinitely small, that is to say, ultimately, when P and Q coincide (*in figura indefinite parva QRPT lineola QR . . . est . . . ut quadratum temporis*, Theorem 3; see above p. 31).

2. What holds for a polygonal surface composed of a finite number of triangles

can be extended to an infinite number of triangles that compose a curvilinear surface (*triangula numero infinita et infinite parva*, Theorem 1; see above p. 22).

3. It is permitted, in relations of proportionality, to substitute some lines for others, because they differ from them very little when certain points approach one another so as to coincide (*Loquor de spatiis . . . minutissimis, inque infinitum diminuendis, sic ut pro . . . scribere liceat*, Theorem 2, p. 26 above; *punctis P et Q coeuntibus scribatur SP pro RL*, Problem 1, p. 33 above; *scribe QR pro PV . . . punctis Q et P coeuntibus*, Problem 2, p. 36 above; *punctis Q et P coeuntibus, rationes . . . fiunt aequalitatis*, Problem 3, p. 38 above).

4. The parts of area generated at each moment of time can be compared and added so as to compose the total areas (*partes arearum singulis momentis genitae . . . per numerum momentorum multiplicatae*, Theorem 4, p. 41 above).

5. The ellipses constructed on the same axis can be deformed to the point of merging with the axis, while certain properties are preserved (*minuatur latitudo illa in infinitum*, Problem 5, p. 53 above).

This mathematical machinery is not reducible to one or two fundamental procedures, as is the case for "fluxions," where Newton subsumes everything under two problems: (a) knowing a relation between two variables, to find the relation between their velocities of increase; (b) knowing a relation between the velocities, to find the relation between the variables (*NMP*, 3:70; see below, p. 210). The second fluxional problem agrees rigorously with the fourth of the cases enumerated, in which the comparison of parts of areas simultaneously swept out makes possible the comparison of the periods (one passes from the increments of the magnitudes to the magnitudes themselves).

The other cases are more intractable. What sort of principle must underlie these reasonings? A simple extension of classical geometry would suffice for the third case: certain relations are enunciated with respect to finite, immobile figures, and the configurations are then deformed in such a way as to make possible the replacement of one line by another line from which it differs very little. Proportions and relations are thus enriched by authorizing the substitution of one magnitude for another if the two are infinitely close to equality.

The most audacious procedure is the first one in the list: how to justify enunciations that apply exclusively to magnitudes in a nascent state? It is on this point that Newton will make the first effort at introducing greater rigor in the manuscripts in which he refines the earliest version of *De motu* (see below, p. 163).

In all the cases it is necessary to presuppose a rather vague kind of "postulate of continuity": the relations must be preserved while the figures are deformed to an ultimate configuration. These procedures all presuppose motion and time: points approach one another (*coeunt*), areas are generated (*genitae*) by the rays, magnitudes diminish continually (*perpetuo*), one area remains always (*semper*) proportional to some quantity, and each instant of time (*singulis temporis momentis*) is associated with the magnitudes that are then produced.

These aspects of the mathematical reasonings of the *De motu* are not an isolated curiosity. The intervention of time in geometry, although not in conformity with the most orthodox Greek tradition, characterizes one entire aspect of the mathematics of the seventeenth century. Other examples of these kinematic reasonings can be found in Newton's work, and also in that of the Italian disciples or contemporaries of Galileo, in Roberval's, and Barrow's. On the basis of these examples, a clear image and well-founded idea of the diverse methods practised between about 1630 and 1680 can be constructed. The procedures of the *De motu* and of the *Principia* can be placed in the context of a plurality of possible methods, in particular as they concern the infinitely small and the justification of its use: theories of indivisibles, fluxional methods, "infinitesimal calculus," and ultimate ratios. Chapter III will be devoted to these new kinds of mathematics.

First, however, chapter II will address the context of the discussion about force. What is a "force" for a philosopher or physicist living in the years around 1680? Is it normal or scandalous that forces should be supposed in the heavens? Have the forces exerted by weights or machines something in common with the new "centripetal force"? Is it necessary to count force among the number of physical realities?

CHAPTER II

ASPECTS OF FORCE BEFORE THE *PRINCIPIA*

INTERLUDE: A RIDDLE CONCERNING "FORCE" IN THE TRADITION

A collection of exercises that had been designed for the use of students will give an idea of what was commonly understood by force at the time of the *De motu*. Below is an example taken almost at random from one of these collections—it leaves the author's identity to be guessed.

> *Problem 7*. Being given the forces [*viribus*] of several agents, to determine the time in which they will together produce a given effect d.
>
> Let us suppose that the forces of the agents A, B, C are such that they produce respectively the effects a, b, c, in the times e, f, g. In the time x, they will produce the effects ax/e, bx/f, cx/g. Consequently, $ax/e + bx/f + cx/g = d$, and by reducing this,
>
> $$x = \frac{d}{a/e + b/f + c/g}.$$
>
> *Example*. Three hirelings can accomplish a certain job [*opus*] in a certain time, namely A once in three weeks, B three times in eight weeks, and C five times in twelve weeks. In how much time can they complete the job by working together?
>
> The forces [*vires*] of the agents A, B, C, are such that they would produce respectively the effects 1, 3, 5 in the times 3, 8, 12; and we seek the time in which they will complete the effect 1.
>
> Consequently, in place of $a, b, c; d; e, f, g$, write 1, 3, 5, 1, 3, 8, 12; we obtain:
>
> $$x = \frac{1}{1/3 + 3/8 + 5/12},$$
>
> that is, 8/9 of a week, or 6 days, 5⅓ hours, which is the time in which they will finish the whole task.

The usual absurdity of such a problem is ignored, and it is assumed that the workers when united have the exact sum of the forces of work that they possessed when separated (Aristotle had criticized this oversimplification long before[1]).

What is a force in this problem? It is that which permits the accomplishing of a certain result in a certain time. If A produces the effect a in the time e, and B the effect b in the time f, the forces of A and B can be represented by the quotients a/e and b/f. It can also be said that the force A, during the time e, produces the effect a:

$$A \cdot e = a \text{ (in Latin, } \textit{vis} \cdot \textit{tempus} = \textit{effectus}\text{).}$$

During another time x, the same force would produce an effect $A \cdot x$ or $(a/e) \cdot x$.

The nature of the effect matters little. It can be a pit to be dug, a load to be moved along a path, etc. Problem 6 in the same collection of exercises proposes an example of "effect" that is very remarkable (above all from the pen of an author who has left heaps of manuscripts):

> If a scribe in 8 days can write 15 folio pages, how many scribes of the same sort will be necessary to write 405 folio pages in 9 days?

Now who wrote these problems? I point out that they were probably used in teaching at Cambridge around 1673–75, then assembled into a book around 1683, and published in 1707. The name of the author is then easy to guess: it is Isaac Newton himself.[2]

The ambiguity of this notion of force is very great. According to whether the concept is defined by one "effect" or another, very different specifications of "force" result, some of them incompatible with the modern principles of mechanics. If the effect is a motion, and the force of the mover that is the cause of this motion is to be evaluated, what should be measured? The choice that is the most obvious, the most conformable to ordinary intuition, is that of the distance traversed in a given time.

This was the measure adopted in the Aristotelian tradition of the Middle Ages on the basis of a passage of the *Physics* (7.5):

> In an equal time, a force A will move the half of B through the double of [the distance] C. . . . And if the same force moves the same body in a certain time through a certain distance, it will move it through half the distance in half the time.[3] (Aristotle, *Physics* 7.5.250a1–6)

The *force* which moves, the *weight* moved, the *distance*, and the *time*, must be capable of entering into relations of simple proportionality.

But it is not necessary to hide behind the authority of Aristotle to affirm that force is measured by its effect, and in particular, by the distance traversed. Kepler, for instance, when analyzing the motions of planets, employed this traditional notion of force as an entity well known, although at times he found it necessary to clarify what the parameters were that needed to be considered in one or another particular case (see below, p. 79ff.). Thus, according to Kepler, if solid orbs cause the planets to revolve, it will be necessary to determine for each orb its motive force, which is to be evaluated according to the size of the orb to be moved and the time of revolution assigned to it; in this manner will be measured "the motive force that is sufficient to the round orb, and from whose vigor and constant strength arises the time of revolution."[4] The force is evaluated by means of the distance traversed and the time elapsed.

This manner of assessing the force that moves a celestial orb may seem acceptable enough as long as it remains general and vague. But when Kepler came to quantitatively analyze the diffusion of the "motor virtue" issuing from the sun and its action on the planets, he presupposed that the planet would be moved so much the more rapidly as it receives more of the solar virtue. The velocity of the planet thus depends directly on the force that it receives. It is not acceleration, but velocity, that is the sign and measure of force. That concept is contrary to the principle of inertia, which assumes the continuation of uniform rectilinear motion without exterior cause. Velocity no longer has anything to do with force; it is change of velocity that corresponds to force. May it no longer be said that the effect is the measure of force? The statement remains true, but the physicist of the twentieth century defines the effect differently: if uniform rectilinear motion is only a certain state, the only real effect will be change of motion and not motion itself.

It is not in this form that Newton will enunciate the principle of inertia. For him, the perpetuation of motion results from a force, which is the force inherent in the moving body (*vis insita, vis corporis*). For him, therefore, force is not at all correlative to acceleration: even uniform motion presupposes a certain kind of force.

Thus, in Newton's work, there are superimposed different strata of a centuries-long meditation on force. The most traditional layer is represented by the algebraic exercise transcribed above. A more elaborate conception is found in a critique of the Cartesian conception of motion:

> Force is the causal principle of motion and rest. (Hall, 114)

The mention of rest is the most original feature of this definition: the return of the body to a state of rest presupposes also a causal principle. A little earlier Newton had rather maintained a thesis Cartesian in inspiration:

> Force is the pressure or crouding of one body upon another. (Herivel, 138)

Finally, there is the new sort of force invented by Newton, the "centripetal force," which will be ranged in the *Principia* under the more general rubric of "impressed forces," in contrast to "inherent forces." It is this impressed force that will become after Euler's time the force that is now familiar, that is defined and measured by the product $m \cdot dv/dt$.

Sketches of Three Doctrines of Force

These different definitions of force in Newton's writings are a sort of condensed image of the entire evolution of the concept, including the successive refinements that it had undergone. A systematic exposition of the evolution of the notion of force would go beyond the scope of this book, whose aim is only to achieve a better understanding of Book I of the *Principia* through a very exact and detailed exegesis of its procedures of reasoning. This will be impossible without first taking a step backwards. Some brief sketches will suffice to present the most striking features of three of the seventeenth-century doctrines.

First, what does the Magnetic Philosophy say about the diffusion of the solar force? Is a calculus of forces possible? The answer can be found in Kepler's *Epitome*: once it is admitted that there can be forces in the heavens, how can the parameters that play a role be analyzed? The analogy with certain simple situations—the diffusion of light, heat, and magnetic attraction, and the action of running water—permitted Kepler to weigh several significant factors—the intensity of the active source, loss in the ambient medium, receptivity of the body that is moved, and ascendancy of the motive action. Kepler even claimed to deduce the quantitative laws of motion of the stars from this analysis of forces and claimed to prove a priori, for instance, that the planets move in ellipses. In brief, he had already responded to Halley's query, and without turning it wrongside out.

After Kepler, Galileo's legacy will be considered. Centripetal force is an extension of weight and generalizes the law that Galileo had announced as applying to falling bodies: the space traversed is proportional to the square of the time. In Newton's eyes, Galileo's fundamental idea is that of a regular acceleration ordered in accordance with the flow of time: weight generates *a new motion at each instant*. It is this role of time that invites attention: how did Galileo and his school arrive at their conception of the role of time in the action of forces?

The final sketch concerning the doctrine of force in the seventeenth century will be devoted to the discussions of Descartes and Huygens concerning gravity. Their physical study of the fall of bodies and of weight was founded on an analysis of effort (*conatus*). These philosophers spoke, not of forces, but of efforts, of tendencies to move, that is, of *impeded motion*. It was in such terms that Descartes, and then Huygens, described circular motion and "centrifugal force," with Huygens ultimately arriving at a formula of evaluation identical with the one in Theorem 2 of Newton's *De motu*. The Cartesians sought to do without force insofar as possible. After the initial jolt at the creation, the only cause of motion was motion itself under the name of "force of bodies in motion." Everything must be explained by impact, including weight, which would be a sort of repercussion derived from the motion of bodies and their mutual impedance of one another.

Made vigilant by these scruples of the Cartesians, these questions should be kept in mind: What is the mode of being of force? What reality should be conceded to it? If force is the cause of motion or of change of motion, it must surely participate in the reality of motion itself. But this is a weak assurance; the being of motion is hardly yet firm or obvious. More than any other cause, force is the imaginary point where fantasies are refracted: it is impalpable, unseizable, and yet so rich, pregnant with future effect. Leonardo da Vinci spoke of *forza* as an "invisible power," a "spiritual virtue," and Torricelli saw in it a "subtle abstraction," a "spiritual quintessence." It is spiritual because the spirit is the very exemplar of impalpable realities. Material bodies can only receive, conserve, or give up force; they are not of the same nature as it. It is spiritual above all because force is of the order of the virtual or of the future: it is by spirit or mind alone that one encroaches on what is not.

Descartes was embarrassed by that inactuality. He reckoned effort (*conatus*) among the features that make possible the delineation of his world, but effort is neither figure nor motion: it is impeded motion, hence a motion that does not take place. Should one make of it the tension of a spirit, the decision of a will? This would be to ruin the whole Cartesian edifice and fill in the abyss that separates extension from thought. Kepler was more at ease, since according to him the sun and even the planets have an animal life, and force is another name for soul or the vital power.

If force is defined and measured by the effect that it can produce, its entire being is virtual. Thus Simon Stevin refused to explain equilibrium in terms of motions that do not take place, and Galileo hesitated to account for present effects in terms of a proportion "that is not, but is yet to be" (EN 8:438). But Galileo's creative boldness depended finally on this: that he had fully accepted the role of time in the evaluation of forces. *Impeto*, *momento*, *energia*, *acceleratio* imply the consideration of time in a privileged mode. Impetus (*impeto*) is what remains of past motion, and it promises a motion to come. The thrust (*momento*) of a body that falls is made up of an accumulation of similar thrusts acquired instant by instant and conserved in the body, and their total intensity is measured by a virtual ascent: a body that has fallen through a certain height has acquired a *something* enabling it to ascend again to the same height (by whatever path). Matter receives force instant by instant and conserves it in order to release it either little by little or all at once, in the instant of an impact or collision.

The consideration of time was already present in the traditional analysis of machines as the time during which the force is dispensed. In the seventeenth century, a new sort of time was introduced: elementary time, the ever-repeated instant punctuating the action of force. Force itself is nothing durable; weight that is exercised permanently is nothing substantial or available to be used up;

it is only the effect of surface or the result of the ever-repeated action of gravity. The only real entity is the incessant and ever-renewed repetition of the "thrust" or "push."

Newton will neglect a part of this heritage: he will admit as a first principle the reproduction and accumulation of the thrust, instant after instant, "in the mode of gravity," but he will avoid all reasoning based on a notion of energy and will not rely on ideas of conservation, even though he knows how to present the mathematical skeleton of such reasoning magisterially (Prop. 39 of the *Principia*; see below, p. 250).

THE DIFFUSION OF THE SOLAR VIRTUE: KEPLER AND THE CALCULATION OF FORCES

Kepler inaugurated an entirely new astronomy: he investigated the physical causes of motions and evaluated the forces acting in the heavens. The Copernican system had turned the universe upside down, but it had not at all changed the way of doing astronomy. The description of the celestial motions remained purely kinematic—without any causal explanation, without dynamical theory. The sun was at the center of the world, but it did not act on the planets. All the motions were still to be explained in principle by uniform rotation of transparent spheres, each planet being embedded in the thickness of one of these spheres. With Kepler, the sun became the cause of the motions due to the force or virtue emanating from it.

The medievals had already discussed the thickness of the celestial spheres, and they had also asked what attributes must be attributed to a mover capable of moving eternally. More recently, J. C. Scaliger had proposed a doctrine of celestial intelligences. But these speculations remained isolated from technical astronomy. Kepler, by contrast, took up the whole treasury of the traditional mathematical astronomy, enriched by the observations of Tycho Brahe, and he reformulated it, making of it a system founded on dynamic and harmonic considerations.

From within this system, much later, were extracted the famous "three laws":

> The planets follow elliptical trajectories, with the sun occupying one of the foci.
>
> The radius connecting the planet to the sun sweeps out equal areas in equal times.
>
> The cubes of the major axes are proportional to the squares of the periods.

It is Kepler's dynamics that is of principal interest. It is woven of analogies, images, speculations that are sometimes strange—it is not difficult to lose the thread of inquiry in the meanders of this work. Often ignored is the more mathematical part of this study of forces: Kepler developed his dynamical considerations to the point of giving precise quantitative evaluations, and he undertook calculations concerning the diffusion of the solar force and its effects. It will be shown how he dealt with the geometric laws of this diffusion, rejecting an inverse-square law, and how he demonstrated the proportion between the distances and the times (Second and Third Laws) and deduced the ellipticity of the orbits.[5]

This quick presentation of Kepler's dynamics will, in general, follow the *Epitome astronomiae Copernicanae*,[6] which is more synthetic, more complete, and perhaps less known than the *Astronomia nova*—at least today, for no doubt this "manual" of Copernican and Keplerian astronomy was in its own day more often read than the *Astronomia nova*, which contemporaries found too disconcerting and too confusing.

The Destruction of the Solid Spheres

Kepler's enterprise presupposes the destruction of the ancient world, at least in two of its most fundamental and emotionally charged features:

1. The Earth is no longer the immobile center of the universe.
2. The solid, transparent spheres that carry the planets have been demolished by the observations and reasonings of Tycho.

The privileged role of the sun will be discussed later. As to the second point, in a passage at the beginning of the *Epitome*, where he discusses the order of the elements in the cosmos, Kepler himself explains why there can no longer be celestial spheres:

> The Ancients gave the first place, in the region of the elements, to the sphere of fire; they divided the region of the aether into numerous contiguous solid spheres, one contained in the next. Do you have arguments opposing this?
>
> 1. Tycho Brahe observed that the trajectories of certain comets passed from one place to another in the region where there were believed to be solid orbs.
> 2. If outside the surface of the air there were, higher up, other surfaces of orbs that touched one another, then there would be various reflections, as in mirrors placed against one another. But this appearance is not produced.
> 3. The sphere of fire would be more tenuous than the sphere of air, the celestial spheres more tenuous than the sphere of fire, and higher celestial spheres than lower, for the heaven is more tenuous than the elements. If therefore the rays from the stars must pass through so many interposed media of different densities, they would be frequently refracted when they strike obliquely (as that happens necessarily with eccentric orbs and epicycles) before arriving at the surface of the air; hence the visible stars, by virtue of the refracted rays, would appear outside of their true places. . . . Yet the places of the stars agree with themselves perfectly

> without supposing any refraction down to the surface of the air. Therefore there are no orbs differing in degree of density down to the sphere of the air. (*GW*, 7:54)

Later in the same book there are other, different lists of arguments against the solid orbs. Among these arguments is another borrowed from Tycho that applies peculiarly to the Tychonic system:

> The third reason is drawn from Tycho's own principles, according to which (as also according to the principles of Copernicus), Mars is sometimes closer to the Earth than is the Sun. Yet Brahe did not believe this permutation possible if there were solid orbs, since then the orb of Mars must intersect the orb of the Sun. (*GW*, 7:261)[7]

The arguments concerning luminous rays are the newest and most remarkable. What boldness to thus evaluate the refraction of the celestial spheres and calculate the path of light beyond the moon as if the media were terrestrial! The difference in nature between the heavens and the world here below would preclude reasonings of this kind. This is also why there could not be "celestial mechanics"—the very term would seem a barbarism or an absurdity.[8]

Kepler attributes to the celestial bodies a matter analogous to that on earth and hence an "inertia," that is, a powerlessness for self-movement and a tendency to remain at rest (ibid., 7:296). In order to move the stars or the orbs—if in fact there still are orbs—there is therefore necessary a "power" stronger than their inertia. The principles of terrestrial dynamics can then be applied: the terms *forces* and *resistances* will be used, and, for example, the times of revolution of the planets will be evaluated by means of the determinate relation between their inertia and the force that moves them (ibid., 7:297).

Kepler's Realism

Ptolemy, the ancient, had replied long before to several of these arguments by invoking the radical difference between things terrestrial and things celestial. Kepler cites and discusses the following, often referred-to passage from the *Almagest*:

> Let no one judge too difficult the intertwinings of circles that we assume, seeing that it would be very confusing to make a manual imitation of them.[9] For it is not legitimate to equate our human condition to the immortal gods and to seek an assurance concerning the most sublime things from examples of things so dissimilar. . . .
>
> Even if the various intertwinings of circles, because of different motions, and the insertion of the ones in the others, are very difficult to realize in models [*exemplis*] of the theories constructed by the hand of man, and succeed badly in avoiding the obstructing of one motion by another, nonetheless we see in the

> heavens that no obstacle is caused to any of the motions by such a complex combination. Rather, we must not decide what is simple in the heavens from models [*exemplis*] of these things that appear simple to us; for even here the same thing is not equally simple everywhere on Earth. To anyone wishing to judge in this manner of the celestial things, it will easily come to this, that he recognizes nothing that occurs in the heavens as simple, not even the invariable constancy of the first motion, for it is difficult and even impossible to find a similar thing among us (that is to say, a thing that behaves always in the same way). Our judgment must therefore be formed not from terrestrial things [*ex rebus nostratis*] but from the nature of celestial things and from the unchanging tenor of their motion. In this way we shall consider all the motions as simple, and even much more simple than those that seem such to us; since we can suppose no labor [*laborem*] or difficulty involved in their revolutions. Such is what Ptolemy wrote. (*GW*, 7:291–92 = *Ptolemy's Almagest*, 13.2; Ptolemy, *Syntaxis*, 532–34)

According to this passage, the laws of motion of the stars have nothing in common with what is known here below. In the heavens there is nothing of labor or resistance, hence no real force, since force is what overcomes a resistance. The very notion of simplicity loses its meaning: what appears complex and intertwined to human eyes, could be, in the heavens, quite simple and direct.

This text excuses in advance all physical and dynamical absurdities in the representation of the heavens. Ptolemy encouraged a sort of indifferentism: since the heavens are so inaccessible and of a nature so different, why wish that astronomical discourse should "bite" into the reality? The most varied hypotheses can render account of the heavenly motions without claiming to be an exact description.

Here is what the "magister" of the *Epitome* responds to a question from his pupil (the *Epitome* is written in the form of a pedagogical dialogue):

> What do you say to this opinion of Ptolemy?
>
> Even if it is true that the ease of celestial motions should not, for many reasons, be assessed according to the difficulty of the motions of the elements, yet it does not follow that there is on Earth no model approximating [*exempla propinqua*] to the celestial motions. Ptolemy seems to extend this excuse too far, to the point of undermining astronomical reasoning [*rationem astronomicam*]. . . .
>
> This puts all hypotheses under a suspicion of falsity, in making so sharp a distinction between celestial and terrestrial things, to the point that reason itself will be supposed mistaken when it makes a decision as to what is geometrically simple. (*GW*, 7:292)

If Ptolemy was right, it is astronomy itself that is emptied of substance. Kepler battled vigorously against the conventionalists or positivists of his time—those who wished to return to a simple catalogue of celestial appearances or who devalued the hypotheses of astronomy to the rank of pure fictions devoid of physical significance (Ursus, Osiander, Patrizzi).[10]

Ptolemy refused to enter into a truly physical discussion. He merely decreed that the nature of the celestial bodies permits them to move without resistance and even to interpenetrate in their rotations. Kepler demands much more:

> To philosophers it is not sufficient that the matter of the celestial bodies should be fluid and permeable by the globes and therefore not resist the motion of the globes that traverse it. For they want to know by what the globe itself is moved in a circle, above all if it is established that the matter of the globes resists motion. They seek the force by which the mover moves the body from one place to another, since there is no immobile subjacent terrain and a spherical body possesses neither feet nor wings to do it service, so that by moving them as do animals, it might transport its body across the aether, in the manner of birds which traverse the air by supporting themselves on this medium and receiving from it a contrary thrust.
>
> They ask what light of the mind, and what means, permit a mover to discern and to trace the centers of the circles and the orbits that encircle them.
>
> Finally, neither theology nor the nature of things can allow that Ptolemy, imbued as he was with pagan superstition, should make of the stars visible gods (in ascribing to them an immortal life because of their eternal motion) and should attribute to them more than the God Creator himself possesses: that is to say that for them, geometrical relations that are in reality composite are simple, while God has willed that the understanding of these relations should be common to himself and to man his image. (*GW*, 7:293)

According to Kepler, it does not suffice to say that the heaven is made of a peculiar matter, totally penetrable. Philosophers, that is to say, physicists, want forces, points of support; they want deliberations and decisions and instruments for putting them into action. If the motions of the heavens are to be explained, which is the purpose of astronomy, physicists must be able to render an account of these particular motions in the same way that they render account of any motion.

The radical division of this world from the heavens is the action of a pagan who worships the stars. For Kepler the Christian, astronomy is fully possible because the stars are not gods. His philosophy of creation teaches that the only truly inaccessible thing is God himself. All that is created is open and accessible to man.

Ptolemy claimed that what appears complex here below can be simple above. He thus broke the principal link of human understanding with the cosmos and with the Creator. For Kepler, on the contrary, God made humans participants in mathematical truths, without reserve and without degradation. In their understanding of geometrical relations, humans are the equals of God.[11]

Motive Souls and Intelligences

If there are no longer crystalline orbs to sustain and guide the planets, it is necessary to find out what supports and carries them along. And since it is permitted and even required of a Christian to explore boldly and to contemplate the celestial realities, the inquiry can make use of a constant analogy with motions that are known, beginning with the "animal" manner of being moved.

Under what conditions could intelligences and souls preside over the motions of the planets? Here is an agenda for discussion that could lead to indefinite developments as in the manner of medieval disputes. In fact, Kepler dismisses the question very quickly: in the end it is material necessity that gives rise to, determines, and guides the motions of the planets. The only animate force is in the sun; the other motions proceed from natural and material powers. To be sure, the earth, for example, has a life and a soul, as is attested by the tides, the volcanic and meteorological phenomena, the secretions that it produces, and the astral influences to which it reacts.[12] But the soul of the earth does not intervene in any way in its motion around the sun.

A soul, moreover, is not the same thing as a mind. The soul (*anima*) is a simple vital principle, the force or cause that animates living beings, causing them to nourish themselves, breathe, perceive, and move. The mind (*mens*) presupposes the soul, but it is nobler: by virtue of an intelligence (*intellectus*), it is capable of deliberating and making decisions, that is, it gives or withholds its assent (*nutus*). In the case of a displacement, as for the planets, the soul would be rather of the order of the motive force, while the intelligence must indicate the direction and guide the motion.

Kepler examines in detail the functions of the intelligence and the soul in order to decide whether the motions of the stars can be explained by means of them. The question is: what is the condition of minds? Whatever a mind is, it would not be allowable to people the universe indefinitely with minds, because

> A mind is related to a body, which gives it its *situs* and its individuality.
>
> A mind must receive information about the exterior by means of certain organs.
>
> A mind must be able to transmit orders to its members, by certain intermediary materials (and not at a distance).

> What seat shall we give to a mind, in order that it can measure a circle or an elliptical orbit in the liquid fields of the aether? Shall we place it at the center? This would be to place it in the aethereal medium, in a point that differs not at all from all the remaining space of the world, since the orbit of the planet is eccentric in relation to the body of the Sun. But that is absurd, since the principle of individuation of souls comes down to matter and to body. . . . The soul and the mind receive position only in virtue of the body that they inform. . . .
>
> Let us even grant to a mind the faculty of seeing from its seat at the center; how

> then will it cause the planet, at such a great distance from itself, to arrange its orbit about this center? . . .
>
> But if one places the mover-mind outside the center of the orbit, its condition will be yet worse. Either, on the one hand, it will be in the body that occupies the center of the world, and thus all the minds will be in the same body, and we shall find ourselves again in the difficulties already cited, about keeping the planet to its orbit and making it find its way along this orbit.
>
> Or else the mind will be in the globe of the planet itself, and we then ask, in the two cases, by means of what intermediaries the mind knows where the center is around which the planet's orbit must be ordered, and at what distance from this point are both it and its globe. Avicenna thought rightly that the mover of the planet, if it is a mind, needs to recognize the center and the distance by which it is separated from it. (*GW*, 7:295–96)

Does the form itself of the planetary trajectories accord with an intelligent choice? It seems very improbable. The proportions of the periodic times are irrational (*GW*, 7:297); and if it were a mind that made the decisions and gave the guidance, the motion would instead be circular:

> The orbit of the planet is not a perfect circle. Yet if it were the work of a mind, it would have been ordered in a perfect circle, having the beauty and perfection accorded to a mind. On the contrary, the elliptical figure of the planetary paths, and the laws of motion whence this figure results, lead one to think rather of the nature of the lever [*sapiunt potius naturam staterae*] or of material necessity, than of the conception and determination of a mind. (*GW*, 7:295)

Ellipses present a less complete uniformity than do circles; they have about them something of matter, of the deformed; their nature or flavor makes one think of the lever, which is the very exemplar of exclusively material modes of control. Kepler lets it be understood in this passage that the ellipse can be deduced as the outcome of a certain number of material laws of motion. This is the goal toward which he is proceeding: to construct the elliptical orbits and their properties starting from the diffusion of the solar virtue—here, as later, a philosopher is trying to respond to Halley's question.

In the end, the intelligences and the souls can be eliminated almost completely from the causal description of the celestial motions. The sun itself is animated, as shown by its light and the variations of its appearance (*GW*, 7:298–99; Koyré *RA*, 297), but it is not directly by this soul that it moves the planets:

> The prehension of the bodies of the planets, which the Sun drives in circles by turning about itself, is a corporeal virtue, and not animal or mental. (*GW*, 7:299)

As to the planets, their motion is explained by "natural powers that are inherent in their bodies" (ibid., 7:295).

In sum:

> It is not necessary to introduce a mind [*mentem*] which would lead the globes in circles by a decree of reason and as though by its assent [*nutu*], nor a soul that would preside over their circular motion and would imprint its action on the globes. . . . It is only to the unique solar body, placed in the middle of the whole universe, that the circumsolar motion of the primary planets can be attributed. (*GW*, 7:297; Koyré *RA*, 292)

The path leading toward material necessity that Kepler induces his reader to follow is not a rhetorical or pedagogical fiction but rather one that he himself has previously traversed. His first speculations, those prior to 1600, were more clearly animist. Later he wrote that what he had maintained in the *Mysterium* of 1596 remains true provided *soul* is replaced by *force*:

> If in place of the word soul we substitute the word force, we shall have the principle from which the physical force of the heavens was elaborated in the *Commentaries on Mars* [= the *Astronomia nova*] and developed in the *Epitome of Astronomy*. For previously I believed that the cause that moves the planet was a soul, absolutely speaking, because I was filled with the dogmas of J. C. Scaliger concerning the motive intelligences.
>
> But in dwelling on this fact that the motive cause in question weakens with distance, and that the light of the Sun also attenuates with distance from the Sun, I was led to the conclusion that this force must be something corporeal, if not in the proper sense, at least equivocally, just as we say that light is something corporeal, that is to say, a species issuing from body, though immaterial. (*Mysterium*, *GW*, 8:113; see the English translation by A. M. Duncan in Kepler, *The Secret of the Universe*, 202)

Already the young Kepler had modified the traditional animist image: in place of one soul per planet, he believed to be necessary only the one soul in the sun.[13] Then, reflecting on the manner in which the animating force of the sun is diffused, Kepler ended by considering it as a quasi-material reality.

The Role of the Sun

The role of the sun thus comes to be reinforced. The planets only submit to—while resisting—the action that it impresses on them. Copernicus had not gone so far in his praise of the sun. Not only is this great luminary placed at the center of all as the leader and mainspring of the choir of stars, it is necessary to grant to it as well the role of hearth (*focus*), that which maintains life and motion. Light, heat, motion, and harmony find in it their source: it is the lamp that illumines and beautifies, the hearth that dispenses heat, the first mover and the first cause of the motions of the planets, the place where all harmony becomes apparent (*GW*, 7:259–60; Koyré *RA*, 287–88).

The Copernican system had revealed a marvellous harmony among the motions of the several planets: the periods were ranged in increasing order, the slowest planets being most distant. Kepler had long meditated on this relation between velocities and solar distances and claimed that the same thing is true of each planet by itself at different points along its orbit: the farther it is from the sun, the more slowly it goes (*moras esse ut distantias*).

He had recently discovered (in the *Harmonice mundi* of 1618) the exact law of dependence between the periods and the dimensions of the orbits, namely the "sesquiplicate" proportion: the periods are as the cubes of the square roots of the radii. And, he had conjectured that, as the periods become shorter and shorter toward the center, the most rapid motion must belong to the sun itself, which must complete a rotation in an interval shorter than the period of Mercury. This speculation was marvellously corroborated through the observations of Galileo: the displacement of sunspots confirmed a rotation of the sun in some two dozen days. Thence was explained the revolution of the planets round the sun: the sun, by rotating, caused them to revolve about itself, but with a certain retardation.

Three powerful analogical reasons thus prompted Kepler to make the sun the cause of planetary motion:

> What causes led us to think that the Sun is the moving cause or the source of the motion of the planets?
>
> 1. Because it is evident that, to the degree that one planet is more distant from the Sun than another, it advances more slowly, in such a way that the ratio of the periods of their motions is the sesquiplicate ratio of their distances from the Sun. Whence we draw an argument for making the Sun the source of the motion.
>
> 2. The same thing happens for each planet individually, as we shall see below: the closer a planet comes to the Sun, at whatever time, the more the velocity with which it is moved increases, precisely in the duplicate ratio.
>
> 3. This is not contrary to the dignity and proper character of the solar body, which is the most beautiful, and of a perfect rotundity, and the greatest, and is the source of the light and heat whence all life flows to plants; indeed, heat and light could almost be considered as the instruments suitable to the Sun for conferring motion on the planets.
>
> 4. Above all, what gives the greatest degree of probability is the rotation of the Sun within its own proper space around an immobile axis, in the direction followed by the planets and with a period shorter than that of the planet that is nearest and most rapid, Mercury. For this fact is revealed nowadays by the telescope and can be seen any day: the body of the Sun teems with spots that traverse the Sun's disk in its lower hemisphere in 12, 13, or 14 days, slowly at first and at the end, more rapidly in the middle (which allows us to assert that they adhere to the

> surface of the Sun and rotate with it). I demonstrated the necessity of this rotation, by reasons drawn from the motion of the planets, in the *Commentaries on Mars* [= the *Astronomia nova*], chapter 34, long before it was established by means of the sunspots. (*GW*, 7:298)

The Diffusion of the Solar Virtue

The virtue of the sun is carried to the planets by a *species*. This Latin term is from medieval philosophical jargon. For want of a better substitute, it will be translated simply by the English word *species*. The species is the material or quasi-material image diffused from the surface of the object into the ambient space, carrying afar the appearance of its source and transporting also certain active traits of the emitting object.

According to Kepler, the sun does not have need of hands to move the planets:

> In place of hands it has the virtue of its own body, which is emitted in a straight line into the whole amplitude of the world, and which, because it is the species of a body [*species corporis*], turns with the same motion as the body of the Sun, in the manner of a very rapid vortex, sweeping across the entire amplitude of its circuit, as far as it reaches, with the same rapidity as the Sun turning round its center in its own very reduced space. (*GW*, 7:299; Koyré *RA*, 298)

Since the species of the sun transports the appearance and the traits of the body that emits it, it must rotate at the same speed as the body. By this motion the species entrains the planets.

But another species from the sun is already familiar: the rays of light that it sheds and which transport colors and heat (*GW*, 7:303). Kepler asks whether it is necessary to distinguish between light and the species that carries the moving force. Is not light the carrier of the solar virtue? Why distinguish between light and another species? The *Epitome* gives six arguments, of which the last four are retained (*GW*, 7:304):

> "The species of light flows out from the surface of the luminous body," while "the prehensive force must necessarily descend from the body itself, since it is a moving cause analogous to its mobile object."
>
> The resistance that bodies offer to light has not the same character as that which they present to the moving virtue. The body that lets itself be shined upon is immediately illumined over all its surface, and that which resists remains definitively opaque. By contrast the struggle [*contentio*] between matter and force occurs in time: the planet cannot resist invincibly, but it does not capitulate either instantaneously or totally.
>
> Light is arrested by opaque surfaces, while the solar virtue penetrates the bodies that it seizes upon and is not at all eclipsed when another body has intervened.

> Finally, there exist well-known cases, like that of magnetism, in which a motion similar to the celestial motions takes place without the accompaniment of any light.

Nevertheless, although there is a distinction between light and the species that carries force, there is also a great resemblance between them:

> The similarity is quite absolute when it comes to the genesis and conditions of these two species: their descent from the luminous body occurs in a moment, both traverse a medium whether great or small without loss or levy; nothing is dispersed between the source and the illuminable or mobile body.
>
> Both flows are therefore immaterial, in contrast to odors whose substance diminishes, or the heat coming from a red-hot furnace. Whereas odors and heat fill the intermediary space, this species is not in any place, except in the body that is opposed to it and is an obstacle to it—the light indeed on the opaque surface, but the motive virtue in the whole corpulence. In the intermediary space between the Sun and the surface, the species is not, but has been.
>
> But if either of these two solar species should encounter the concave spherical surface of an opaque body, it would disperse itself in this concavity with all the abundance it had when it left the solar body, so that there would be just as much in a vast, far-distant sphere of this kind as in a small, nearby one.
>
> Yet since the ratio of convex orbs is the duplicate of the ratio of their diameters, in unequal orbs the species would be doubly more tenuous in proportion to the distance. Moreover, since circles have between them the simple ratio of their diameters, the species is therefore as much more tenuous in length as it is farther from the source. (*GW*, 7:304)

How to imagine the mode of subsistence, so mysterious and fleeting, of the solar virtue? In the interval between the sun and the body that receives it, it has truly no reality, and it is necessary to say that *it is not but has been* (certain passages of Aristotle's *Physics*, bk. 6, contain similar plays on being and having been).

The resemblance between light and the motive virtue could lead to certain conclusions as to the geometric properties of diffusion. But Kepler does not here appear to decide between the two possible laws:

> If the diffusion is spherical (that is to say, in three dimensions), the proportion of the decrease will be that of the square of the radii.
>
> If the diffusion is circular (plane, or "in length"), the proportion will be simply that of the radii. (*GW*, 7:304–5)

In reality, though, the decision has already been made based on other motives:

> Since one and the same planet . . . on equal portions of its eccentric circle, but at unequal distances from the Sun, makes unequal delays, and this in the very proportion of the intervals between it and the Sun, it follows that the motive virtue

attenuates in length in the same proportion as light attenuates in length, that is to say, according to the proportion of the amplitude of the circles that have these intervals or radii.

[In figure 2.1,] let S be the Sun, with one planet CA being closer, and another FD farther away; and let DH and AI be equal parts of the eccentric [be it understood: in opposite positions on the eccentric], namely, DH more distant and AI nearer. Then, as SD is to SA, so are the delays [*mora*] of the planet on DH and on AI. It follows equally that as SD is to SA, so inversely are the densities of light as to length, of CA the nearer, to FD the more distant.

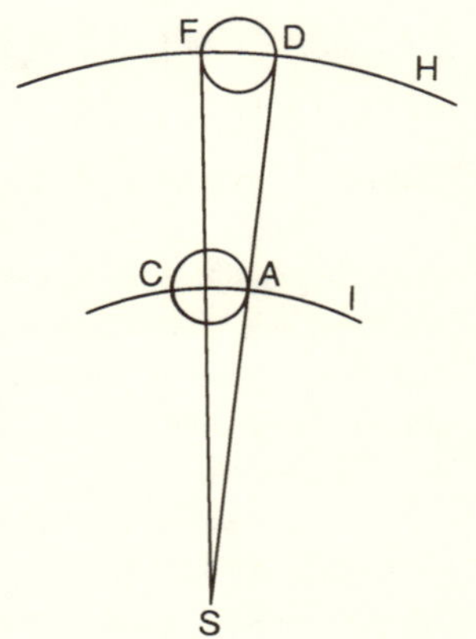

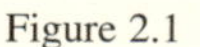
Figure 2.1

Yet light attenuates in the duplicate proportion of the intervals, that is, in the proportion of the surfaces. Why does not the motive virtue also weaken in the duplicate rather than in the simple proportion?

Because the motive virtue has for substrate [*subjectum*] the species of the solar body, not simply as body, but as a body endowed with a motion of revolution around its axis and immobile poles.

Therefore even if the species of the solar body attenuates both in length and breadth, not less than does light, yet this attenuation contributes to the weakening of the motive virtue only in length [*tandummodo causa longitudinis*]. For the local motion that the Sun brings to the planets is produced only in the length along which the parts of the Sun are also mobile, and not in width, in the direction of the poles of the body, with respect to which the Sun is immobile. (*GW*, 7:305)

Thus is a grand occasion missed. Kepler concludes that the species which is emitted by the sun and serves as vehicle to the motive force is diffused in the same manner as the luminous species, but its progressive weakening makes itself felt only in the direction of the line of motion of rotation. The force does not obey an inverse-square law.

The Law of Areas

Kepler admires the order and harmony that reign in the heavens. The cortege of planets obeys the leader, the sun, in following more or less faithfully its rotation, transmitted by the solar body's species, which rotates at great velocity across the cosmos. The farther a planet is from the sun, the slower is its motion. That is true of the planets compared with each other and also of each individual planet at different points of its path. This relation should make it possible to calculate the times required to traverse different distances.

To this end classical astronomy employed certain procedures of regularization or of "equation" of the motion. For Kepler, the law of areas takes the place of these traditional procedures. The word *aequatio* has no algebraic connotation: to "equate a motion" (*aequare motum*) is to equalize it, to redistribute it uniformly. Irregular periodic motion is replaced by a regular or mean motion to which a periodic correction is made by addition or subtraction (the *prostaphaeresis* or addition-subtraction). So it becomes possible to construct convenient tables of the motions of the stars—the ultimate end of the astronomer's activity.

The uniform motion of revolution of the planet must be effected around the "equant point" (*punctum aequans*). In a greatly simplified account, the schema would be the following. The planet's motion can be viewed from three distinct centers. It circulates on an "eccentric" circle with center C, and it is perceived by a terrestrial observer situated at T, which is a little offset in relation to the center C of the motion (if it is the sun that is at the center of the world, the sun here replaces the earth). Finally, the motion is uniform about a third point A (the equant point): seen from A, the planet P appears to circulate uniformly.

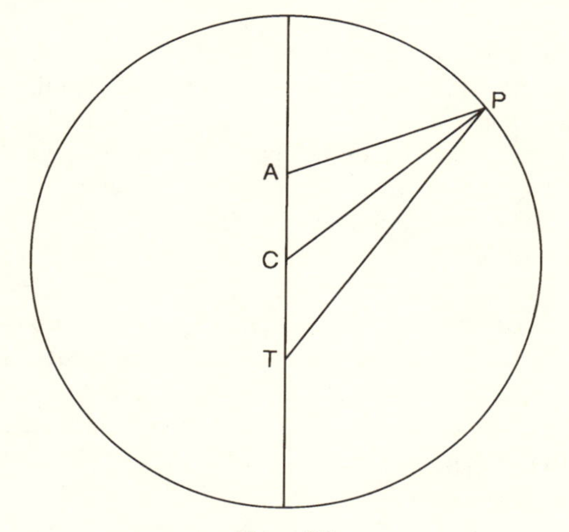

Figure 2.2

Copernicus had rejected the equant point as a fiction incompatible with the principles of a physically correct astronomy (*De Revolutionibus*, bk. 4, chap. 2 and bk. 5, chap. 2, in *Oeuvres complètes*): he thought that all the motions must be circular and uniform around their true centers. Kepler, on the contrary, regrets this severity on Copernicus's part and appreciates the "physical excellence" of the equant (*GW*, 3:73–77; *GW*, 7:380).

But the equation that Kepler proposes has a meaning very different from the traditional equation. It no longer amounts to assigning arbitrarily a fictive point round which the planet would have a constant angular velocity; henceforth it is from the body of the sun itself that the uniformity must be observable. This uniformity is of a dynamical nature and no longer kinematic. The diffusion of the force must be taken as the invariant fundamental, the regularity to which the deviations are referred.

The principle of the entire reasoning is the proportionality between velocity and distance: the farther the planet is from the sun, the more time it takes to traverse an arc of given length. In order to evaluate the difference in the times required to traverse these parts, it would thus suffice to compare the distances separating the parts of the trajectory from the sun. If this calculation is made for very small arcs cut off successively on a part AB of the orbit, the following proportion is obtained:

$$\frac{\text{sum of distances between the arc AB and the sun}}{\text{time to traverse AB}} = \frac{\text{sum total of the distances along the orbit}}{\text{total period}}$$

Kepler imagines cutting the orbit into sections of one degree each in order to calculate the successive distances to the sun. But this procedure is very tedious. Since all the distances between points of the orbit and the sun are contained in the corresponding sector, why not substitute the measure of the area of the sector for the sum of the distances? It is not exactly the same thing if the summit of the sectors is not the center of the circle, but the method will be much more convenient. It is in chapter 40 of the *Astronomia nova* that Kepler explains this connection of ideas. The title of the chapter, "Imperfect Method for Calculating the Equations from the Physical Hypothesis" (*GW*, 3:263), indicates clearly the nature and significance of the result; it is not a demonstrated proposition, but rather a convenient procedure of calculation.

It is indeed the "law of areas," but its status is very different from what it will be in Newton. Proposition 1 of the *De motu* and of the *Principia* could be entitled "first fundamental theorem on the action of any central force." For Kepler, by contrast, not only is the proposition "imperfect," but it presupposes a particular hypothesis as to the variation of the force issuing from the sun.

In the *Epitome*, the presentation of the law of areas is very long and confusing[14] and is more clearly sketched in the *Astronomia nova*. There it is the first mention of this law in Kepler's work and precedes, as well, the announcement of the discovery of the ellipticity of the orbits:

> Since the delays of the planet on equal parts of the eccentric are proportional to the distances of these parts [from the sun], while this distance changes from point to point along the whole semicircle of the eccentric, it was no easy task that I took upon myself, in inquiring how the sums of the individual distances could be obtained. For without obtaining the sum of all—and they are infinite in number—we could not say what the delay would be in each part, and so the equation would

remain unknown. For as the whole sum of the distances is to the whole periodic time, so is any part of the sum of the distances to its time.

This is why I began by dividing the eccentric into 360 parts, as if these were least particles, and I posited that within one of these parts the distance did not change. I sought the distances at the commencement of these parts or degrees, . . . and I took the sum of them. Then for the time of revolution, although it consists of 365 days and 6 hours, I took a different and rounded value, and set it equal to 360 degrees or a complete circle, which is the mean anomaly [or mean motion] of the astronomers. Thus I posited that the sum of the distances is to the sum of the times as any particular distance to the time that belongs to it. Finally I added up the times for individual degrees; and comparing these times, or degrees of mean anomaly, with the degrees of eccentric anomaly, that is, with the number of parts up to the last for which I had calculated the distance, I obtained the physical equation. . . .

But as this method is mechanical and tedious, and does not permit calculating the equation for any degree separately, without taking account of the others, I looked for other methods.

As I knew that on the eccentric there is an infinity of points, and thus an infinity of distances; it occurred to me that all these distances are contained in the plane of the eccentric. For I remembered that long ago Archimedes, when seeking the ratio of the circumference to the diameter, dissected the circle into an infinity of triangles. For in this lies the hidden force of his demonstration, which proceeds by reduction to the absurd. Thus since I had previously cut the circumference into 360 parts, I cut the plane of the eccentric circle into as many parts by drawing the lines from the point [A] whence is computed the eccentricity. . . .

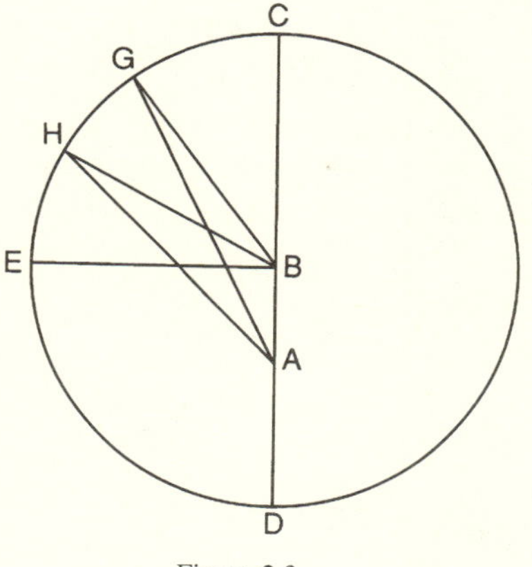

Figure 2.3

Just as the straight lines drawn from B to the infinity of parts of the circumference are all contained in the area of the semicircle CDE, and the straight lines from B to the infinity of parts of the arc CH are all contained in the area CBH, so

> also the straight lines issuing from A to the same parts of the circumference or arc are similarly contained. Since finally both the straight lines issuing from B and those issuing from A fill one and the same semicircle CDE, and since those drawn from A are the distances of which the sum is sought, it seemed to me possible to conclude that by calculating CAH or CAE one would have the sum of the infinity of distances on CH or on CE. Not that an infinity can be gone through, but I considered that the measure of the power by which the collected distances are able to accumulate delays was in this area, and that we could obtain this measure in knowing the area, without the enumeration of least parts. . . . Thus CGA will be the measure of the time of mean anomaly. (*GW*, 3:263–65)

The idea of using the surface as equivalent to the lines therein contained will be taken up again by the practitioners of "indivisibles." But here the employment of the infinitely small is very free and not well founded. Cavalieri will cite these pages of Kepler as an example of the experiments of precursors: "Kepler was dreaming [*hallucinari*] when he believed that the surface was equivalent to the sum of the distances" (*Geometria indivisibilibus*, last page of the preface). The essential difference in Cavalieri's eyes has to do with Kepler's using infinitely small elements that were concentric—while Cavalieri's method is restricted to parallel indivisibles. The Keplerian procedure was exposed to all sorts of errors.

Kepler himself knows that the reasoning is seriously in error, and in the *Epitome* he will recognize that the result contradicts one of the premises: if the area measures the time, for an equal area the base and the height must be in an inverse ratio; the velocity at a point must therefore be inversely proportional to the *height* of the triangle, and not to the *distance*. For arcs that are very askew in relation to the center of forces, the difference can be important.

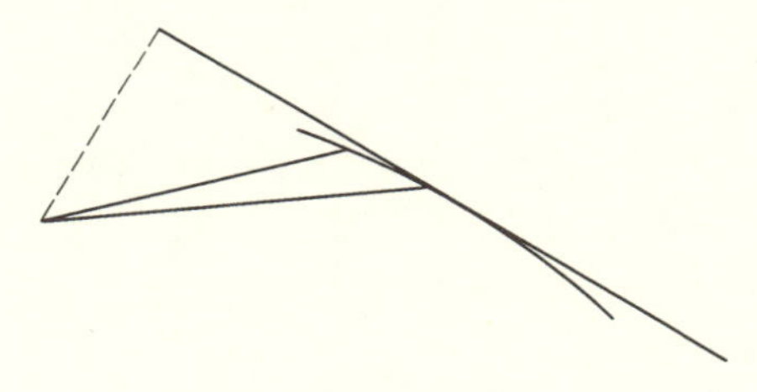

Figure 2.4

The argument thus has the following following steps: velocity inversely proportional to distance; time proportional to area; velocity inversely proportional to height . . . There must therefore be a fault in the reasoning! Kepler will sidestep this difficulty in the *Epitome*, it seems, by distinguishing the motion of rotation around the sun from the motion of approach and recession (the "libration"; see Koyré *RA*, 318–19, 344).

THE "CAUSES OF THE PERIODIC TIME" AND THE THIRD LAW

The analysis of dynamic factors is not far developed in the texts remarked on up to now. The law of areas—in the form that it takes in the *Astronomia nova*—presupposes only a very simple relation between velocity and distance. Much richer discussions on the action of forces can be drawn from Books 4 and 5 of the *Epitome*. Thus the determination of the periods of the planets requires, according to Kepler, consideration of four distinct factors:

> Four causes concur to determine the length of the periodic time:
>
> The first is the length of the path.
> The second is the weight or abundance of matter to be transported.
> The third is the vigor of the motive virtue.
> The fourth is the mass or space in which is spread out the matter to be moved. (*GW*, 7:306)

The first three "causes" are easy to admit and call to mind the Aristotelian analysis in *Physics*, 7.5: the space to be traversed (*longitudo itineris*), the weight to be moved (*pondus seu copia materiae transportandae*), and the force that must move it (*fortitudo virtutis motricis*). The fourth "cause" is less obvious; it is, one might say, the *hold* of the motive virtue, the more or less of surface that the body presents to the power that entrains it (*moles seu spatium in quo explicatur materia vehenda*). Kepler gives an analogical justification of it:

> As in the mill whose wheel is incited to turn by the impetus of the river, the wider and longer the vanes or blades that have been affixed to the wheel, the greater the force that can be drawn from the river by the machine, . . . so also, in the vortex of the solar species that is hurled out in a circle and causes the motion, the more the body occupies of space, that is, the more it occupies of the moving virtue in width and depth . . . the faster also will the body be carried forward, all other things being equal. (*GW*, 7:306–7)

How is this analysis applied in the case of the planets? It is necessary to evaluate each of the four "causes" as a function of the distances of the planets from the sun and their respective sizes. The last two causes counterbalance each other: the force diminishes as the radius, as shown above, while the *hold* offered by the planets increases as the solar distance (because the volumes of the planets, Kepler believes, are proportional to the radii of the orbits).

It remains therefore only to evaluate the contribution of the first two factors, the quantity of the matter to be moved and the length of the path. For the same period and same angle, the length varies as the radius, while the quantity of matter increases as the square root of the radius (Kepler calls this the subduplicate or halved proportion). It is therefore necessary to combine two proportionalities, a proportionality with distance and a proportionality with the square root of the distance:

> Yet the simple proportion and the subduplicate proportion of the intervals together constitute the sesquialterate proportion: therefore the periodic times are in the sesquialterate proportion [as the 3/2 power] of the intervals. (*GW*, 7:307)

The "third law" is thus deduced from a rudimentary quantitative study of the action of the sun on the planets.

The Double Action of the Sun

All that precedes is valid for circular orbits. But whence do the elliptical orbits arise? It is necessary to understand that the sun has a double action on the planets: not only does it make them revolve about itself in virtue of the species that it emits across the immense spaces of the cosmos, but it also draws the planets toward itself in the manner of a magnet.[15]

If the magnetic action were the only one, the planets would be irresistibly attracted and end by falling into the sun—at least those that present to the sun their "friendly" face. Those that turn towards it their "unfriendly" face would be pushed toward the stars (*GW*, 7:301; Koyré *RA*, 300). Each planet, in fact, is traversed by magnetic fibers that have one end friendly and the other end inimical to the sun (*GW*, 7:333). These fibers remain constantly oriented in a fixed direction whatever the trajectory of the planet. Hence they present alternately their friendly and unfriendly faces in accordance with the successive positions of the planet in its orbit. By turns the planet is thus attracted and repelled, and in certain positions the two influences balance one another. The alternate motions of approach and recession of the planet constitute what Kepler calls the *libration*; it combines with the revolution about the sun. The elliptic trajectory results from the superimposition of these two actions.

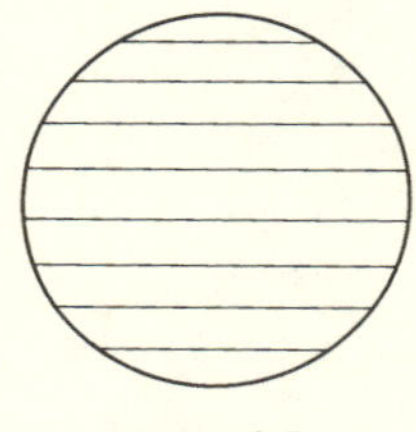

Figure 2.5

The most formidable task of the new dynamic astronomy consisted in evaluating the quantity of the libration experienced by a planet during any given portion of its path. In the first place, the time during which the planet "feels" the motive force must be taken into account; if it takes more time to traverse an arc, it will receive more of the force. The slowest planets must therefore feel more of the "libratory" force. But this factor is exactly offset by the attenuation of the force with distance. The quantity of libratory force is therefore exactly the same on equal parts of the eccentric circle (*GW*, 7:366).

It would be also necessary to evaluate the relative intensity of the magnetic action with the changing angle between the magnetic fibers and the direction to the sun: the greater inclination of angle weakens the magnetic action. In what measure?

A comparison with heat can assist in the calculation of the attenuation of the force with obliquity:

> It is thus that a ray of the Sun, considered in respect to its function of heating, heats very strongly when it strikes a plane surface at right angles. But if it strikes obliquely, it heats less, and this in the measure in which the perpendicular let fall from the Sun onto the plane extended is smaller than the oblique ray. (*GW*, 7:367)

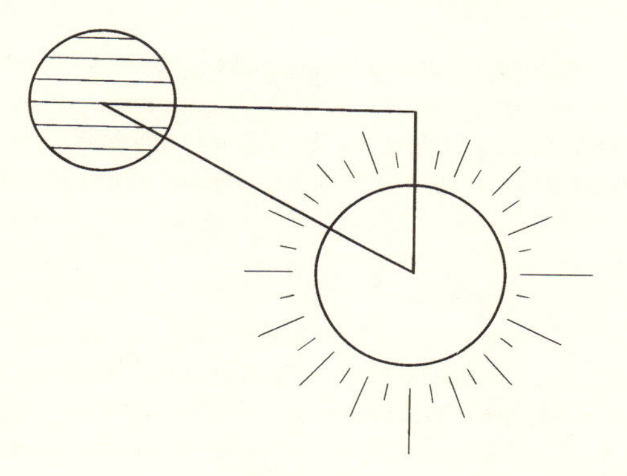

Figure 2.6

Kepler gives here not just a comparison, but another aspect of the same reality: always solar rays, whether taken with respect to their action of heating or their virtue of attracting.

Another analogy can be useful: when the beam of a balance is inclined, the obliquity weakens the tendency to motion in a certain proportion that is also given by the cosine of the angle. Here, once more, one has to do with more than an analogy because the two phenomena are aspects of the same reality:

> For as the Sun attracts [*trahit*] the planet, so does the Earth attract bodies, which are called heavy because of this traction. (*GW*, 7:367)

It is thus very plausible that the ratio of the oblique forces to the direct forces should be that of the cosine (Kepler called it "the sine of the complement"):

> It is therefore established that as EG and IH are to LC and KF, thus will be the total modulus of the forces issuing from the Sun and present in I and E, in relation to the portion that the planet admits when its fibers are in the situations EG and IH. (*GW*, 7:367)

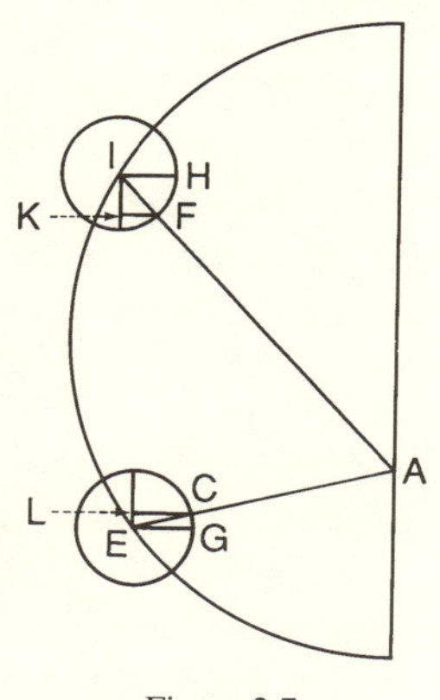

Figure 2.7

In each point of the trajectory, the planet is subject to a force that causes libration and is proportional to the cosine of the angle between the direction of the sun and the orientation of the magnetic fibers of the planet. This force can thus change sign and become repulsive.

The Deduction of the Ellipse

The next step consists in adding up all these actions of libration for a given arc. The libratory force for each point is known, and it must be calculated for an entire arc.

By means of a division of the spherical surface into least parts, Kepler demonstrates that the sum of the cosines (or sines of the complements) for a whole arc is equal, independent of sign, to the arrow (*sagitta*) or versed sine (*sinus versus*).[16] In terms of the infinitesimal calculus, this result corresponds to:

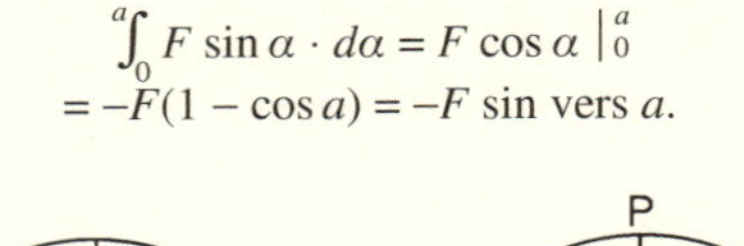

$$\int_0^a F \sin \alpha \cdot d\alpha = F \cos \alpha \Big|_0^a$$
$$= -F(1 - \cos a) = -F \text{ sin vers } a.$$

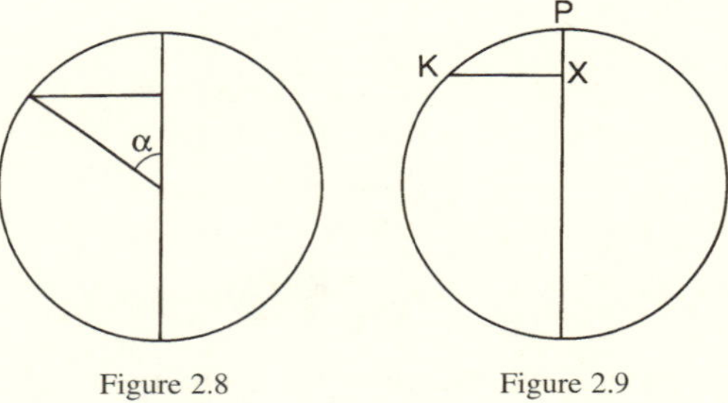

Figure 2.8 Figure 2.9

The libratory force accumulated along the arc PK is therefore proportional to the *sagitta* PX of the arc.

The libration for a semirevolution can be evaluated: in orbiting about the sun placed in A, the planet approaches it through the distance AP – AR (see fig. 2.10). On the quarter of a circle the libration is equal to the half of this difference, that is, to the length AB (in fig. 2.11), which separates the sun from the center of the eccentric circle. (At the end of the reasoning, it is understood that this distance AB is the eccentricity of an ellipse.)

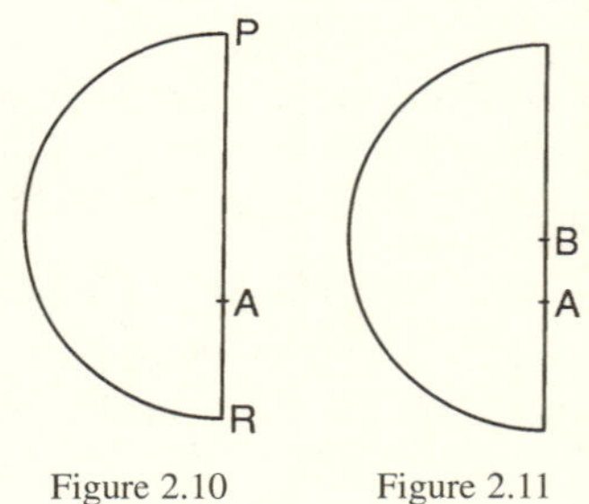

Figure 2.10 Figure 2.11

To know where the planet is after a quarter of a circle, it suffices to understand that it has approached the sun by this distance AB. If a circle is described (fig. 2.12) with center A and radius AP – AB, the intersection of this circle with the radius BD will be the position of the planet. For this particular position, the distance AE between the planet and the sun is equal to the radius BD of the circle.[17]

For an arbitrary position of the planet, for instance, after the traversal of the arc PG on the eccentric, what will be the total libration? The distance between the planet and the sun will be equal to AP diminished by the libration for the arc PG, and this libration is to the libration for a quadrant in the ratio of the corresponding *sagittae* or versed sines to one another:

$$\text{Sun} - \text{planet distance} = (\text{maximal distance}) - (\text{maximal libration}) \times (\text{vers.sin arc PG})/(\text{vers.sin } 90°),$$

or $AH = AP - BA \cdot FP/BP$.

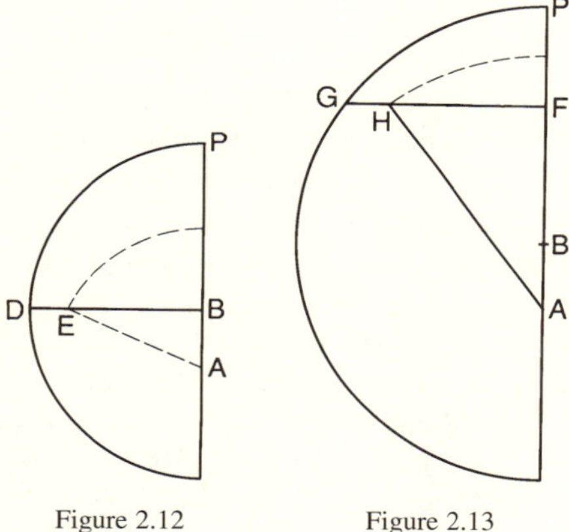

Figure 2.12 Figure 2.13

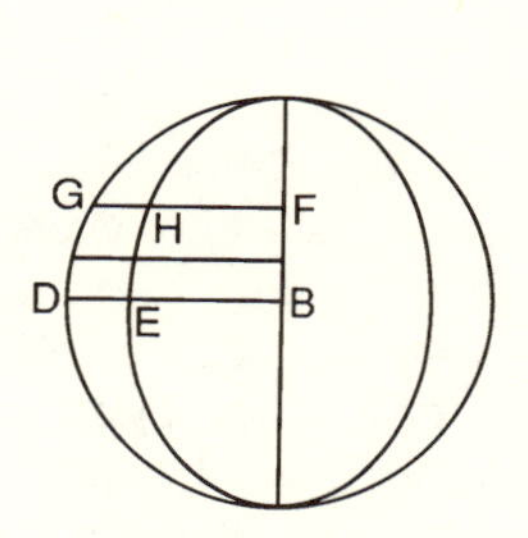

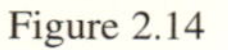
Figure 2.14

Kepler shows that these successive positions of the planet sketch an ellipse of which the sun occupies the focus A.[18] The property that Kepler considered to be fundamental in defining the ellipse is the constant affine relation between the segments HF and EB cut off by the ellipse and the corresponding segments GF and DB cut off by the circle. The ellipse, it may be said, is a circle flattened in a determinate ratio.

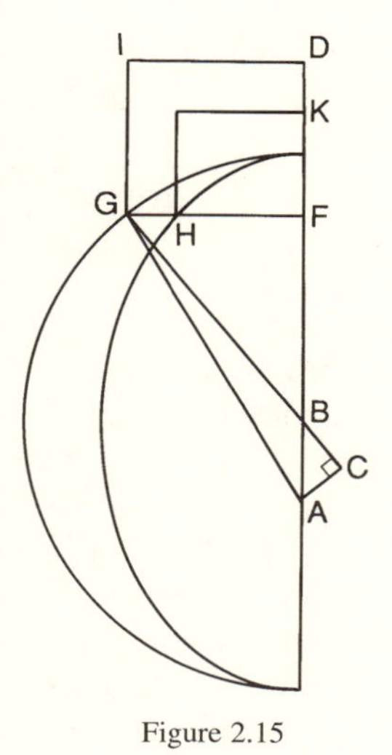

Figure 2.15

Here, schematically, are the last steps of the reasoning. Consider two auxiliary magnitudes: the segment AC let fall perpendicularly from the focus A onto the extended radius GBC; and the area in the form of a gnomon, which is the difference between the square constructed on GF and the square constructed on HF. Kepler demonstrated that this difference HIK is equal to the square on AC; then that

AC/GF is constant when G traverses the circle (it is the "eccentricity"). Therefore (GF^2/gnomon HIK) is also constant, and so also is $GF^2/(GF^2 - HIK)$, that is, GF^2/HF^2. Hence GF and HF are in a constant ratio as G traverses the circle. Consequently point H is on an ellipse (*GW*, 7:372–73).

One of the difficulties of the procedure comes from the fact that the planet does not actually traverse the eccentric circle, yet the libration received is to be evaluated as if the planet traversed the arcs of the circle. Kepler applies, so to speak, a method of "false position" in geometry. Later, Newton will be in a similar predicament (see below, p. 151): it is necessary to evaluate the force received and therefore to assume a certain trajectory, although the force in its turn modifies the trajectory.

Poor Welcome for a Dynamical Theory

The doubtfulness of this last reasoning arises from the mathematical difficulty of the problem previously called "the inverse problem": to pass from the forces to the trajectories. But this is not the burden weighing most heavily on Kepler's theory. In his attempts to evaluate the effects of the solar virtue and to derive from it the properties of the planetary motions, the great innovator did not persuade his contemporaries.

Kepler's renown in the seventeenth century was due to his works on optics and to the practical success of his *Tabulae Rudolphinae* of 1627, as recognized, for instance, by Huygens: "One finds that in general the *Rudolphines* [tables] are those that approach the heavens most closely" (1666 or 1667; *HO*, 19:261). While allowing that the Copernican system had been "perfected" by Kepler (*HO*, 21:357), Huygens accepted only the third of the "Three Laws" and proposed a circular astronomy in his plans for a planetarium (*HO*, vol. 21). He denied all action to the sun and denied, too, the Keplerian relation between velocities and distances:

> In explaining my inequality of the motion of the planets, I shall speak of the false conclusion of Kepler, who would have it that the Sun moves them, and this unequally according to the distances. (*Pensées mêlées*, *HO*, 21:350)

Even those who believed in elliptical orbits did not generally accept the celestial dynamics of the *Epitome* or the *Astronomia nova*. Ismaël Boulliau, for example, expounded a purely geometrical elliptical astronomy of Pythagorean inspiration. To the question of "whether the Sun moves the planets" (*Astronomia Philolaica*, chap. 12, 21), Boulliau replied that "the planets and the celestial bodies are moved by their proper form" (*per propriam formam moveri*, *a forma propria habent motum*, ibid.). In place of an action by the sun, he preferred to suppose "an active faculty that emanates from their form" (*agendi facultas a forma emanans*, ibid.). Boulliau challenged Kepler's arguments in detail; they were "the fictions of a too ingenious mind" (ibid., 26).[19] It is true that the planets follow elliptical paths, but not at all that the sun attracts them or acts on them.[20]

The action of the sun, according to Kepler, is multiple: it moves the planets both by the gyration of its "species" and by its quasi-magnetic attraction. To evaluate the effects, it is necessary to take account of the diffusion of each of these powers across the celestial spaces, of the relative efficacy of their actions on the planets, and of the combination of the two actions. For that, there is needed a rich panoply of analogies: it is necessary to compare the modes of diffusion and of action of light, heat, odors, and magnetic effluvia, and it is necessary to represent with clarity the action of fluids.

A cultivated reader of the seventeenth century trained in the mathematical disciplines would yet stumble and balk at many points along the thread of Kepler's argumentation. If readers were astronomers, they were used to rigorous geometrical reasonings articulated on the basis of empirical facts that were indeed only approximate, but of which the degree of approximation was well controlled. If they were experts on the science of weights and machines (the "mechanics" of the ancients), they knew how to reason quantitatively concerning weights and resistances, but not at all concerning actions at a distance or the diffusion of "species" or the ascendency of a moving medium over an obstacle.[21]

The dynamics of Newton was not to be the descendent of the Keplerian theory of forces. The latter was too conjectural and analogical. The notion of active force that served as the basis of the *De motu* does not have the richness or multiplicity of aspects of the solar virtue; it is, on the contrary, a very simple and precisely delimited notion: the weight of which Galileo speaks, the unknown but clearly determined force that causes bodies to fall to the surface of the earth.

WEIGHT AND ACCELERATION: FORCE AND TIME IN THE GALILEAN TRADITION

Newton as an Interpreter of Galileo

Whom does Newton recognize as predecessors? The only precursor that he names in relation to the first two laws is Galileo:

> Hitherto I have set down [*tradidi*[22]] principles accepted by mathematicians and confirmed by much experience. By the first two laws and first two corollaries Galileo discovered that the descent of heavy bodies is in the duplicate ratio of the times, and that the motion of projectiles is made in a parabola; experience agreeing, except insofar as those motions are a little retarded by the resistance of the air. (*Princ.*, 21)

The principles expounded at the beginning of the *Principia* are here asserted to be in the direct line of what is already accepted in "mathematics," and moreover their soundness is confirmed by experience.[23] Thus the third law (equality of action and reaction) will be considered as guaranteed, or at least as rendered plausible, by results obtained for impacts (Wren, Wallis, Huygens; *Princ.*, 22–25) and simple machines (*Princ.*, 26–27). The work of Galileo, a mathematician and natural philosopher attentive to experience, is guarantee for the truth of the first two laws (the principle of inertia and the proportionality between force impressed and change of motion).

This interpretation may elicit surprise. Newton supposes that Galileo relied on these two laws in elaborating his doctrine of weight (it was "by" [*per*] them that he "discovered" [*invenit*] . . .). This may be a too striking and overly condensed account or a purely retrospective interpretation. The possible connection between Galileo's results and the Newtonian notion of impressed force is clarified in the following addition to the second edition of the *Principia*:

> When a body falls, its uniform gravity [*gravitas uniformis*],[24] acting equally in the equal single particles of time, impresses in the body equal forces and generates equal velocities; and in the total time it impresses a total force and generates a total velocity proportional to the time. (*Princ.*, 21)

It is not said that gravity or weight is a force, but that it is a source of forces, that it impresses forces. This impression is measured by the time that elapses: at each instant there is added a new force (weight acts "equally" in each "particle" of time). The addition of force is a modification of the velocity (in accordance with the second law of motion): weight adds or subtracts velocity according to whether the fall is or is not in the same direction as the velocity already acquired. All the elementary forces associated with the instants of time together make a total force because the impressed forces are conserved (by the first law of motion) and because the elementary forces collected together compose a "total" force, just as the particles of time compose a total time.

Besides the two laws of motion, it is necessary to lay down a fundamental hypothesis that Newton does not always make explicit: force acts in time; it is regulated according to the time. This is the same assumption that Newton had already adopted in his correspondence with Hooke: to study the action of weight, he said, it was necessary to consider

> ye innumerable converging motions successively generated by ye impresses of gravity in every moment of it's passage . . . The innumerable & infinitly little motions . . . continually generated by gravity. (*Corresp.*, 2:307–8)

Otherwise put:

> These motions are proportional to ye time they are generated in. (*Corresp.*, 2:308; see below, p. 151ff.)

Newton says: "weight impresses forces" (*gravitas imprimit vires*). For Galileo, weight is a source of *momenti* produced one after another in time. These momenti generate at each instant a new velocity, which comes to be added to the velocity already acquired, or to diminish it if it is in the opposite sense. Galileo's *momento* or (in Latin passages) *momentum* is not to be identified either with Newton's "force" or with "momentum" in its modern usage as mass × velocity. The term contains a fertile ambiguity that should be maintained and which no English term is able to translate. In the following discussion, therefore, *momento* will stand for itself.

That the momenti are regulated according to time is not self-evident. Galileo himself had at first believed in a proportionality between velocity and space traversed, and then he had discovered that the acceleration of fall occurs according to the increase in the time. What the course of his thought had been between 1604 (Galileo to Sarpi, EN, 10:115) and 1632 (EN, 7:254ff.) and what motives led him to transform his theory, no one appears to be able to say with certainty. Too many steps are missing in the journey to permit reconstructing this itinerary and to reply to a question that presents itself: is the momento that explains the effects of machines "the same" as the elementary momento of weight associated with each instant? An in-depth response to this question is beyond the scope of this book: it would require a complete picture of the Galilean theory of motion (along the lines of those limned by A. Koyré, M. Clavelin, W. Wisan, and P. Galluzi).

This more modest presentation of the Galilean doctrine has as its goal a better understanding of Newton, and it also aims to point out certain possible connections between the ancient science of machines and the dynamics of the seventeenth century. The quantitative study of forces in antiquity and the Middle Ages was undertaken in order to explain and reduce to law the surprising effects of "machines." Galileo's intellectual enterprise might be defined, briefly, as the attempt to extend to new domains the concepts and modes of reasoning of the science of machines. Weight is only one aspect of this extension, and much is gained by situating the theory of weight in the larger framework that includes, in particular, the action of fluids and percussion.

Galileo's successors understood this better than did Galileo himself, and it will be shown that in his writings, Torricelli extracted from the teaching of the master a conception of the momento of weight that was more explicit, clearer, and simpler than Galileo's.

Newton's theory presents itself as a return to and generalization of Galileo's theory. But other great thinkers did not accept the latter so forthrightly: Huygens attempted to avoid the Galilean definition of weight as regulated by time. Others, in contrast, gave a justification of Galileo's results with new mathematical tools: Torricelli and Barrow demonstrated the fundamental theorems in a framework that presupposed the language of indivisibles.

The Metamorphoses of Force in Simple Machines

The evaluation of forces and their effects was gradually clarified over the centuries in the study of simple machines: the lever, the balance, windlass, wedge, and so forth. The sources of force were principally of two kinds: weight and the muscular vigor of man or animal. The motive power of running water, for instance, was reduced to weight. (What of the force of the wind?) The marvellous power of gunpowder, or as Galileo terms it, "the vigor of fire" (EN, 8:323), is excluded from mechanical operations because it seems to escape from all determination and proportion. By contrast, the force of weight—and of all that is reduced to it—can enter into quantitative determinations and finite relations. Muscular vigor itself must always be able to be translated and measured by a certain weight, as Leonardo da Vinci wrote in a picturesque aphorism:

> The greatest force that a man can deploy . . . is that which he will obtain by putting his feet on the pan of a balance, then supporting his shoulders against something solid; he will thus raise in the other pan of the balance as much weight as he weighs himself, and will carry on his shoulders as much weight as he has of force. (Leonardo da Vinci, *Les Manuscrits*, MS A, fol. 30v)

It is ultimately the notion of weight that is the basis of the theory.

But weight by itself explains nothing. How is it that unequal weights can be made to equilibrate at the two ends of the lever, if not because weight acts not in isolation but combines with other factors? Several ways of reasoning are possible. One can view the problem geometrically and choose as a fundamental combination the product of the weight and the distance to the point of support. Or one can choose a more "dynamic" mode of analysis.[25]

For instance, the *Mechanical Problems* [Aristotle?], which had its origin in the entourage of Aristotle and was rediscovered in the Renaissance, bases all explanation of mechanical effects on the study of the force's path. A weight placed at the end of a lever traverses a certain arc of a circle, and the form of this circular arc modifies the effect of the weight. The trajectory is more or less incurved, according to whether the radius of the circle is smaller or greater, and the force is thus more or less turned aside. The unknown author of the *Mechanical Problems* takes it as a fundamental axiom that the same force (*ischus*) must generate a greater displacement if it is less hindered or turned aside. The weight will therefore have more effect as the circle on which it moves is less incurved. It results from this that the weight suspended on the greater arm of the lever generates a greater or more rapid displacement.[26] On a balance of unequal arms, a small weight can thus equilibrate a large weight and traverse a greater circular arc because the efficacy of the small weight is less hampered by curvature. Applied initially to the lever and the balance, this explication can be extended gradually to other mechanical effects: windlasses, gears, pliers or tongs, distribution of the motive action on sails or oars, and so forth.

This mode of reasoning is dynamic in inspiration and contrasts with the Archimedean manner of presenting the same phenomena: in *The Equilibrium of Planes*, Archimedes developed a type of geometry of weight and of centers of gravity that posed as a starting point for his demonstrations that "equal weights at equal distances balance." Archimedes' treatise inaugurated a more geometrical tradition in the study of simple machines, which may be less interesting from the perspective of this book since the very notion of force disappears, and perhaps also less fertile in consequences since it is limited to certain phenomena. The product of the weight and the distance to the point of support cannot serve, for instance, to account for the inclined plane.

The dynamical study of the effects of machines considers the transmission and conservation of force. Force undergoes various modifications when it enters into gearwork and the articulations of a system of machines, but it must remain "the same" force, in a sense that needs to be clarified. What, exactly, is the invariant, the magnitude conserved through these transformations? According to Hero of Alexandria and Leonardo da Vinci, that gain in power does not occur without slowing down:

> The more a force is extended from [gear]wheel to [gear]wheel, from lever to lever, or from screw to screw, the more powerful and slow [*potente e tarda*] it becomes. (Leonardo da Vinci, *Les Manuscrits*, MS A, fol. 35v)

> If a [gear]wheel moves a machine, it is impossible that, without doubling the time, it should move two [machines]; therefore, [it is impossible that] it do as much work [*tanta operatione*] in an hour as two other machines, also in an hour; thus the same wheel can cause to turn an infinity of machines, but in a very long time they will not do more work than the first machine in an hour. (Ibid., fol. 30v)

From this observation Galileo will draw a general principle of great import: the velocity of displacement compensates for the weakness of the force.

Treated in this way, the science of machines is not at all "statics"; it is not limited to the study of mechanical systems in equilibrium. Velocity and displacement are among the primitive terms of the explanatory theory; displacement is not solely virtual displacement. Simple machines like the lever and the pulley are then a particular case of the couple *force-resistance*, and the same principles that are applied to the study of the lever can also render an account of the action of a mover on a body to be moved or of the motion of a body against a resisting medium. Thus in the little treatise *Mechanical Problems*, alongside such traditional problems as the balance with unequal arms, there is a discussion of projectiles, vortices, and different sorts of gyratory motions.

The Deposition of Force

The idea of a metamorphosis of force can be applied outside the narrow framework of simple machines and can be extended to the whole of nature. Conservation, transmission, accumulation, and loss—none of these avatars of force are limited to mechanical effects. The force of a machine can be transformed into the force of a projectile:

> With as much force as you have drawn your crossbow, with just so much will the arrow be launched by it. (Ibid., fol. 30r)

> Every crossbow charged in the same time with the same force, however varied its size, length or weight, will always cast the same weight to the same distance. (Ibid., fol. 32v)

It is in the operations of machines that the laws of the transmission of force are demonstrated, but the whole of nature is the theater of the *forza* of which Leonardo speaks, and which gives life to material bodies by being modified and consumed:

> Force is a spiritual virtue, an invisible power that . . . is introduced and infused into bodies, which thus find themselves drawn and turned aside from their natural habit; it gives them an active life of a marvellous power, it constrains all created things to change form and place, it runs with fury to the death it desires, and as it goes diversifies itself in accordance with causes. Slowness makes it great and velocity makes it feeble; it is born through violence and dies through liberty. (Ibid., fol. 34v)

Force cannot be seen or touched, yet in its way it subsists and can be conserved in bodies. Certain astonishing effects are explained when one considers the force that is stored in the course of time. For example, Torricelli (in a lecture given in 1642) asks how it is possible that a small boat striking against the wharf with great speed makes much less of an effect than a large ship striking the wharf with a very small speed.

> Let us imagine in a pond or extremely calm port a large galleon, separated from the wharf by, say, ten paces, and a man draws it by means of a rope with all his force.
>
> For my part, I believe that the vessel, however slowly it moves, when it comes to strike, will give such a blow to the wharf that it could cause a tower to tremble.
>
> If the same man, from the same distance, with the same force and on the same tranquil water, drew a small felucca or rather a very light plank of fir, this too on reaching the wharf will strike it, and with a much greater velocity than the galleon; yet I believe it will not have a thousandth part of the effect made by the large vessel.

Let us seek the cause of this diversity of effect. Here the force of the blow does not come from the velocity, since the plank strikes with a much greater velocity than the ship; the power that drew the one and the other was the same; and yet the greater mass made the greater effect.

Does it remain then, could one say, that the cause is to be attributed to the matter? I am of opinion that neither does the matter do anything here. It is certain that the matter by itself is dead, and serves only to impede and to resist the virtue that operates. The matter is nothing else than an enchanted vase of Circe, that serves as a receptacle for the force and the momenti. Yet the force and the momenti are abstractions so subtle, they are quintessences so spiritual that they cannot be enclosed in other vials than in the intimate corpulence of natural solids.

Here, then, is my opinion: the force of the man who pulls is what acts and gives the blow. I do not mean the force that he exerts in the instant of time in which the wood comes to give the blow, but all that he had earlier exerted from the beginning until the end of the motion.

If we ask during what length of time the strain endured while he pulled the galleon, the response might be that to move this great machine over a space of twenty paces would require a half hour of time and continual strain. By contrast, in drawing the small piece of wood, one would need to exert effort through less than four musical beats.

Yet the force that continually during the space of a half hour sprang forth as from a live source from the arms and nerves of this hauler, has not at all vanished in smoke nor flown away in the air. It would have vanished if the galleon had not been able to move, and it would have been entirely extinguished by a reef or an obstacle that impeded the motion.

In reality it was all impressed in the viscera of the beams and panels of which the vessel was composed and with which it was laden, and there within it was conserved and accumulated, except for the little that the resistance of the water may have taken away.

What wonder is there if this blow, which carries with it the momenti accumulated during a half hour, has a greater effect than that which carries in itself only the forces and momenti accumulated in four musical beats? (*Academic Lectures*, *TO*, 2:27–28)

This lovely text of Torricelli is, as it were, a résumé of Galileo's teaching: no one has understood anything of force who has not taken account of the time. Why did the great galleon on colliding cause the wharf to tremble so, while the much more rapid plank produced hardly any effect? Because the plank had been pulled to the wharf in a few minutes, while a half hour was required to haul the ship. During all this time the hauler did not cease to impress force on the enormous mass. The key to the phenomenon is in this difference in duration: the ship has accumulated the momenti of the hauler during a much longer time. At the instant of impact, the momenti accumulated during the preceding time are liberated. The motive power has deposited or impressed momento moment after moment, and these momenti are expressed by an increase in velocity that is different in bodies of different mass.

The force itself can be transmitted, can change state, and can dissipate, but it exists always in a body. This aspect interested Maxwell—he cites this text of Torricelli in the last paragraph of his *Treatise on Electricity and Magnetism*: energy must always reside in a material substance, even when it appears to be transmitted to a distance without intermediary medium. Even what is called dissipation is not really such: if a force is "extinguished" by an obstacle, it is because it passes into the obstacle; the reef absorbs the momenti of the ship.

Torricelli's analysis could serve as a guide for clarifying the Galilean theory of weight. The hauler, the human source whence "gushes forth" the force, can be replaced by another mysterious and inexhaustible source: weight, which is also "a fountain of momenti," as Torricelli calls it in another lecture, which will be mentioned later. Weight is an invisible reservoir of successive momenti, and time is its regulator.

But there is a distance yet to traverse across Galilean terrain before it can be seen how the accumulation of successive momenti may serve as key in the analysis of weight.

Velocity and Momento in the *Mechanics* of Galileo

The supreme principle that guides Galileo in his study of mechanical effects is that nature does not let herself be vanquished or deceived: a force can never overcome a superior resistance (EN, 2:155; 8:328, 572, etc.)—at least if it is clearly seen what is meant by force and resistance and none of the factors (e.g., space and time) are neglected. If one takes account only of the most apparent resistance (the weight to be moved) and of the force that must overcome it, one will understand nothing of machines—one will believe that a small weight can move another that is much greater.

But if one pays attention as well to space and to time, that is, to velocity, the paradox is clarified: the small force overcomes a weight ten times greater than itself because it has employed ten times more of time or distance. Everything happens as if the small force traversed the path traversed by the large resistance not once but ten times, transporting each time the tenth part of the resistance. It is in this way that Galileo reasons in his small treatise *Mechanics*, probably written in 1593 (according to Viviani) for his students in Padua:

> I have seen all the mechanicians go wrong in wishing to apply machines to operations that are impossible by their very nature. . . . Of these errors it seems to me that the principal reason is the faith these men have had and continue to have in the art of being able with little force to move and raise very great weights; outwitting Nature in some way with their machines. But Nature's instinct and

well-established constitution is such that no resistance can be overcome by a force that is not stronger than it. . . .

At the beginning of our considerations there are four things to take into account: the first is the weight to be transported from one place to another; the second is the force or power that must move it; the third, the distance between the two termini of motion; the fourth concerns the time in which this change is to take place, and this time reduces to the same thing as the swiftness in velocity of motion, since a motion is determined as swifter than another if in a shorter time it traverses an equal distance.[27]

Yet, if one has assigned a certain resistance, no matter what, and determined upon any force and designated any distance, there is no doubt that the given force will drive the given weight over the designated distance, because even if the force is very small, the weight may be divided into many particles, each of which is less than the force, and transporting them one at a time, the force will in the end have driven the whole weight over the designated distance; and at the end of the operation one cannot rightly say that the great weight has been moved and transported by a force much smaller than it, but rather that the force has several times recommenced its motion and several times traversed the space which was traversed a single time by the entire weight.

Whence it appears that the velocity of the force has been as many times greater than the resistance of the weight, as this weight is greater than the force, since in the time in which the moving force has measured many times the interval between the termini of motion, the weight has passed over it only once; and one must not then say that a great resistance has been overcome by a small force, contrary to the constitution of Nature. One could say that the constitution of Nature had been vanquished only if the small force, in transporting the greater resistance, had moved with a velocity equal to that with which the resistance was moved, a thing that we affirm to be absolutely impossible, in any machine that one can imagine.

But because it could happen that, having only a small force, one needed to move a great weight all together, without dividing it into pieces, it will then be necessary to have recourse to a machine, by means of which one will transport the proposed weight over the space assigned, with the given force. But one cannot prevent that the same force has to traverse a great deal of distance, measuring this same space again and again, as many times as it is exceeded by the weight.

So that at the end of the action we shall not have received other advantage from the machine than that of having transported the said weight with the given force to the given distance, but all together. The which weight, divided into parts and without other machine, would have been transported by the same force within the same time and over the same interval.

And this is one of the benefits that one derives from machines: because it often happens that, having but a slight amount of force, but much time, one moves a great weight while keeping it intact. But if someone should wish and attempt, by means of machines, to obtain the same effect without increasing the slowness of the body moved, he would find himself mistaken, and would show that he did not understand the force of mechanical instruments and the reasons for their effects. (EN, 2:155–57)

A machine is thus a procedure for avoiding the many successive voyages that the small force would have to make and that are impossible if the resistance is to remain intact and undivided. The machine collects this succession of journeys into a single displacement, but the "reason" of the operation is the same: the small force must traverse three times the distance that the resistance traverses if it is three times weaker than the resistance.

This inequality of velocities compensates for the difference in weight and makes it possible to explain, for example, what happens on a lever or balance of unequal arms:

Let us consider the balance AB divided into unequal parts at the point C, and let the weights suspended at the points A and B have the same proportion as the distances CB and CA; it is manifest that the one will counterbalance the other, and hence if one were to add to one of the two a very small momento of weight [*un*

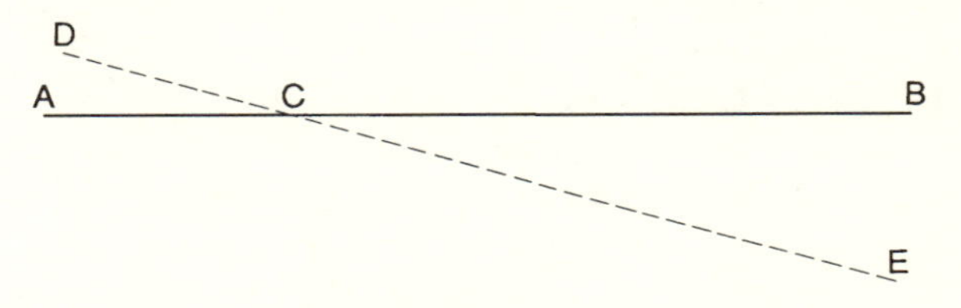

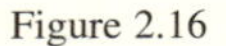

Figure 2.16

minimo momento di gravitá], it would move downward, causing the other to rise; so that if one adds an insensible weight to the heavy body B, the balance will move, descending from the point B towards E, and the other end will mount from A towards D. And because, in order to cause B to descend, it suffices to add to it no matter how small a weight, we shall not take account of this insensible quantity, and we shall not distinguish between the fact that one weight can be sustained by another, and the fact that it can move it.

Let us now consider the motion that the heavy body B makes in descending to E, and that which the other, A, makes in mounting to D. We find without doubt that the space BE is greater than the space AD in the same proportion as the distance BC is greater than the distance CA, since at the center C there are two equal angles at the summit, namely DCA and ECB, and consequently two similar circular arcs, AD and BE, and these two have the same proportion as the radii BC and CA with which they are described.

Hence the velocity of the weight B which descends is found to be greater than the velocity of the weight A which ascends, in the proportion in which the weight A exceeds the weight B; the weight A not being able to rise to D except slowly, while the other weight B moves rapidly to E.

> This is not, then, a marvel, nor something foreign to the natural constitution, that the velocity of motion of the weight B compensates for the greater resistance of the weight A, while the latter moves slowly to D and the former descends rapidly to E. And equally in reverse, if the weight A is placed in the point D, and the weight B in the point E, it will not be unreasonable that the first can, by descending slowly to A, raise the other rapidly to B, restoring by its weight what it loses by the slowness of its motion.
>
> And this discourse can show us how the velocity of motion has the power to increase the momento in the moved body. (EN, 2:163–64)

The Italian term *momento* can designate very different aspects of weight and the tendency to move: at the beginning of the text, it is the minimal impulse that can disturb the equilibrium and is not to be included in the evaluation because it is "insensible"; in the last sentence, on the contrary, the momento is the force or impulsion proper to a heavy body, and increase in velocity increases it.

Momento is not exactly the same thing as weight. Galileo decided to designate by momento the significant magnitude intervening in mechanical effects and in natural operations that obey laws analogous to mechanical laws.[28] This magnitude is not weight simply, but a modification or modulation of weight. Momento is "the propensity to go downward," which derives not only from weight, but also from the respective situations of different bodies. This compound propensity is the heir of the ῥοπή or ῥεπειν of the Greek authors (Aristotle, Archimedes, Eutocius): each heavy body exercises a tendency to motion, a pressure or inclination (ῥοπή), which can depend on factors other than simple weight. For Archimedes this tendency depended on the situation in relation to the point of support.

Galileo did not adopt Archimedes' point of view. Rather than the geometrical *situs*, it is the velocity that modifies the momento of the heavy body. Galileo distanced himself from a strictly geometrical method by admitting velocity and time as the decisive parameters of the analysis. It is the comparison of velocities that makes it possible to understand why the small force overcomes the large resistance. The large velocity "compensates for" the small weight—it "restores" or "equalizes" it. The true magnitude to be taken into account in mechanical effects and in whatever resembles them, the magnitude that is conserved through levers and gearwheels and which renders the marvels of the art of machines "natural" and intelligible, is the combination of weight and velocity.

The Critique of Virtual Entities

Yet the introduction of velocity into the analysis of equilibrium did not proceed without difficulty. Why explain by a displacement the equality between two magnitudes that are perfectly motionless?

Stevin had already criticized this procedure in a syllogism:

> The reason why equal weights, suspended from equal radii, are in equilibrium, is known from a common maxim, but not the cause of the equilibration of unequal weights from unequal radii, proportional to them [the weights]: this cause having been recognized by the Ancients, they judged that it was hidden under the description of the circumferences described by the ends of the radii, as is seen in Aristotle in his *Mechanics*, and in his followers. Which we deny for this reason:
>
> E. What remains motionless, being suspended, describes no circumference;
> A. Two weights suspended in equilibrium are motionless;
> E. Therefore, two weights suspended in equilibrium do not describe any circumference. (Stevin, *Oeuvres mathématiques*, 501)

In a manuscript written toward the end of his life, and probably intended for the *Discorsi*, Galileo himself criticizes the recourse to virtual entities:

> Sagredo: Let a balance with unequal arms be supported at the point C, with AC the greater arm and BC the smaller; we seek the cause why it is that, if two equal weights A, B are placed at the extremities, the balance does not remain at rest and in equilibrium, but inclines to the side of the greater arm, moving into [the position] EF.

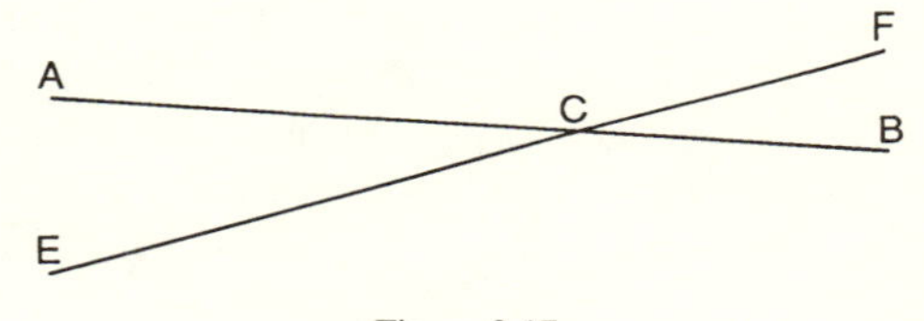

Figure 2.17

> The reason usually assigned for this is that the velocity of the weight A, during its descent, would be greater than the velocity of the weight B, since the distance CA is greater than CB; whence it follows that the body A, which is equal to B in weight, surpasses it in the momento of the velocity, and therefore prevails over it and descends in causing it to rise.
>
> One doubts the value of such a reason, which does not appear to have a conclusive force: for though it is true that the momento of a heavy body increases if one adds velocity to it, and that this momento surpasses the momento of an equally heavy body that is at rest; yet if both are at rest, that is, where there is neither motion nor velocity of one greater than the other, to claim that this inequality which is not, but has yet to be, could produce a present effect, is hard to understand, and indeed I see a difficulty there.
>
> Salviati: You are right to put that in doubt. I too have been discontented with such an explanation; but by a different route I have found one that satisfies me. (Fragments, EN, 8:438)

The Generalization of the Mechanical Momento

This hesitation attests to a philosophic carefulness in Galileo, but it is rather marginal in the development of his ideas on nature and forces. Having brought to light an invariant in the case of mechanical effects, Galileo generalizes it in extending it to the motion of fluids. Thus he prefaces his study of floating bodies (*Discorso intorno alle cose che stanno in su l'acqua*) with the following principles and explanations:

> I borrow two principles from the science of machines. The first is that weights that are absolutely equal and moved with equal velocities, have equal forces and momenti in their operation.
>
> Momento among the mechanicians signifies this virtue, this force, this efficacy with which the mover moves the resisting body; which virtue depends not only on simple weight, but on velocity of motion, and on the diverse inclinations of the paths along which the motion occurs, because a heavy body that descends along a very inclined path has more impetus than one descending along a less inclined path. . . .
>
> The second principle is that the momento and force of the weight increases with the velocity of motion, so that weights that are absolutely equal, but conjoined with unequal velocities, are of unequal force, momento, and virtue, the more rapid being the more powerful in the proportion of its velocity to that of the other. . . .
>
> Such a compensation between gravity and velocity is found in all the mechanical instruments, and was considered as a principle by Aristotle in his Mechanical Questions [= *Mechanics*]. (EN, 4:68–69)

The idea of a compensation between weight and velocity is a thread running through the whole of Galileo's career, and in a text that is probably one of the last by the old master—the scholium that he dictated to Viviani for insertion in the Third Day of the *Discorsi*, for a new, posthumous edition—there is this passage:

> Thus we may assert and affirm that when equilibrium (that is, rest) is to prevail between two mobile bodies, their momento, their velocity or their propensity to motion—that is, the spaces they would traverse in the same time—must correspond reciprocally to their weights, exactly as is demonstrated in all cases of mechanical motions. (EN, 8:217)

In this particular passage, momento designates only the velocity (real or virtual) and not the combination weight-velocity. But the idea remains the same, and with the help of this principle of compensation, Galileo demonstrates the laws of motion on an inclined plane. He shows first how the vertical momento of a heavy body, which he calls its "total momento" (*momento totale*), is diminished by the inclined plane, and then derives the proportion between the oblique, partial momento (*momento parziale*) along the inclined plane and the total momento.

Pendulum and Percussion as Extensions of Mechanical Effects

The adequate measure of force in natural effects is therefore not weight but rather a compound magnitude—the momento in which velocity comes to modify or modulate weight. A small weight can move a greater, provided that their velocities are inversely proportional. It is as if the small power went faster in order to condense into the same interval of time the actions that it would have spread out over a longer time. In face of a very heavy and imposing load, the most natural behavior would be to seek to divide it, in ten for instance, so as to move it a little at a time. But if it cannot be divided into pieces, the ten trips into which the action would have been divided will have to be carried out concurrently in one move. Since the job cannot be divided into ten trips, the force must travel ten times faster. Machines are, so to speak, artifices for shortening time.

This idea of an amassing of successive actions, a concentration of operations in time, much resembles a pedagogical or rhetorical fiction. But Galileo extends it to new phenomena in which the accumulation of motive actions—momenti—is not at all fictive. The explanations of percussion and the pendulum presuppose the concentrating of momenti accumulated gradually and successively into a single result—global momento—and are illustrated in a short Galilean piece about machines, which has its picturesqueness.

Galileo, head philosopher and mathematician at the Medici court, was to give his advice on an extraordinary machine that an inventor had proposed to the grand duke. It was a question of a bizarre, composite instrument that combined the effects of the lever, the winch, the pendulum, and percussion. The analysis of this apparatus gives Galileo the pretext for a small tract explaining the different kinds of machines. After a general discussion on the equality that nature always maintains between force and resistance, Galileo retraces the grand lines of his explanation of mechanical effects:

> We should do well to consider wherein this adjustment between Art and Nature consists. The calculation and the reason for it are very easy and clear, since all is equalized by virtue of the speed and slowness of the motion, or let me say the slowness and length of time. It is true that a single man whose force has a momento sufficient for one hundred pounds, will raise and pull along the ground ten thousand pounds of weight, but if we pay attention to the length of the trip the man makes and to that made by the column [to be transported], we shall find that when the latter has been displaced a *braccia*, the mover has traversed a hundred, which

> reduces to saying that the mover is displaced a hundred times more rapidly than the column.
>
> Whence one sees that by equalizing the parts, if this stone had been divided into one hundred equal parts, each would have been of one hundred pounds, and therefore equivalent to the force of the mover, who in one hundred trips of one braccia each would have transported the hundred pieces of stone the distance of a braccia, moving at the same velocity, that is to say, in the same time.
>
> The advantage of the windlass is not therefore that it diminishes the fatigue or the time, but that the column is transported whole and not in pieces that one could then reattach and unite into one as required for our usage. Whence one sees that if the weight to be transported were a vessel of water of one hundred barrels, it would bring me little or no advantage to be able to transport with the windlass the entire large container, filled, in one trip, using the force of the one man, rather than transport it by means of the same man, in just as much time, but barrel by barrel in one hundred trips, given that the water reunites together and becomes again a single mass.
>
> There are two other ways, in appearance different from this one, that Art has found for overcoming very great resistances with very little force. One is impact [*urto*], or shall I say, a blow or percussion, to which, it seems, there is no resistance that does not give way.
>
> The other consists in bringing about, so to speak, a conservation or amassing together of forces. This occurs if I impress my force, which is, say, of ten degrees, on a mobile body that conserves it, and then again I impress my force, so that joined with the first ten degrees there are now twenty in the body, which conserves them; and continuing, yet more times, to impress ten degrees and again ten, one will unite in this reservoir a hundred, two hundred, or a thousand degrees of a virtue powerful enough to overcome very great resistances, against which my simple virtue of ten degrees would be without effect.
>
> For such a conservation of force, we have an appropriate example in this very heavy pendulum that you have adapted to this lever; by receiving impulsions from the very weak force and putting them into reserve, it collects them and makes, so to speak, a very great capital, that it can then distribute freely and apply to the overcoming of resistances that the first force was far from being able to overcome. (Fragments, EN, 8:573–75)

Thus two new phenomena come to be ranged under the rubric of mechanical effects: the pendulum and percussion. In both cases a small force becomes capable of overcoming a much greater resistance.

The pendulum is not here a means of measuring time or studying motion and isochronism; it is considered as a machine for accumulating forces. At each return a new impulsion is impressed on the suspended body. Maintaining and reinforcing the oscillations puts in reserve the successive impulsions. This way of utilizing the pendulum is after all the most usual one. Consider, for instance, the bell ringer who gives progressively more amplitude to the motion of the bell, each impulsion being added to the previous one. Or the besiegers who, having suspended a battering ram on ropes, impress on it a to-and-fro motion before striking the enemy's gate. In both these cases, moreover, the effect of the pendulum is finally realized in the form of percussion. The deposition of force takes place in a discontinuous manner by discrete impulsions, but its principle does not differ from the one explained by Torricelli in connection with the ship and the small plank: the momenti are "capitalized" in matter, which will deliver them all at once if there is an impact.

The Amassing of Momenti in Percussion

Galileo returns to this idea in his posthumous text on percussion, which describes the deposition of force in a heavy door and in a bell. In the first of these two cases, the force is exercised continuously, and in the second, by means of successive impulsions:

> He who closes the bronze doors of San Giovanni would attempt in vain to close them with a single, unique and simple pressure; but with a continual impulsion, he impresses on this very heavy body a force such that, when the door comes to strike against the doorsill, it causes the whole church to tremble. By this one sees how one can impress on bodies—including the heaviest—and multiply and conserve in them the force that is in a certain time communicated to them.
>
> We can see a similar effect in a great bell. It is not by pulling the chord a single time, nor even four or six times, that the bell is put into vigorous and impetuous motion, but by pulling it a great many times, and reiterating the pull over a long interval; the last tugs add force to the force acquired by means of all the preceding tugs, back to the first. The larger and heavier the bell, the more it will acquire of force and impetus, which must be communicated to it over a longer time and by virtue of a greater number of tugs than would be necessary for a small bell. The latter, by contrast, acquires its impetus rapidly, but also loses it more quickly, not being impregnated (so to speak) with as great a force as the larger bell. (Sixth Day, EN, 8:345–46)

The two phenomena that Galileo had evoked previously—the pendulum and percussion—are here treated exactly in the same way. The tugs given successively to the bell are of the same nature as the continual impulsion (*impulso continuato*) that it was necessary to give to the bronze door to close it. (In Newton's *De motu*, examples of a more uncommon nature illustrate this parallel treatment of successive impulsions and continuous force: the continuous case is approached as a limit of a series of impulsions; see above, pp. 22–26).

The analysis that Galileo proposes is subject to a severe restriction: it applies only to "violent" percussion, that which is the result of a force "exterior" to the body itself. The force is impressed on the door or in the bell by a person and not by the weight of the object itself. (It is indeed the *forza* of which Leonardo speaks, that which gives life to bodies by being deposited in them and being freed at the moment of impact.)

Could "natural" percussion, that which results from weight, be analyzed along the same lines? Galileo's same posthumous text treats similarly the "natural" or "internal" force of weight and the external violence that ends in impact.[29] Both deposit in the body, instant after instant, new momenti, that are expressed by new degrees of velocity:

> The momento of a heavy body, in the act of percussion, is nothing else than the composite and aggregate of an infinity of momenti, each equal to a single momento, whether it is the internal and natural momento of the body itself (which is the momento of the body's own absolute weight, exerted eternally when it is placed atop any body that resists it), or the external and violent momento which is that of motive force.
>
> Such momenti, during the time in which the heavy body is moved, are accumulated from instant to instant with a uniform augmentation, and are conserved in the body, precisely in the way that the velocity of a heavy body that falls is increased. Just as in the infinity of instants of even a very short time, a heavy body passes always through new and equal degrees of velocity while always retaining those that it has acquired in the time already elapsed, so also in the moved body are conserved and compounded from instant to instant these momenti, whether they be natural or violent, conferred by nature or by art. (Sixth Day, EN, 8:344)

The same body can receive two sorts of momenti, either "violent," coming from an exterior force, or natural, due to its own weight. This weight is in reality linked to time. The body exercises it "eternally" against the horizontal support that holds it up; once the body is freed from the support, the momenti do not disappear, but are accumulated "instant after instant." This internal and natural momento (*momento interno e naturale*) obeys the same laws as the external or violent momento (*momento estrinseco o violente*). Like the momenti given by the porter of the church of San Giovanni on the heavy bronze door, the momenti of weight are deposited in the body, instant after instant, and accumulate in it.

Yet Galileo does not draw from this phenomenon the consequence that seems to follow: What is the connection between the momento and the degree of velocity? Should not the acceleration of a body in free fall be exactly this accumulation in time of the elementary momenti of weight? Galileo maintains a fragile distinction between the momenti that accumulate and the degrees of new velocity that are added to those which have been retained previously.[30]

The momenti are accumulated "precisely in the way" that the velocity of fall increases, but they are not declared the causes of the increase in velocity. "Just as" the body passes through an infinity of degrees of velocity, "so also" the successive momenti are compounded in it.

Acceleration and the Momento of Weight

It is not Galileo himself, but rather his genial disciple Torricelli, who provides the clearest expression of the passage from the momento of weight to acceleration. To be sure, Torricelli writes as if his theory were already firm and explicit in the work of his old teacher, as if Galileo had clearly announced the link between the variation of velocity and the renewal of the momento. In brief, Torricelli reads Galileo in the manner of Newton: the law of fall is the consequence of a deeper law linking acceleration to the persistent action of force.

In the second of his *Academic Lectures* of 1642, Torricelli seeks to render intelligible the phenomenon of percussion. He finds the key to it in the Galilean theory of fall. The initial question is: why can a weight striking a table of marble break it, although when merely laid upon it, it seems "to do" nothing?[31] In the end, this question will be replaced by another: what is meant when it is said that a weight rests on a support?

> Let us submit to our contemplation the marble table, and the fact that to break it without percussion, it would be necessary to place on it, at rest, a heavy body of at least a thousand pounds. If another heavy body, weighing only one hundred pounds, is placed at rest on this table, it will certainly not have a force sufficient to break it; since for this effect not a hundred but a thousand pounds are required, as we supposed. It is therefore evident that the momento or, as one might call it, the activity of such a heavy body, for breaking the plane surface beneath it, will by itself be as nothing.
>
> No one will deny that the momento of such a body is of one hundred pounds, as it is in reality, and that when multiplied it can break the table. We even assert that it is of one hundred pounds, and that with this momento of one hundred pounds it exerts its weight not only now, but it will always exert it uniformly on the plane that is placed below it, in such a way that in each of the instants of time—time that runs continuously—the body applies against the marble table its violence of just one hundred pounds at a time.
>
> To confirm the truth of it, we can consider the same heavy body placed on the balance; I believe that everyone will grant me that, in whatever twinkling of an eye I look at the body, in this instant it exerts its gravity with its total force of one hundred pounds, neither more nor less. And if anyone placed it on his own hand, he would perceive that no single instant of time ever passes without the heavy body's (so to speak) generating on it a pressure with a force of one hundred pounds, directed towards the center of the Earth. But on the other side, the marble

placed beneath, in each of the instants of time that pass, is continuously responding to the heavy body that presses it, with a *momento* of resistance [*momento di resistenza*] not as one hundred but as a thousand pounds. It results that, if in imagination we assign any instant in the time that passes, we shall find that in this assigned instant there is an unequal contrast between a force of one hundred and a repugnance of one thousand, and therefore although the heavy body rests and presses eternally on the marble, it will never do anything toward breaking it that goes beyond what it did in the first point of time in which it was placed on top. (*Academic Lectures*, *TO*, 2:6–7; *Opere* [1975], 556)

Torricelli is no longer opposing weight and force, that is, the natural momento of bodies and that which they receive from outside: here weight itself exercises a violence (it "applies against the marble table its violence of just one hundred pounds," "it exerts its gravity with its total force of one hundred pounds"). Force henceforth designates both the action of weight and that of an external agent.

Weight is itself a sort of "activity." The heaviness of the body is not an entity in itself, a durable reality; it is the repetition, instant after instant, of a certain momento. One will feel it, says Torricelli, if one sustains the body with the hand: the arm must furnish at each instant a new resistance.

Torricelli has a very lively sense of the flow of time ("time that runs"; further on he speaks of the time that produces the momenti). Constant and durable weight is only an appearance; it results from the renewal of the force. If weight seems to be a static entity that can be measured on a balance or against an obstacle, it is because the momento comes to die or be extinguished instant after instant in the support that resists it. At each twinkling of the eye there is reproduced this battle of opposed momenti.

This passage might be viewed as the birth of dynamics. Torricelli unifies machines, percussion, and free fall. What he renders possible is not the passage from static weight or "mechanics" to the acceleration of fall, but rather the inverse: the instantaneous momento is first and makes it possible to understand what weight is. In slightly anachronistic terms, it could be said that Torricelli bases static weight on acceleration: the momento of weight is nothing without the time that creates and renews it.

The explanation of percussion and of the acceleration of free fall are then easy to give: if the momento that is renewed at each instant is not extinguished each time by a support or an obstacle, it will accumulate in the body which falls and produce an increase in velocity. Torricelli thus renders comprehensible the marvellous effects of the body that hits the table and breaks it, although that body was incapable of doing so if merely placed on top of it:

But let us return to the heavy body at rest, and let this be, for example, a ball which with a force of one hundred pounds presses continuously on the marble table placed underneath. Although by itself the momento of the heavy ball, which is of one hundred pounds, acting always in an isolated way and without being multiplied, does not suffice to overcome the obstacle of the table, which is as one thousand, even in an infinite time, if we take ten balls equal to the first, all together, or if instead we can enclose in a single ball all the virtue and all the activity of the ten balls in question, we would have a force of a thousand pounds united together, and it would be precisely such that if it were placed on the marble (the resistance of which we suppose can be overcome by one thousand pounds), the marble would break.

Now without multiplying the matter, I believe that by multiplying the time, the producer of these momenti, and by finding also any means whatever of conserving these momenti produced by time, we would have the same effect and the same increase of force.

Let me explain myself with an example. I need one hundred bottles of water from a certain fountain, but I become aware that this fountain gives only a single bottle of water per hour; must I then lose all hope of ever being able to obtain the hundred bottles of water from this fountain? Certainly not. Let one wait for one hundred hours and conserve the water that pours forth continuously; thus one will have the hundred bottles of water that one desires.

The heaviness of natural bodies is a fountain whence continuously pour forth momenti. The heavy body of which we have spoken produces at each instant of time a force of one hundred pounds, and therefore in ten instants, or better said, in ten extremely brief times, it will produce ten of these forces of one hundred pounds each, provided that one can conserve them.

But insofar as it rests on the body that sustains it, it will never be possible to have as we desire the aggregate of the forces all together, because immediately that the second force or momento is born, the preceding has already vanished, or has been so to speak extinguished by a contrary repugnance, that of the underlying plane which in the very time that the momenti are born, kills them successively one after the other.

But without more wearisome prolixity, it is the very definition that Galileo gives of naturally accelerated motion, that suffices to reveal these mysteries of Nature concerning the force of percussion. Let the outpouring source of gravity be opened. Let the heavy ball be raised to such a height that, in falling down, it can remain in the air ten seconds of time, and consequently generate ten of these momenti that are proper to it. I say that these momenti are conserved and aggregated together. That is manifest by the continual experience of heavy bodies that fall and of accelerated motion; we see that heavy bodies after their fall have a greater force than they had when at rest. Reason also persuades us of it, since if the subjacent obstacle, with the continuous repugnance of its unfriendly contact, extinguishes all the preceding momenti, when the obstacle is removed, the disappearance of the cause must entrain that of the effect as well.

And when the heavy body after falling comes to strike, it will no longer apply as previously the simple force of one hundred pounds, which was the daughter of a single instant, but the multiplied forces, daughters of ten instants, which are

> equivalent to one thousand pounds: precisely as much force united and applied at once as was required in order for the marble to be broken and vanquished. (*Academic Lectures*, *TO*, 2:7–9; *Opere* [1975], 558–59)

Torricelli interprets Galileo's theory in simple and clear terms: in each instant the heaviness generates a new momento, which is not other than the weight of the body.[32] The difference between impeded weight and percussive weight comes from a difference in time: the successive momenti of the weight in free fall have not been absorbed instant after instant but have been accumulated and discharge themselves all at once in the impact.

Still to be resolved are the difficulties having to do with infinity (Galileo had already announced that impact is infinite in relation to weight): in every interval of time, no matter how small, one can "denumerate" an infinity of instants, hence the forces that are "daughters" of these instants must form an infinite aggregate. Torricelli recognizes this difference, which could be called dimensional, between weight and weight multiplied by time.

There exists yet another disconcerting point: an exterior force that is deposited in a body generates more or less velocity depending on the mass of the body, as Torricelli himself had illustrated with the example of the ship and the plank. How does it happen that heaviness or gravity generates only equal velocities in all bodies? Torricelli's response (see *TO*, 3:253) is that there is in each body exactly as much "moving virtue" or "momento" as there is of "matter" or "mass" (*moles*).

Static Momento and Acceleration in Galileo's Works

It was perhaps this last difficulty that kept Galileo from asserting so direct a link between momento and the increase of velocity in the course of time.[33] Yet Galileo had given formulations closely akin to those of Torricelli, particularly in the scholium added to the *Discorsi*, the so-called "Viviani scholium," which was mentioned above (p. 97). This text is decisive for the present inquiry because in it Galileo shows how to deduce from the proportion between static momenti the relation between velocities acquired after a certain time. Salviati here explains how the "total momento" of a body suspended vertically differs from its "partial momento" when retained on an inclined plane; the principles of the science of machines allow him to derive the correct proportion between these two momenti.

But how to pass from this proportion between momenti to the determination of the velocities after a certain time? Here is the key phrase of this passage, ever enigmatic for the translators:

> Quali furono gl'impeti nella prima mossa, tali proporzionalmente saranno i gradi della velocità guadagnati nell'istesso tempo, poiche e questi, e quelli crescono colla medisima proporzione nel medesimo tempo. (EN, 8:218)

Vertically the body weighs with all its weight, it exercises what Galileo calls its "total momento," while on the inclined plane it exercises only a "partial momento" because "the impetus [*l'impeto*], the power [*il talento*], the energy [*l'energia*], or the momentum [*il momento*] of descent of the moving body is diminished by the subjacent plane" (ibid., 215). If the plane were maximally inclined, that is, horizontal, the push or impetus would be entirely annihilated—"extinguished" as Galileo puts it (a formulation found in Torricelli, see p. 91 above).

What happens if the body on the inclined plane is released, and another equivalent body is let to drop in free vertical fall? The relation between the two impeded momenti (the static weights) is known. How to derive from it the evaluation of the velocities after a certain time as each body continues on its trajectory? It is to this question that the passage of interest replies. Galileo claims quite simply that the ratio between the velocities will always remain the same as the ratio of the initial momenti. A possible translation of this passage is:

> Whatever were the impetuses at the very beginning of the motion, such will be, proportionally, the degrees of velocity acquired in the same time, since both [impetuses and speeds] increase in the same ratio in the same time.

A Latin manuscript offers an invaluable parallel:

> Velocitates mobilium quae inaequali momento incipiunt motum, sunt semper inter se in eadem proportione ac si aequabili motu progrederentur. (Fragments, EN, 8:386; *Discorsi* (1958), 590; see Galluzzi, *Momento*, 293)

That is to say: "the velocities of bodies that commence their motion with unequal momenti, are always between them in the same ratio as if they advanced with uniform motion."

The Latin fragment uses *momentum*, where the Italian employs *impeti nella prima mossa* [impetuses in the first setting into motion]. In both cases it is the initial tendency to motion: both the weight, such as a balance would measure it, and the accelerative momento that disturbs the bodies at the start of their motion. For two equal bodies in different situations on planes more or less inclined, for example, the relative intensity of the momenti is known by virtue of the science of machines.

Galileo asserts that as they increase, the velocities always maintain the same ratio. After a certain time, each body will have gained velocity, but always in the same ratio as that of the initial momenti. Thus, in his last text, the old master is on the verge of saying what his disciple Torricelli announces explicitly: the acceleration of the falling body is nothing else than the renewal of its weight through successive instants of time. But Galileo never expresses himself so directly; he maintains a kind of parallelism (as seen in connection with percussion) between the relation of the momenti and that of the additions of velocity. Thus he declares, for example:

> We lay it down that the velocity of motion is augmented or diminished in the same proportion as are augmented or diminished the momenti of heaviness *[Ponatur igitur augeri vel imminui motus velocitatem secundum proportionem qua augentur vel minuuntur gravitatis momenta].* (Fragments, EN, 8:379; *Discorsi* (1958), 478; see Galluzzi, *Momento,* 305)

The degrees of velocity increase "as" the momenti of heaviness, but the causal link remains hidden.

The Demonstration of the Law of Fall in Galileo's *Discorsi*

These ambiguities and hesitations of Galileo as to the nature of force and acceleration, and as to the connections between static weight and time, should call attention to the text that was the foundation of the "new science" of the seventeenth century—the passage of the *Discorsi* that demonstrates the proportionality between spaces traversed and the square of the time for a body in "naturally accelerated" motion.

The most decisive part is that preceding the law:

> *Theorem 1, Proposition 1.* The time in which a certain space is traversed by a body in uniformly accelerated motion starting from rest, is equal to the time in which the same space would be traversed by the same body carried in uniform motion whose degree of velocity would be one-half the greatest and final degree of velocity in the preceding uniformly accelerated motion.
>
> Let line AB represent the time in which the space CD is traversed by a body in uniformly accelerated translation starting from rest in C. And let EB, erected in any way on AB, represent the greatest and last among the degrees of velocity added in the instants of time AB. With AE joined, all the lines drawn from each of the points of the line AB parallel to BE, will represent the increasing degrees of velocity after the instant A. Having then divided BE at its midpoint F, and drawn the parallels FG and AG to BA and BF, we shall have constructed the parallelogram AGFB, which is equal to the triangle AEB, for with its side GF it divides AE at its midpoint I.

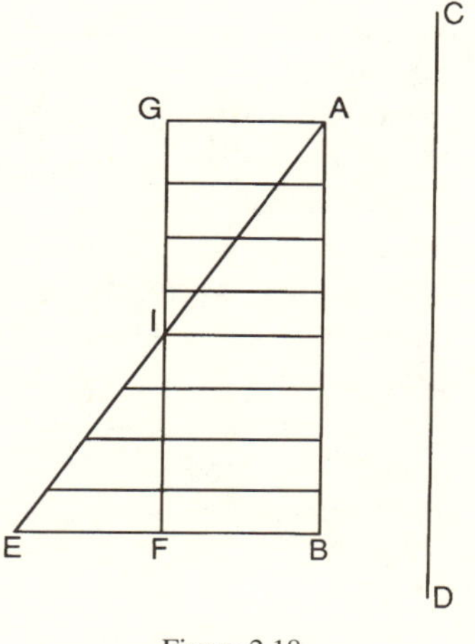

Figure 2.18

> Now if the parallels of the triangle AEB are prolonged as far as GIF, we shall have the aggregate of all the parallels contained in the quadrilateral equal to the aggregate of those included in the triangle AEB; for those in triangle IEF are equal

> to those contained in triangle GIA, and those found in the trapezoid AIFB are common.
>
> But since to each and all instants of time AB correspond each and all points of the line AB, from which points the parallels drawn and included in the triangle AEB represent the increasing degrees of added velocity, and the parallels contained in the parallelogram represent in the same way just as many degrees of velocity not increasing but uniform, it appears that as many momenti of velocity have been expended in the accelerated motion according to the increasing parallels of the triangle AEB, as in the uniform motion according to the parallels of the parallelogram GB. For what is lacking in the way of momenti in the first half of the accelerated motion (the momenti represented by the parallels of the triangle AGI) is made up by the momenti represented by the parallels of the triangle IEF.
>
> It is therefore clear that equal spaces will be traversed in the same time by two bodies, of which one is moved with a motion uniformly accelerated starting from rest, and the other with an equable motion having a momento which is one-half the momento of the maximum velocity of the accelerated motion. Which [proposition] it was required to demonstrate. (EN, 8:208–9)

This theorem makes it possible to avoid dealing directly with accelerated motions: one can henceforth replace a uniformly accelerated motion by a uniform motion whose constant velocity is half of the final velocity of the accelerated motion. It is by virtue of this substitution that Theorem 2, the law of fall, is demonstrated immediately afterward: the spaces traversed are proportional to the squares of the times (since for uniform motions the spaces traversed are in the compound proportion of the velocities and the times, and here the velocity is proportional to the time).

What are the mathematical springs of this reasoning? Galileo first proves that the two areas, triangular and rectangular, are equal; then he considers the segments traced or that could be traced in the interior of these figures; finally he attributes to these segments their physical significance (in several steps: degrees of velocity, momenti, spaces traversed).

The second stage can appear to be infinitesimal or "indivisibilist" in nature: "all" the parallel segments enclosed within one given contour and "all" the segments enclosed within another contour are compared taken as wholes (this procedure will be contrasted with Cavalieri's processes on p. 176ff. below). The manipulation of infinite aggregates is not in fact utilized to demonstrate the equality of areas.[34] The equality in area of the two figures results very simply from Euclid's *Elements* (in virtue of the equality of the triangles EIF and GIA). Galileo has recourse to the infinity of lines only at the moment in which he passes from surfaces to the segments contained in them. The infinitist content of the passage is very subtle: Galileo asserts that to each instant corresponds a point, to each point a parallel, and to each parallel a degree of velocity; finally he declares that there are "as many" momenti in one motion as in the other. There are indeed an infinity of instants, of points, and of degrees, but no sum or composite is made except when it is a matter of "momenti." Here is the only respect in which this passage is close to the procedures of Cavalieri: "as many" momenti were dispensed in one motion as in the other;[35] in both motions they were infinite in number, but nonetheless it is permissible to compare these infinites by comparing the finite magnitudes (the surfaces) in which these two infinites are respectively contained.

What, from the physical point of view, does this comparison of sums of momenti signify? It substitutes for a reasoning that would concern degrees of velocity: the parallels represent degrees, but one does not sum the degrees. The notion of momentum is convenient and flexible: a momento can be a sum of momenti, and by accumulating an infinity of momenti one can obtain a single resultant momento.

The passage from the sum of momenti to distance traversed is, Galileo's text declares, evident.[36] Should it be said that the area "represents" the distance traversed? Doing so would be to go beyond what the demonstration expressly says and to enter upon a path that was yet to be traced, that of the Galilean tradition.

Infinitesimal Demonstrations of the Law of Fall (Barrow, Torricelli)

At the beginning of his *Lectiones geometricae* of 1670, Barrow proposed a demonstration of the law of fall; Newton probably read it, if he did not hear it presented in a lecture. The work is a treatise on geometry, but its way of proceeding is (at least to modern eyes) somewhat peculiar: curves, surfaces, and volumes are generated by different motions; and before studying the geometrical properties of objects by analyzing the motions that generate them, Barrow sets forth with a certain generality the properties of motions and forces:

> After these preliminaries concerning time, we come to consider the efficient force of motion [*vim motus effectivam*], which (whatever its nature or the origin whence it arises, for we leave this discussion to physicists) is also conceived with good right as a quantum, and like other quanta as subject to calculation.
>
> Experience, in fact, shows clearly that, when two bodies are moved from the same starting-point and along the same path, one of the two often gains on the other, or traverses a greater space in the same time. And that cannot arise from anything else than a greater force or motive power [*a majori vi seu potentia motiva*], by virtue of which the one of the moving bodies prevails over the other and is said to be more rapid.
>
> Nothing evidently prevents that this excess of space traversed should occur according to all sorts of proportions, and as a consequence this force will be conceived as divisible into parts of any size whatever (it is permissible and customary to call them degrees, as the intensive parts of each quality . . .), hence into

an infinite or indefinite number of parts. What relates and separates them, their common term, or (according to the supposition that the quanta consist of an infinity of atoms) their absolutely smallest part, will be called rest, that is, the supreme slowness or least velocity. By the successive increase in this part, or its continuous intensification, we may conceive that there is aggregated or produced any degree of velocity, however great, in the same manner as we imagine a line as generated by the apposition or motion of points, and time by the succession or the flow of instants.

Consequently, considering for itself the reality in which one will be able to present correctly to the mind or imagination the quantity of this force, it will suffice to offer, in place of the force, any regular magnitude (that is to say, a magnitude in whose parts we come to grasp clearly and rapidly any difference and any proportion whatever). For reasons of simplicity and therefore of clarity it is the straight line that agrees most exactly with the representation of any degree.

Yet this force, considered in itself as to its generation and absolutely, does not imply time, and can therefore be conceived separately from it; for we can conceive that a body may be endowed with it in any instant of time and during any interval of time. Nonetheless, insofar as it is calculable [*computabilis*] and subject to mathematical evaluation, velocity is specified by the joint designation of a space and a time. . . .

To each instant of time, or to every indefinitely small particle of time (I say instant or indefinite particle, for just as it matters little whether one supposes the line composed of innumerable points or of indefinitely small lines [*lineolis*], so it amounts to the same to suppose time constituted [*conflatum*] of instants or of innumerable very small times [*tempusculis*]; also, for the sake of brevity, let us not fear to employ instants or points in place of indefinitely short times), to any moment of time, I say, there corresponds a certain degree of velocity which one must conceive the body then to possess. To this degree corresponds a certain length of space traversed. . . .

Yet since the moments of time do not in reality depend the one on the other, one can suppose that in the instant following [*proximo*] the body receives another degree of velocity, . . . to which corresponds another length of space, which will have to the first the proportion that this degree of velocity has to the preceding.

As the instants of time are all equal to each other, the ratio of spatial lengths will depend only on the ratio of velocities; the first ratio will be equal or similar to the second. . . . If through all the moments [*per omnia . . . momenta*] of any time one assigns the degrees of velocity that correspond to them, there will be aggregated out of them a certain quantum, of which the respective parts, that is to say, the particles corresponding to these same times, are proportioned to the parts of spaces traversed, and thus the magnitude representing the quantum composed of these parts can also represent space traversed. . . . Since the degrees of velocity, always different, whether equal or unequal, which are possessed in the different instants of time, are expressed by straight lines, as previously stated, and these always different degrees traverse, each of them, moments of time, independently of one another and without mixing, . . . the plane surface that results (determined according to the quantity of the time and the ratio of the degrees of velocity that have been supposed) will present very exactly the aggregate of the degrees of velocity, and the parts of this surface will be proportional, as previously stated, to the parts of space traversed. . . . This surface we shall henceforth call, for brevity's sake, the aggregate velocity or velocity representative of the space.

But let no one be offended . . . if we have said that in each of the instants of time there is completed a certain length, as if I should affirm that there could be an instantaneous motion.

For if one has posited that the times are composed of moments, the lines will also be composed of points; if, on the other hand, unequal lines are composed of an infinity of points, equal in number, it follows necessarily that the points of the lines will be unequal in the same proportion as the lines themselves. Therefore, by lengths traversed in these equitemporal moments, it will be necessary to understand unequal points of this sort, from which the total length traversed is, so to speak, composed. (Barrow, *Lectiones*, 7–9)

Boldness is here combined with the most scholastic ponderousness. On the notion of force, there is only the very banal: force is the cause of motion, and if one body traverses more space than another in the same time, it is by virtue of its greater "force or motive power." Thus, uniform rectilinear motion evidently presupposes a force; it even provides the most obvious way of characterizing the "efficient force of motion": by the space traversed in a given time. (The discussion of the nature and origin of forces, to be sure, is left to the physicist.)

Force, like other magnitudes or quanta, is divisible to infinity, or more precisely, there can be distinguished parts endowed with magnitude, which are "related or separated" by their common boundary, just as segments cut from a line are at once "related and separated" by the point-boundary (it is the Aristotelian conception, see *Physics*, bk. 6). Alternatively, the points can also be considered as "least parts" which compose the line, in agreement with the doctrine of the atomists (and of Galileo). A little later in the passage, the same duality of viewpoints is permitted with respect to time, which is composed either of very small times endowed with size or of punctual instants.

The atom of force is then rest, the least velocity. How to conceive that velocity or force is composed of a sum of rest? Barrow is not very clear on this point. He is content with juxtaposing two possible modes of composition: successive accretion of infinitely small parts and "intensification" of a single infinitesimal part—two modes that correspond, for the line, to the apposition of points and to the generation by the point in motion. (What is to be understood by this intensification of rest, which generates velocity, in the manner of the flux of a point? Can rest, in entering into a certain sort of transformation, generate velocity?)

Finally, leaving aside these somewhat confused speculations, Barrow takes up again the linear figuration of force (that is to say, of velocity). A straight line will adequately represent force. He goes on to say that with each instant of the interval of time there is associated a degree of velocity, and to each degree of velocity there corresponds a length of space traversed, these lengths having between them the proportion of the successive degrees.

Can one sum these degrees? Barrow avoids saying so directly: "there will be aggregated out of them a certain quantum." Of what nature is this "quantum"? For each instant the component particle is proportioned to a space traversed, so that "the magnitude representing the quantum . . . can also represent space traversed."[37] By being expressed in a more geometrical mode, which is supported by the linear figuration of time and velocity, the plane surface that "results" from all these degrees presents or suggests the aggregate of the degrees.

It is easier to see now on what conditions it was possible to demonstrate that the area under the curve of velocities "represents" the space traversed. Note that here the area is not decomposed into a sum of infinitesimal rectangles *v.dt*. Barrow recurs to the idea of a distance associated with each instant: each point-instant brings with it a distance represented by a line, of which the aggregate is a surface (if the line is a distance, how can the surface also be a distance?).

These modes of reasoning imply a strange and audacious conception: with each instant is associated a distance traversed, "as if I should affirm that there could be an instantaneous motion." (The *Physics* of Aristotle challenged this idea and demonstrated, for example, that the vacuum is impossible because motion in a vacuum would be instantaneous. Galileo himself uses the impossibility of instantaneous motion to refute his own "ancient error" [EN, 8:204–5].) Barrow displaces the difficulty by transferring it into pure geometry: the "lines" (that is to say, the segments) are composed of points, and two unequal lines have the same number of points, since there are an infinity in each of them; it is therefore necessary that these points be unequal, more or less "long" according to the size of the segment in which they are found. Consequently it is not so astonishing to suppose a motion in an instant.

This but replaces one oddity by another: how can one speak of points that are unequal to one another? The idea probably came from Torricelli, for whom it was the basis of a profound and original mathematical method.[38] It was Torricelli who imagined giving a size to points, a size depending on the length of certain segments of the figure studied. This passage of the *Lectiones geometricae* is therefore a very strong indication (among others[39]) of indirect contacts between Barrow and the Torricellian milieu. Yet it is unfortunate that, owing to the premature death of the young Torricelli, the transmission was indirect and that contemporaries, instead of encountering the ideas in the somewhat turgid presentation of Barrow, did not have access to the clear and explicit expositions that are found in the manuscript work of Torricelli and that remained unpublished until 1919!

Torricelli enunciated in different passages of his manuscripts, this very astonishing thesis of his: it is necessary to give a variable thickness to surfaces, a variable width to lines, and a variable size to points. (This is the point of departure for a theory of indivisibles radically different from Cavalieri's; see below, p. 188.) Here is how these new tools can be used in the study of accelerated motions:

> Let BA be the time, and during this time BA let a body traverse the lines GF and OH—on the one hand, GF with a uniform motion having a constant degree of velocity AV, and on the other hand, OH with a nonuniform motion having degrees of velocity homologous to the lines AC or ME.

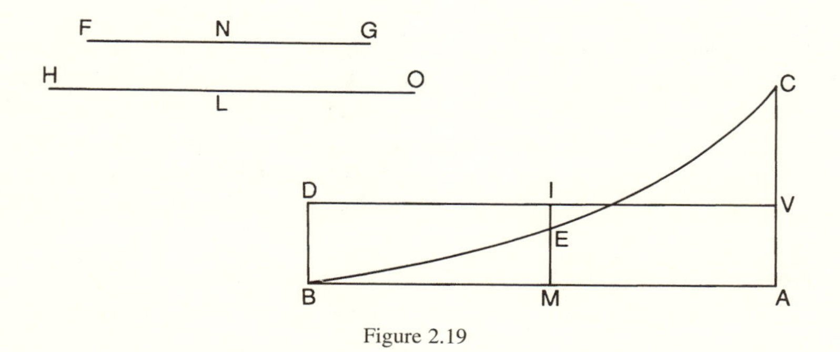

Figure 2.19

> I say that the spaces traversed GF and OH are between them as the figures BECA and BAVD.
>
> For there are as many points in the space GF as in the space OH, namely, as many as there are instants in the same time, but these points are unequal.
>
> Now let us take any instant of this time, for example M, and let N and L be the points traversed during this instant [*puncta peracta hoc instanti*]. As the lines MI and ME are to one another, that is to say, as the impetuses [*impeti*], so will be the spaces N and L, and this in all cases.
>
> But the antecedents [the MI on the one hand, the N on the other] are all equal: consequently, as BAVD is to BECA, so will be the quantity of all the points of GF to the quantity of all the points of OH, which are the same in number, or in other words so will be GF to OH. (*TO*, vol. 1, pt. 2, 259; figure modified)

Torricelli compared two motions, one uniform and the other accelerated. His reasoning is valid even for the most general case of a variable acceleration. The fruitful innovation resides in using indivisibles of length (which did not exist for Cavalieri): to each instant, that is to say, to each point of BA, corresponds a point on GF and a point on OH. These points represent distances traversed during this common instant, and each of the distances GF and OH will be the "sum," so to speak, of these points traversed during the series of instants.

But the points on GF are all equal, while the points that make up OH are of variable length, proportional to the velocity in the instant considered (the points are "as the impetuses"), that is to say, proportional to the variable length of the segments such as ME. All the points of GF (constituting GF) are to all the points of OH (constituting OH) as all the segments of the surface BAVD to all the segments of the surface BECA, or in other words as the surfaces themselves. The area under the curve of the velocities is thus found to be the measure of distance traversed.

In contrast to Barrow, Torricelli did not say that there are as many points on one segment as the other "since there is an infinity in both." Such a statement would have been vague and would have led to a pure and simple contradiction. Torricelli argued differently: there are as many points in the two distances, *because there are as many instants*. Time serves as the fundamental variable: having assigned a particular instant, one considers the velocity of each of the two moving bodies in this instant and associates with this instant a length traversed. It is thus that Newton also reasons in many cases: all the variations are functions of time, which is itself divided into unassignable or infinitesimal instants, and one compares the increases of the several variations during this instant.

Huygens' Demonstration of Galileo's Law

In the eyes of Newton and Torricelli, Galileo's law of fall presupposed a renewal of the momento or force and its accumulation instant by instant. But not all the seventeenth-century authors who were inspired by Galileo reasoned in this way. Huygens believed it possible to demonstrate Galileo's theorem without explicitly supposing the dependence between time and the variation of velocity—he even believed it possible to derive demonstratively the fundamental property of heaviness, that at each equal interval of time there comes to be added an equal velocity.

In his *Horologium oscillatorium* of 1673, Huygens' sequence of theorems in the second part is based on three initial principles. The first corresponds to the principle of inertia (under a restrictive form); the second is a principle of composition of motions; and the third is a sort of principle of relativity:

Hypotheses

I. If gravity did not exist, and if the air did not impede the motion of bodies, then any body, once having received a motion, would continue it with a uniform velocity in a straight line.

> II. But by the action of gravity—whatever its source may be—it comes about that bodies move with a motion compounded of the uniform motion they possess in one or another direction, and of the downward motion arising from gravity.
>
> III. And each of these motions can be considered in isolation, and neither is impeded by the other. (*HO*, 18:125)

These hypotheses are adapted to the terrestrial situation and require only making abstractions of the weight of the body and the resistance of the air. Nothing here announces the possibility of a general dynamics applicable to celestial bodies. The mystery of gravity is left intact with a sort of axiomatic neutrality; "the downward motion arising from gravity" is not otherwise specified. It is the following argumentation of the book that says more about the properties of such a motion. The hypotheses are followed by a commentary:

> Let C be a heavy body that, starting from rest, traverses in a certain time F the space CB by virtue of the force of gravity.
>
> And let us also suppose that the same body had received from elsewhere a motion by which, had there not been gravity, it would have traversed in the same time F, with uniform motion, the straight line CD. Since the force of gravity is added, the body will not go from C to D during the time F, but will arrive at a point E placed in a straight line under the point D, in such a way that the space DE will always be equal to the space CB; that is to say, in such a way that the uniform motion and the motion due to gravity will each accomplish its journey without impeding the other. (*HO*, 18:125)

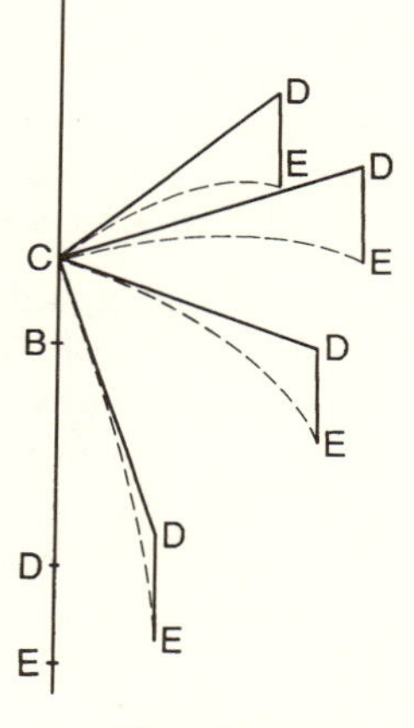

Figure 2.20

Huygens is careful to leave indeterminate the component of the motion that is due to gravity.[40] The motion received "from elsewhere" is uniform, but of the law governing "the downward motion arising from gravity," nothing is known. It is known only that in a time F the body would traverse the space CB or another vertical space always equal to CB.

Following this last passage, Huygens makes clear that there can be a composition of motions whatever the direction of the uniform motion received at the start. The figure (fig. 2.20) shows several possibilities, vertical and oblique. The segment CD that would be described by the uniform motion alone is always compounded with a segment DE which represents the vertical trajectory due to gravity during the same interval of time. The diagram shows stippled paths CE with noticeable curvature. What are their exact forms? They cannot be determined yet because the law according to which DE is traversed remains unknown.

In the two extreme cases in which the body is launched vertically, either upward or downward, the component DE appears as a diminution or an addition to the uniform motion CD. Huygens is confident that, starting from these two cases of vertical ascent or descent, he can deduce the law of motion of fall and even its cause:

> If in each of these two cases we consider the two motions separately, as we said, and if we suppose that neither is impeded by the other, then it will be possible to deduce therefrom the cause and the laws of acceleration of heavy bodies. (*HO*, 18:127)

The culmination of the demonstrative reasoning is reached in Proposition 3, in which Huygens shows that the spaces traversed in free fall are proportional to the squares of the times. But the decisive step is taken in Proposition 1, in which the fundamental property of acceleration appears as a demonstrated result:

> *Proposition 1.* In equal times, equal parts of velocity come to be added to the heavy body, and the spaces traversed in equal times, starting from the commencement of the descent, continually increase by an equal amount.
>
> Suppose a heavy body, starting from rest in A, falls during the first interval through the space AB, and that when it arrives in B, it has acquired a velocity by which, moving uniformly in the second interval, it could then traverse a certain distance BD.
>
> We know therefore that the space that will be traversed in the second interval will be greater than the space BD, because even if all action of gravity had ceased in B, the body would traverse the space BD.
>
> Yet it will be borne by a motion composed of the uniform motion by which it would have traversed the space BD, and of the motion of heavy bodies which fall, by which necessarily it would descend through a space equal to AB. That is why, if to BD we add DE equal to AB, we know that in the second time the body would arrive in E. 
>
> A
> B
> D
> E
>
> Figure 2.21
>
> If now we ask what velocity it will have in E, at the end of the second interval, we shall find that it must be the double of the velocity that it had in B at the end of the first interval. We have said, in fact, that it moves with a motion composed of the uniform motion with the velocity acquired in B, and the motion produced by gravity, which, since it is exactly the same in the second interval as in the first, must have conferred on the heavy body, during the course of the second interval, a velocity equal to that which it had at the end of the first.
>
> Also, since the body has conserved all the velocity that it had acquired at the end of the first interval, it is clear that it will possess at the end of the second interval twice the velocity that it had acquired at the end of the first interval, that is, the double velocity. (*HO*, 18:127–29)

At the end of each interval of time it is necessary to add to the space traversed in uniform motion (by virtue of the velocity already acquired at the beginning of the interval) a space corresponding to the initial fall AB, that is, the space traversed by the body falling without initial velocity. During each interval the body traverses a space and acquires a velocity, and this acquired velocity transforms itself into space traversed during the following time. The space traversed in each interval of time is therefore made of two elements: one corresponding to the uniform motion, that is to say, to the velocity already acquired at the beginning of the interval of time, and the other generated by the new action of gravity during this interval.

But how does one know that this new action of gravity will produce a result—a space traversed—equal to that which it produced in each of the preceding intervals? "The motion produced by gravity" is "the same in the second interval as in the first"; it therefore always confers the same addition to the velocity: such is the premise that Huygens introduces in his demonstration. But is it not precisely what he is attempting to demonstrate? (Perhaps the subtlety of the argumentation resides in the passage from "motion" to "velocity.") Galileo and his successors had the merit of accepting humbly and explicitly this premise: gravity generates equal velocities in equal times.

GRAVITY AND CENTRIFUGAL FORCE: THE ANALYSIS OF EFFORT IN THE CARTESIAN TRADITION

Gravity remained an enigma in the tradition stemming from Galileo. The cause of gravity was unknown, but it was held as assured that it generated—or withdrew—equal velocities in equal times because it renewed its action at each instant. Whence arose this ever renewed momento? No one knew; it sufficed that its properties could be demonstrated with certainty and that they agreed with "experience" (EN, 8:197, 202–3, 208).

The work of Galileo, taken up repeatedly and reinterpreted by the philosophers of the seventeenth century, was thus the source of a dynamical theory of formal and abstract character. The properties of acceleration were studied in a hypothetico-deductive manner. The claim to truth of the theory as a whole resulted from the agreement of its consequences with the observable motion of heavy bodies.

One of the initial decisions in the construction of the theory is quite singular and remarkable: it is that of according to time the fundamental role. Koyré has insisted on the importance of this choice, showing that Galileo thus went beyond the boundaries of a strict geometrization of gravity and pointing up the very special nature of time ("Space is rational . . . while time is dialectical" [Koyré, *EG*, 97]). It is time that regulates the action of force and the variation of velocity; in the formula of Torricelli, it is *the dispenser of momenti*.

It has been shown how Huygens attempted to do without this first condition and sought vainly to "deduce" the fundamental law relating force and time. It could be said that, like Descartes, he "eliminates the time" (Koyré *EG*, 118 n. 2). Newton, on the contrary, holds to the Galilean dynamic as established; he accepts the "scansion" of impulsions by the flux of time and seeks to generalize it by considering other forces that might act "in the mode of gravity" (*ad modum gravitatis*, Herivel, 193; see pp. 139–40 below).

Descartes' Criticisms of Galileo

For Descartes, Galileo's theory remained abstract and without physical significance: one could always deduce properties from a definition posed arbitrarily, and nothing guaranteed that they applied to gravity. Descartes did not attack Galileo's experience or observation. But others, particularly in France, did so: the measurements they obtained did not agree, said Mersenne and Roberval, with those that Galileo claimed to have found and did not at all corroborate the theory of the *Discorsi*. Descartes took his position on the terrain of natural philosophy: the properties announced by Galileo did not agree with the process whence gravity resulted.

It would be necessary first to know what gravity is. Such is the theme of the remarkable letter in which Descartes criticizes the *Discorsi*:

> Nothing that he says here can be determined without knowing what gravity is. (Descartes to Mersenne, 11 October 1638, AT, 2:238; *Oeuvres et Lettres*, 1026)

Galileo failed to meet the requirements of a correct method, as is attested by the overly lax method of exposition in the *Discorsi*:

> It seems to me that he is very deficient in that he continually makes digressions and does not stop to explain anything completely; which shows that he has not at all examined things in order, but, without having considered the first causes of nature, he has only sought the reasons of some particular effects, and thus he has built without foundation. (AT, 380; *Oeuvres et Lettres*, 1024–25)

Measured against the standard of the Cartesian scientific ideal, Galileo's work was fragile and poorly established. To be sure, Descartes agrees with Galileo in the respect that, like himself, "he seeks to examine physical matters by means of mathematical reasons." But the articulation of natural philosophy with mathematical reasoning was carried out differently by the two thinkers. According to Descartes, before subjecting the phenomenon of gravity to calculation or geometrical speculation, the Italian should have laid down a firmer foundation and established the nature itself of the object studied. "That is to say, he has built everything on air" (AT, 2:288).

The charge of lack of foundation is repeated several times in this letter:

> Everything that he says about the velocity of bodies that descend in a vacuum, etc. . . . is constructed without foundation; for he should have previously determined what gravity is; and if he knew the truth of it, he would know that it is nil in a vacuum. (AT, 2:386)

(It will be shown later why, according to Descartes, gravity would be nil in the vacuum.) Galileo, because he did not know the nature of gravity, spoke of it too superficially and geometrically. The idea of a constant acceleration is a completely abstract conception that does not agree with the reality of bodies that fall:

> He supposes that the velocity of weights that fall increases always equally, which I once believed also, like him; but I now believe that I know by demonstration that it is not true. (AT, 2:386; *Oeuvres et Lettres*, 1029)

Galileo had gambled on the simplicity of nature by supposing that the velocity increased proportionally with the time. Descartes puts this regularity in doubt in the name of an investigation into the cause, or the nature, of gravity. In particular, it could be that the acceleration was greater at the start of the motion:

> The reason that makes me say that bodies that descend are less pushed by the subtle matter at the end of their motion than at the beginning, is just this, that there is less inequality between their velocity and that of this subtle matter. (Descartes to Mersenne, 11 June 1640, AT, 3:79)

Descartes even refuses to admit that a heavy body passes through all degrees of velocity starting from rest. The interlocutors of Salviati, in the *Discorsi*, were repelled by this idea because they had difficulty in conceiving or imagining that a body could go over less than the width of a hand in a thousand years. Descartes does not discuss the mathematical aspect of the question or its plausibility for the imagination; he asks whether this continuous progression of degrees of velocity agrees with the reality of gravity. To Mersenne, who presented him with an argument in favor of the division to infinity of degrees of velocity, he replies:

> As for your example of the inclined plane, it proves indeed that every velocity is divisible to infinity, which I admit; but not that when a body begins to descend it passes through all these divisions. And when one strikes a ball with a mallet, I do not believe you think that this ball, at the beginning of its motion, goes less fast than the mallet; nor finally that any bodies that are pushed by others fail to move, from the first moment, with a velocity proportioned to that of the body that moves them. But according to me, gravity is nothing else than this, that all terrestrial

> bodies are really pushed towards the center of the Earth by the subtle matter, from which you easily see the conclusion. (Descartes to Mersenne, 29 January 1640, AT, 3:9–10)

If gravity results from a series of shocks, there could be a discontinuous change of velocity. The transmission of motion in impact perhaps does not obey a regular law such as Galileo had supposed for gravity.[41]

The Cartesian Theory of Gravity

The objections of Descartes to Galileo's reasoning, like the theory of Huygens that will be studied further on, imply a certain positive or "physical" conception of gravity, which is necessary to discuss briefly.

A knowledge of the mechanism of gravity must be the prerequisite and foundation for a study of the velocity of bodies that fall, as Descartes explains succinctly to de Beaune:

> As for gravity, I imagine simply this, that all the subtle matter from here to the Moon, turning very rapidly around the Earth, drives towards it all the bodies that cannot move so fast. But it drives them with more force when they have not yet begun to fall, than when they are already falling; for finally, if it happens that they descend as rapidly as it moves, it would no longer push them at all, and if they were to descend more rapidly, it would resist them. Whence you see that there are many things to consider before one can determine anything concerning the velocity; and it is this that has always kept me from doing so. (Descartes to de Beaune, 30 April 1639, AT, 2:544; *Oeuvres et Lettres*, 1052–53)

The essential features of the hypothesis are the following: the earth is surrounded by a vortex of "subtle" or "celestial" matter, impalpable and invisible. This very rapid vortex is enclosed within a determinate space because it is limited by other neighboring vortices. The rotational motion gives to this subtle matter a strong centrifugal tendency, which Descartes calls "an effort to recede from the center." This tendency does not generate any motion because the vortex is enclosed. However, if at the heart of the vortex there are portions of a different matter that do not participate in the rotational motion, the subtle matter will have the tendency to push them toward the center in order to take their place. Gravity is this sort of contrary push, a derived effect that is inverse to the centrifugal tendency:

> The force [*vis*] that each of the parts of the celestial matter has for going away from the Earth can have its effect only if these parts in rising push and cause to descend under them some parts of terrestrial matter that give place to them. For, since all spaces around the Earth are occupied either by the particles of terrestrial bodies or by the celestial matter, and all the globules of this celestial matter have an equal propensity to move away from the Earth, none of them would have any

> force to push away and dislodge the others that are similar to it. But as this propensity is less in the particles of the terrestrial bodies than in the globules of celestial matter, whenever the latter have above them some particles of terrestrial matter, they must exercise against them the force of which I have spoken. And thus the heaviness of any one of the terrestrial bodies is not strictly caused by all the celestial matter that circulates round it, but only by that precise part of the celestial matter that, if the body descended, would immediately rise to occupy its place. (AT, 8:213)

It can still be asked why the terrestrial matter does not have the centrifugal tendency of the other matter. Descartes is not very explicit on this difficult point.

Huygens is more so in his *Discours sur la cause de la pesanteur* of 1690.[42] According to him, it is necessary to suppose that the subtle matter is agitated by very rapid motions, which because of the encirclement by and pressure from the surrounding vortices, have ended by assuming an exclusively circular speed (*HO*, 21:455). But these motions of rotation take place in very different directions, as in Brownian motion (Huygens invokes, as an analogy, the vortical motion of a fused bead of metal). The portions of terrestrial matter are therefore struck at each moment by a great multitude of subtle particles in all possible directions. The overall effect is indetectable: the impacts cancel each other because in any instant the body receives them in opposing directions (*HO*, 21:456).

The Analysis of Effort

The key notion of this theory is that of "the effort to recede from the center" [*conatus recedendi a centro*]. Descartes gives an analysis of it in connection with the study of light, in part 3 of the *Principia philosophiae*:

> For it is a law of nature that all bodies moved circularly attempt to recede from the centers around which they revolve. I shall attempt to explain, as exactly as I can, this force by which the globules of the second element, and also the matter of the first element aggregated around the centers S, f, attempt to recede from these centers. For I shall show further on that light consists in this force alone; and many other things depend on the knowledge of it. (AT, 8:108)

Light, according to Descartes, is not a motion but only a force (*vis*), or more exactly, an effort (*conatus*). What does that mean? Effort is related to motion; it is what remains when motion is impeded:

> Yet because often a great number of different causes act on the same body simultaneously, and they impede each other's effects; according as we consider one or another [of these causes], we can say that the body at the same time strives or tends to move [*eodem tempore tendere sive ire conari*] in different directions.

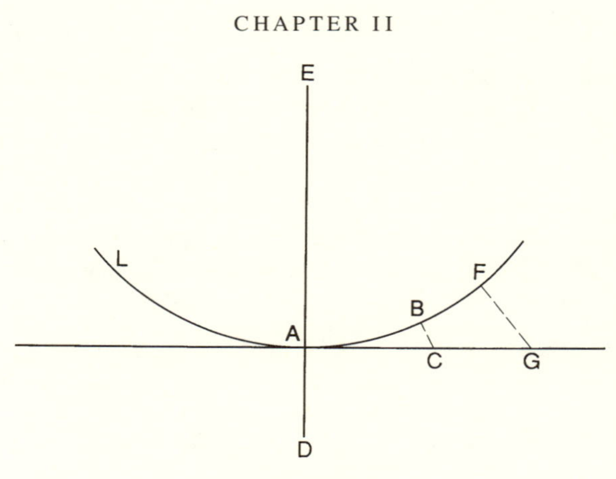

Figure 2.22

> As for example the stone A rotating in the sling EA round the center E certainly tends from A toward B, if one considers all the causes that concur simultaneously to determine its motion, because it is in fact carried there. But if we consider only the force of motion that is in it, in accordance with the law of motion explained previously, the stone when it is at the point A tends towards C, at least if we suppose that AC is a straight line tangent to the circle at the point A. For if the stone left the sling at the moment of time in which, coming from L, it arrived at A, it would in fact continue from A toward C, not toward B. And although the sling impedes this effect, it does not, however, impede the effort. If, finally, we do not consider the whole of the latter force of motion, but only the part that is impeded by the sling, that is, if we distinguish it from the other part that achieves its effect, we shall say that the stone, when it is at the point A, tends only toward D, or that it only attempts to recede from the center E along the line EAD. (AT, 8(1):108–9)

Descartes proposes three different manners of considering the motion of the body at point A and the causes acting on it. If all the causes are taken into account, the body is made to go to B. But if only the "force of motion" of the stone at point A is considered, that is, the impetus belonging to the stone itself in virtue of its inertial tendency (and it is only of this—of the force that a body possesses by its motion or resistance to motion—that Descartes is willing to use the word *force*: see *Principia philosophiae*, pt. 2, paras. 40, 43), it can be said that the stone has a tendency to direct itself toward C. Finally, if the last-named cause is resolved into two portions, namely one that achieves a result and the other that is impeded, the impeded part, Descartes says, has a radial direction along EAD. It is this impeded motion that is called "the effort to recede from the center."

The last assertion of the paragraph is open to two objections. In the first place, why is the impeded motion directed towards D, along the radius? It seems clear from everyday experience that the cord of the sling is stretched radially in the direction EAD at each instant, but the abstract decomposition of the motion does not by any means yield a component in this direction. In A the tendency of the stone is directed toward C, and this tendency is realized only partially, in such sort that the body is finally directed toward B. It is quite unclear how the tendency of A toward C contains the two elements suggested by Descartes: the first, which contributes to the motion toward B, and the second which, being impeded, tends toward D. It was Huygens who posed this question and resolved it, as will be seen shortly.

In the second place, there is a difficulty that is no longer geometrical but may be called ontological: What is an impeded motion? What sort of reality and power should be accorded to a motion that does not occur? The latter is a question that affects Descartes' entire natural philosophy. He recognizes that the observation of actual motions does not suffice for a rational description of the universe; it is necessary also to include virtual motions—tendencies to motion, "efforts." To give an account of the state of a natural body, it is necessary to say what it would do in the next instant if nothing impeded it. The contrary-to-fact conditional (the counterfactual of Goodman: see *Fact, Fiction, and Forecast*) is even more inevitable in the Cartesian context because motion is always impeded from the moment it commences: in the world of the plenum, "in all motion a complete circle of bodies moves simultaneously" (*Principia philosophiae*, pt. 2, para. 33).

The tension of a cord, the pressure on a surface, are the very type of these phenomena that correspond to an effort and allow its detection. But is an effort something real, physically observable, and measurable? How certify the existence of an effort, and how measure its intensity, if one is limited to the effort itself? Yet the mime in the theater succeeds in making the audience believe that he braces himself with all his strength against an imaginary wall.

The *Principia philosophiae* opens itself to such an interrogation:

> When I say that these little globules [of the second element] strive to recede from the centers round which they revolve [Descartes is writing about light, which is the centrifugal effort of the second element], one should not believe that I there attribute to them any thought whence this effort might arise; I mean only that these globules are disposed and incited to motion in such sort that they are on the point of moving [*ituri*] really in this direction, if only they were not impeded therefrom by any other cause. (AT, 8:108)

The notion of effort must be clarified by ridding it of its mental connotations; to admit something of thought into matter would be ruinous for the Cartesian dualism. Efforts (*conatus*) are neither attempts nor endeavors nor preparations. Descartes manages by drawing on the resources of the Latin language, which has at its disposal future infinitives and participles: the small globules are *ituri*, destined to go, ready to go. It is fortunate that French and English do not possess the future participle: one is forced to assume an explicit stance on the future as unavailable and unmanifest.

Perhaps Descartes would have considered that the reality of future motion is guaranteed by the divine immutability: it is God who conserves each thing in the following instant such as it was, insofar as possible; in Him this future state had its reality.

Other solutions are possible. Like Descartes, Hobbes, who relied on different ontological presuppositions, also gives effort a decisive place in his system; but he defines this effort as a commencement of motion, that is to say, as an unassignable or imperceptible motion (*Leviathan*, 1:6; *De corpore*, 177). Thus he also avoids every connotation of the virtual, mental, or future in the description of the universe.

Huygens' Theorems on Centrifugal Force

After Descartes, the theory of centrifugal effort was taken up and developed in a magisterial way by Huygens,[43] who created the very term *centrifugal force* (*vis centrifuga*).[44] Huygens published his results in 1673, without demonstration, as an appendix to his *Horologium oscillatorium*. The detailed analysis and the demonstrations he set forth in a small Latin treatise, *De vi centrifuga* (*HO*, 16:255–301), which remained unpublished until after his death.[45] The importance of the former work lies primarily in the result that he enunciated and demonstrated there for the first time (the essentials of the manuscript go back to 1659). Huygens gave a quantitative evaluation for the centrifugal force as a function of the velocity of rotation and the size of the circle traversed.

For a fixed period of rotation, the centrifugal force is proportional to the radius of the circle; for different constant velocities on a given circle, it is proportional to the square of the velocity. These relations could be written algebraically, with v designating the velocity, r the radius, and ω the angular velocity v/r:

$$\text{Centrifugal force} \propto \omega^2 r \text{ or } v^2/r.$$

From these first two propositions there follow fifteen other theorems, which are often direct consequences of the first two, and which establish, in diverse ways, a relationship between gravity and centrifugal force.

The principle of the demonstrations is the same one that Newton took up in the *De motu* (see above, p. 10): the force is to be represented and measured by the divergence DF between the tangent and the actual trajectory.

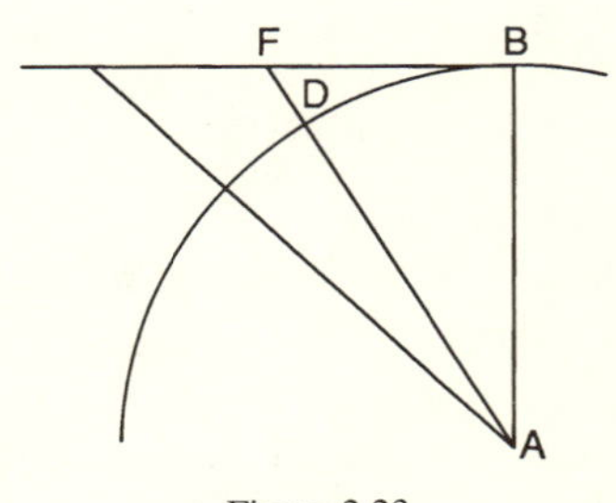

Figure 2.23

This conception could have its roots in the Cartesian notion of impeded motion. The tangent represents positions that have not been reached, although the body, "as much as in it lies," tends to follow this direction. The effort, that is, the impeded motion, can therefore be represented by the segment DF. (Such a measure of force could also be linked to an important text of Galileo, which will be discussed later.)

Here, in Proposition 1, is how Huygens demonstrates the proportionality between centrifugal effort and the size of the circle, for a given angular velocity:

> *Proposition 1.* If two equal bodies in equal times traverse unequal circumferences, the centrifugal force in the greater circle will be to the centrifugal force in the smaller circle as the circumferences or diameters are to one another.
>
> Let there be two circles with radii AB and AC, along which two equal bodies are transported in equal times. On the two circles take very small similar arcs BD and CE, and on the tangents at the points B and C take BF, CG, each equal to its arc. Thus a body transported on the circumference BD possesses an effort to recede from the center in the direction of the [restraining] cord of a motion naturally accelerated, and to traverse in this motion a space DF, in a determinate part of time. On the other hand, the body revolving on the circle CE possesses a similar effort to recede from the center, but by this it would cover in the same part of time the distance EG.

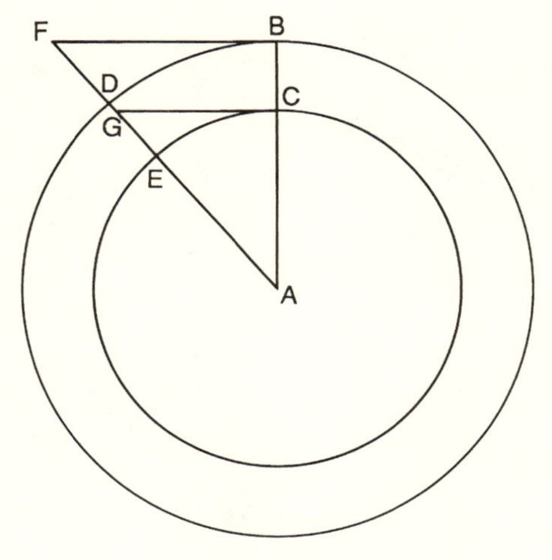

Figure 2.24

> Consequently the force by which the cord is drawn, in the case of the large circle, is to that by which it is drawn in the case of the small one, as the magnitude DF is to the magnitude EG. Yet it is clear that FD is to GE as BF to CG, that is to say, as BA to AC. Therefore the centrifugal force on the large circumference

will be to the force on the smaller as the circumferences themselves, or their diameters. Q.E.D. (*HO*, 16:269)

The moving bodies have the same angular velocity, hence the positions attained after a given interval of time are aligned on the common radius AED. If it is accepted that the centrifugal efforts are to one another as the divergences EG and DF, it is sufficient to say that the two configurations ABDF and ACEG are similar, one being a magnification of the other. The ratio of magnification is that of the radii. Consequently the divergences, and also the centrifugal efforts, are proportional to the radii.

Proposition 2. If equal bodies revolve on identical or equal circles or wheels with unequal velocities, but both with uniform motion, the force of the more rapid for receding from the center will be to the force of the slower in the duplicate ratio of the velocities. That is to say, if the cords that retain them are directed downward while passing through the center of the wheel, and sustain weights which compensate for [*quibus . . . inhibeatur*] the centrifugal forces of the moving bodies and are constantly equal to these forces, the weights will be between them as the squares of the velocities.

Let there be a circle with center A and radius AB, on the circumference of which move a first, slower body with a velocity represented by the line N, and another with a greater velocity which is O. Now if one takes the very small arcs BE, BF, which are to one another as N to O, it is evident that during the same part of time in which the slower body has traversed BE, the more rapid will have traversed BF. On the tangent take BC and BD equal to BE and BF.

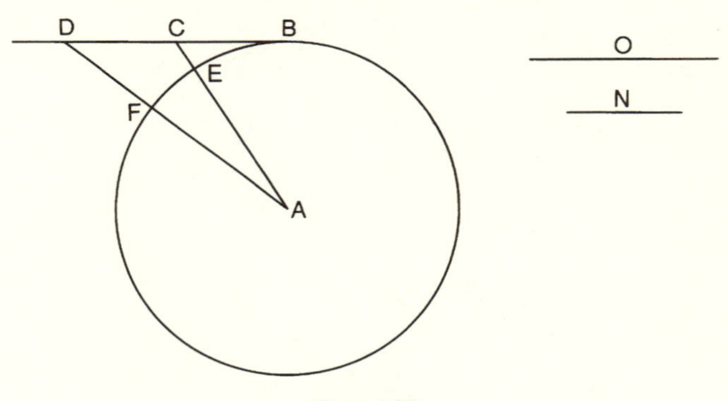

Figure 2.25

It is evident that each of the bodies possesses an effort to recede from the center in naturally accelerated motion in the direction of the cord; but by this motion the body that moves more slowly will recede from the circumference on which it presses through a distance equal to EC; and that which is more rapid will in an equal time recede through the distance FD. Therefore, insofar as DF is greater than CE, by so much is the traction of the more rapid body more vigorous in relation to the traction of the slower.

But since we have taken the arcs BE, BF very small, the ratio of DF to CE must be considered the same as the ratio of the square of DB to the square of CB, as we explained a little previously. And DB is to BC as FB to BE, that is to say as O to N. Therefore the square of O will be to the square of N as FD to EC and consequently as the centrifugal force of the more rapid body to the force of the slower. Q.E.D. (*HO*, 16:269–71)

The velocity of each body is represented by a rectilinear segment. In the same interval of time, the two bodies traverse arcs proportional to these segments, N and O. The more rapid will go farther, and its divergence from the tangent will be greater. Huygens asserts without proof that these divergences are proportional to the squares of the arcs (or of the corresponding segments of the tangent), at least if the arcs are very small. He claims to have explained this "a little previously," that is to say in the preamble, which will be discussed.

Some questions arise from the reading of these demonstrations:

Why is the effort in the direction of the cord?

Why does Huygens speak of "a motion naturally accelerated," caused by the centrifugal effort?

Why does Huygens introduce, in the second phrase of the enunciation of Proposition 2, a mechanism permitting a comparison between centrifugal force and weight by virtue of the weights that compensate for or inhibit the effort to recede?

Which is Prior, Gravity or Centrifugal Force?

The comparison between gravity and centrifugal force runs throughout Huygens' treatise. For instance, Proposition 5 asserts:

> If a body moves on the circumference of a circle with the velocity that it acquired in falling from a height equal to the fourth of the circle's diameter, it will possess an effort to recede from the center equal to its gravity, that is to say, it will pull on the cord that retains it as vigorously [*aeque valide*] as if it were suspended on it. (*HO*, 16:275)[46]

In the enunciation of Proposition 2, as has been shown, the equality between gravity and centrifugal force is materialized in a mechanism of suspension at the center of the wheel: the cord that retains the body in rotation passes through the center of the wheel and supports a weight vertically. Such ways of relating the two forces are not a purely theoretical fiction, and it is known, for example, that at least Hooke and Huygens constructed and studied various sorts of conical pendulums and other procedures for physically comparing centrifugal effort and gravity. Huygens tells how in 1667 he made experiments with a round table pierced by a hole at the center. He attempted also to evaluate the velocity with which the earth must rotate in order that the centrifugal force at the equator should exactly equilibrate gravity.[47]

But what do these comparisons signify? What are they meant to establish? It is not known whether Huygens took the properties of gravity to be already established or if he was seeking to verify them. The status of gravity and centrifugal force in relation to each other does not emerge clearly in a naive reading of the text *De vi centrifuga.*

If this question is put in the context of the previously discussed debate between "Galileans" and "Cartesians," that is, between those who accepted the law of acceleration of the *Discorsi* and those who put it in doubt in the name of a mechanism that would explain gravity, the position of Huygens' text is ambiguous: does the author accept Galileo's theory or does he range himself on the side of Descartes?

In appearance, the works of Galileo are taken as a basis on which Huygens supports himself in order to progress further. The laws of the accelerated motion of fall serve as a first foundation and point of reference. Huygens explicitly accepts Galileo's law of fall: if the resistance of the air is taken into account, the ratio of distances traversed to the squares of the times "agrees with experience" (*HO*, 16:255). Moreover, the text opens with the affirmation:

> Gravity is an effort to descend.

In everything that follows, the properties of the motion of fall form the assured background on which are progressively sketched the characteristics proper to centrifugal force. And finally, in the last lines of the preamble, the two realities are declared to be similar (*HO*, 16:267; *plane similis conatus est*). In the case of the cord stretched vertically by a weight, as in the cord of the sling stretched by rotation, the tension corresponds to an impeded motion by which the body would traverse spaces proportional to the squares of the times. In brief, Huygens, taking as given the properties of fall, shows finally that centrifugal force is comparable to gravity, and of the same nature as it.

Yet the perspective suddenly reverses itself the moment one perceives Huygens' strategy as a whole. His aim is not to develop Galileo's theory, but rather to show that Galileo's law of acceleration follows from the vortical conception inspired by Descartes (at least when considering only the first moment of motion). Gravity is a derived reality, not a firm starting point.

Huygens finally establishes, in his *Discours sur la cause de la pesanteur* (1690), that the laws of centrifugal effort, when applied to the vortex of fluid matter surrounding the earth, justify Galileo's laws:

> One can finally find here the reason of the principle that Galileo has assumed in order to demonstrate the proportion of the acceleration of bodies that fall, which is that their velocity augments equally in equal times. For the celestial bodies are pushed successively by the parts of the matter which endeavors to rise into their

> place, and which, we have seen, acts continually on them with the same force, at least in the descents that come under our observation. It is a necessary consequence that the increase of velocity is proportional to that of the time. (*HO*, 21:461–62)

In the *De vi centrifuga* itself, nothing permits discovery of this logical priority of centrifugal effort over weight. A manuscript sheet, undoubtedly contemporary with the *De vi centrifuga*, is much more explicit as to the general framework in which Huygens reasons:

> The gravity of a body is the same thing as the effort made by an equal [volume of] matter, in very rapid motion, to recede from the center.
>
> That which maintains the body suspended holds back this same matter and keeps it from receding; and that which lets the body fall thereby offers to this matter the possibility of receding from the center along the radius.
>
> Yet since at the beginning it receded from the center according to the progression of odd numbers starting from unity [that is to say, as the successive squares], it could not do it without forcing the heavy body to approach the center in a motion similarly accelerated, so that these two things are necessarily equal at the commencement of the motion: the receding of the matter far from the center, and the approach of the body that falls towards the center.
>
> From that, if we have found how much a body descends in a certain time, for example if it falls in 1‴ through the distance of 3/5 of a line, we shall know also the ascent of the matter that recedes from the center, to wit, equally, 3/5 of a line in 1‴. We shall know in this way the velocity of this matter, given the terrestrial radius. (*HO*, 17:276–77)

The term *matter* here is a discreet abbreviation for "subtle matter." Huygens undoubtedly means the Cartesian fluid that turns at great velocity around the earth, forms another vortex around the sun, and so on.[48] Because this subtle matter is invisible, its motion cannot be discovered except through the inverse motion of the heavy body that comes into its place. The fall of the heavy body is the indirect evidence for the ascent of the "matter" in question.

Huygens deduces from experiments on falling bodies an evaluation of the velocity of rotation of the "fluid matter": since the pendulum that beats seconds measures 3 feet and 8.5 lines, it follows that the velocity of the fluid matter is that of a body that would complete a revolution round the earth in 1 hour and 24.5 minutes (*HO*, 21:460).

The acceleration of gravity is thus a consequence of that of the centrifugal force. An exact and well-founded theory of gravity must in principle begin with centrifugal force. Without doubt it is for pedagogical or heuristic motives that Huygens began with gravity in his tract *De vi centrifuga* and appeared to base his reasoning on Galileo's theory in order to establish his own theory of centrifugal force.

Tension, Effort, and Accelerated Motion: The Generalization of Gravity

The long preamble of the *De vi centrifuga* (*HO*, 16:254–67) clarifies and justifies the theorems reported above. The overall arrangement of this preamble is not apparent at first glance. The text opens with a discussion of gravity, and it is only after three pages that it begins to treat of centrifugal force:

> Let us now see what the nature and intensity are of the effort [*quis quantusque sit conatus*] made by bodies attached to a whirling cord or wheel to recede from the center. (*HO*, 16:259)

In all that precedes this passage, there is no mention of centrifugal effort or of circular motion but only of gravity and the tension of a cord that supports a weight. Huygens' procedure consists in analyzing the tension generated by weight in order to draw from it the foundations of a correct analysis of centrifugal effort. To comprehend his course of reasoning in the first pages of this preamble, one must supply a question and an analogy at the beginning of the text:

> Why, when the stone in the sling revolves, is the cord of the sling under tension?
>
> This tension resembles that in a cord on which a weight is suspended.

There is, therefore, some chance of comprehending centrifugal force if one first studies gravity.

The first notion is that of tension or traction: in the case of the sling, the hand feels the tension of the cord; similarly the hand perceives the tension that is caused by a weight and is transmitted along the cord sustaining the weight. What does this tension signify? How can it be evaluated? This part of the reasoning culminates in a convention stipulated as follows:

> We posit that an equal traction [*attractio*] is felt each time that . . . (*HO*, 16:259)

The formula can seem odd: "we posit that . . . [there is] felt . . ." [*sentiri ponimus*]. Huygens seeks to objectify the feeling or notion of force. Tension as such cannot figure in a correct physical description; the muscular feeling is only an indication for which it is necessary to be able to substitute a definition and an objective evaluation. It will be shown later that, unhappily, this objectification is not yet radical enough, and that the definition of the *direction* of the effort also requires more restrictive stipulations.

For Huygens, all becomes clear when he sees in tension the correlate of the motion that would take place if the body were released. It is the motion toward which the body "tends," towards which it makes an "effort." The concept of effort (*conatus*) is the mean term between tension and motion—the body exerts a tension because it makes an effort to move:

> The cord is stretched because the heavy body makes an effort to recede [*ideo trahitur filum quoniam grave conatur recedere*]. (*HO*, 16:257)

This will also be true of motion of rotation later in the text:

> The reason why the cord is stretched can now be perceived more clearly. (*HO*, 16:261)

It becomes possible to objectify and to measure tension. It suffices to evaluate the motion that would occur if the body were released. It is the case of heavy bodies: gravity is nothing else than an effort to descend (these are the first words of the text: *gravitas est conatus descendendi*; *HO*, 16:255). And this effort is weaker on an inclined plane because the motion that would take place would be less rapid:

> Also one feels here [in the case of the inclined plane] a smaller effort and an effort precisely as much less, in relation to another, perpendicular effort, as the space traversed by the heavy body in the same time would be smaller on the inclined place than on the perpendicular. (*HO*, 16:257)

Of little importance is the direction of the effort, of little importance even its cause; what matters is that the traction of the cord can be defined by means of the virtual motion of the body:

> Whenever two bodies of equal weight are each retained by a cord, from the moment that they manifest an effort to recede in the direction of the cord with the same accelerated motion, by which equal spaces would be traversed in the same time, we posit that there is felt an equal traction in these cords, that they are [equally] pulled downward or upward or in any direction whatever. And it matters little from what cause such an effort arises, from the moment that it exists. (*HO*, 16:259)

The intensity of the tension or effort can be evaluated by estimating the distance traversed in a given time. Very different kinds of tension (for instance, that arising from weight and that originating in circular motion) can thus be compared by investigating the distances that the bodies would traverse if the impediment to motion were removed.

A further condition imposes itself if one wants to make use of this measure of tensions: it is necessary to restrict attention to the first instant of motion. Huygens illustrates this requirement by an example (*HO*, 16:259): a ball suspended on a thread is supported against an incurved surface but in such a way that the surface is vertical where it is in contact with the ball. The effort of the ball is indeed vertical, despite the incurvation of the surface that guides the ball. Once the thread is broken, the ball will deviate obliquely, but it is the commencement of the motion that counts if one wishes to define the effort of the ball. Therefore, the evaluation of

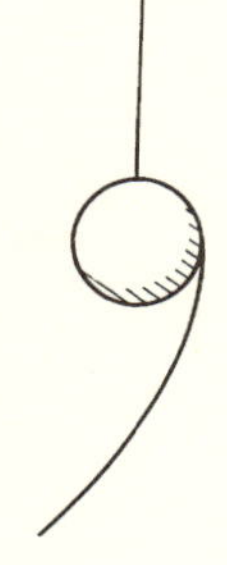

Figure 2.26

distance traversed is restricted to a part of time arbitrarily small (*accepta temporis parte qualibet exigua* or *quamlibet minimam temporis particulam*; *HO*, 16:259). This manner of analyzing effort is reminiscent of what Newton stipulates in the *De motu* as to the spaces traversed at the commencement of motion: Huygens, like Newton, limits the evaluation to nascent motions.

The discussion of this condition consists simply in the clarification of the very notion of effort: what is required is to evaluate a tendency, not an actual motion; all that counts, therefore, is its commencement. The incurved surface does not play a role very different from that of the thread: whether the ball is retained by a thread or is compelled to traverse a surface that causes it to deviate as soon as the motion has begun, these are two obstacles that it is necessary to suppress in thought in evaluating the virtual path. However, the consideration of this surface makes it possible to clarify how direction and change of direction figure in the analysis of motion, and this clarification will be decisive in the case of centrifugal force: while the thread retains the ball without imposing direction on it, the surface causes the motion to deviate, and it is necessary to consider the tangential trajectory in order to evaluate the effort.

Acceleration in Circular Motion

Starting from the paradigm of gravity and having shown how virtual motion explains and measures tension and effort, Huygens applies this result to the tension of the cord of a sling. What is the motion that a revolving body endeavors to make?

Given a situation in which a ball of lead is at the end of a cord held by a man on the rim of a large, horizontal wheel that is rotating very rapidly, the ball tends to escape along the tangent BCD. The effort of this ball can be determined as follows:

> It appears now that when the man arrives at E the lead would be in C if it had been released at the point B, and that it would be in D when the man arrived in F.
>
> If on the other hand the points C and D were on the straight lines AE, AF prolonged, the lead would certainly make an effort to recede from the man along the line issuing from the center and passing through its position. And this in such

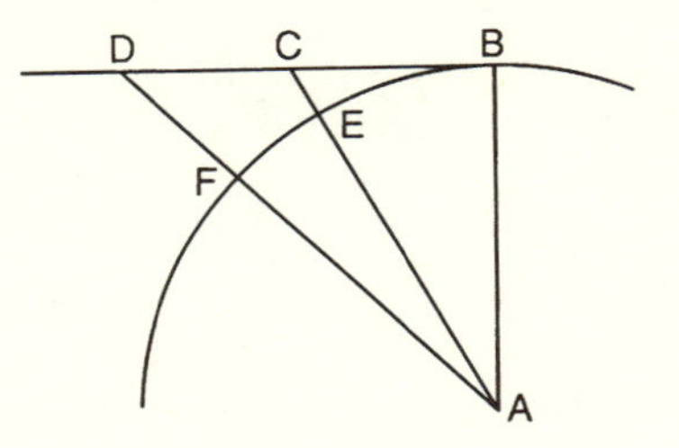

Figure 2.27

a way that in the first part of the time it would recede from it through the distance EC, and in the second part through the distance FD.

Yet these distances EC, FD, and the others in sequence increase as the series of squares starting from unity, 1, 4, 9, 16, etc. . . ., for they agree with this series so much the more accurately as one takes smaller particles BE, EF, and consequently they are to be considered as differing not at all at the beginning. (*HO*, 16:261)

The successive deviations EC and FD between the tangent and the circle measure the effort of the ball to recede from the center of the circle. At this stage of the reasoning, Huygens proceeds as if these deviations take place in the radial direction and are perfectly rectilinear (in the following pages, this approximation will yield to a more accurate analysis). The segments EC and FD, he claims, increase as the squares of the segments BC and BD, at least if they are very small. By what right does he assert this proportionality (for he does not specify the underlying reasoning)? There are at least three ways to arrive at this result. The first is that found in Newton's *De motu* (see fig. 2.28): by Euclid 3.36, $CE \cdot CT = CB^2$, and if CT is nearly equal to the constant diameter ET, $CE \propto CB^2$.

Or again, by reasoning (in fig. 2.29) on the right-angled triangles CEB and CBA (BE being rectilinear and orthogonal to CA), there follows the proportionality $CE : CB = CB : CA$.

Finally, a circle can be approximated in the neighborhood of point B by a parabola in which the segments CE and DF are successive ordinates of this parabola, increasing as the squares of the abscissas BC and BD (fig. 2.30). It is then necessary to suppose that these segments are nearly vertical and not oblique.

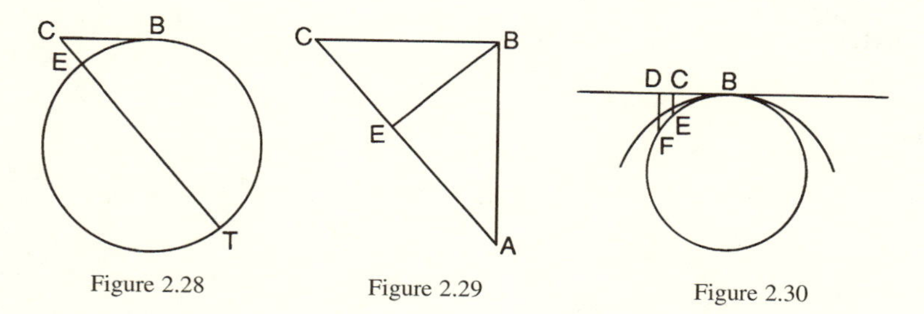

Figure 2.28 Figure 2.29 Figure 2.30

This last manner of proceeding is not the simplest in appearance, but it seems to have been the argument presupposed in Huygens' reasoning. It is in fact the very one used to demonstrate Proposition 16 of the *De vi centrifuga* (concerning the tension in the thread of a pendulum when the ball descends a quarter of a circle):

Let BGE be a parabola of which AB is the semi–latus rectum and B the vertex.

Because the divergences [*recessus*] of the globe B from the circumference BC, when it traverses the straight line BD in uniform motion, are considered at the

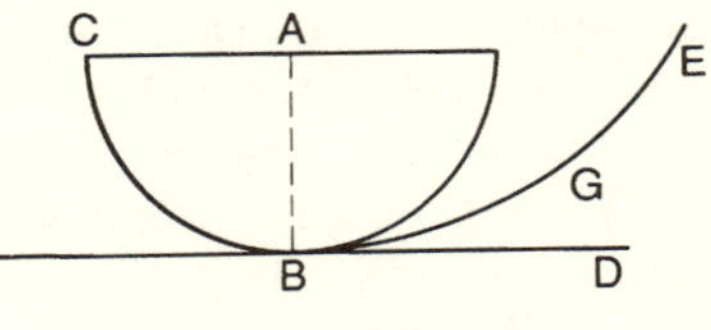

Figure 2.31

commencement near the point B as identical with the divergences from the parabola BGE, it is clear that the centrifugal force that the globe possesses in B, in virtue of its rotation alone, is an effort to recede from the center A or from the circumference BC of a motion accelerated according to the numbers 1, 3, 5, 7, etc. . . . , and that it is therefore similar to the effort by which bodies endeavor to descend and which we call gravity. (*HO*, 16:297; the demonstration seems to be borrowed from a different manuscript, perhaps a vestige of a more explicit version of the preamble.)

Whatever the mode of demonstration, the result is the same: the divergences increase in the same manner as the spaces traversed by a heavy body that falls:

> It is therefore clear that this effort is completely similar to the one experienced when the globe is held suspended by a thread, since in this case also it endeavors to recede along the line of the thread with a motion similarly accelerated, namely with a motion by which after the first part of time it would have traversed the small space [*spatiolum*] 1, and after two parts of time 4 small spaces, after three 9, etc. (*HO*, 16:263)

Since the virtual motions would be the same, the efforts are similar: the tension generated by rotation is of the same nature as that of a thread stretched by a weight.

The Enigma of the Radial Direction: Evolute and Rotating Framework

The enigma in the Cartesian analysis remains: the radial direction is difficult to explain given the virtual motion, which is along the tangent. Huygens poses the question very clearly, beginning on the third page of the preamble:

> What seems on first view difficult to comprehend is why the thread AB is stretched, while the globe endeavors to go along the straight line BH which is perpendicular to AB. (*HO*, 16:259–61)

The response is given a few pages later and presupposes abandoning the rectilinear approximation of the deviations in seeking the true trajectory:

> Because the points C and D depart a little from the straight lines named, in the direction of B, it is found that the globe endeavors to recede from the man not

> along a straight line issuing from A, but along a certain curve which touches this straight line at the place where the man stands. (*HO*, 16:263)

The virtual trajectory, that which the body must follow when released from the thread, is not a straight line, but "a certain curve." This curve, in its point of departure B, is tangent to the radial direction ABV: at the point B it thus merges with the straight line issuing from the radius (see fig. 2.32).

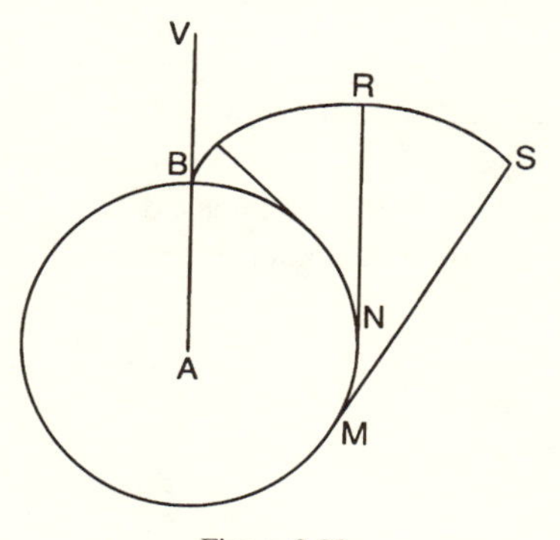

Figure 2.32

> If we wish to describe this curve, it will suffice to unroll the thread on the circumference BNM and to guide its end towards RS in such a way as always to have under tension the part that has left the circumference BNM; in this motion the end of the thread will describe the line BRS in question. But this line will have the following property: if at any point of the circumference, such as N, one draws the tangent to the circumference meeting the curve in R, the straight line NR will be equal to the arc NB. (*HO*, 16:263)

A reader of the *Horologium oscillatorium* will know what curve is here in question: the evolute. It is generated as follows: given a curve with a thread wound on it, if one pulls on the free end of the thread and unrolls ("evolves") it while keeping it always stretched, the end of the thread will describe another curve, which is called the evolute of the first curve. It was Huygens himself who invented and utilized the evolute in the study of the isochronous pendulum (pt. 3 of the *Horologium oscillatorium*: *De linearum curvarum evolutione*).

The relation between the evolute of the circle and the trajectory of the ball released in B is not evident at first reading. (Huygens assumes his reader will have much experience with changes of reference frame and with the geometry of motions.) An explicit presentation of the underlying reasoning presupposes the decomposition of the motion according to two different reference frameworks. First, with a fixed framework in which the wheel is in uniform rotation, a schematic can be made showing the successive positions of the man on the rim of the wheel and of the ball of lead released in B for different instants after the separation of the ball:

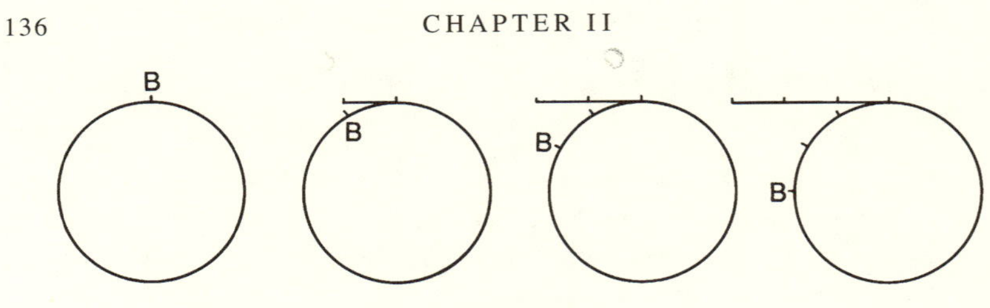

Figure 2.33

If the reference frame is now that of the observer being joined to the wheel and turning with it while maintaining the illusion of immobility, figure 2.34 shows the successive positions occupied by that observer (immobile in O) and by point B, whence the ball was released from the wheel but which is now considered as a point of space exterior to the wheel, so that it appears to turn:

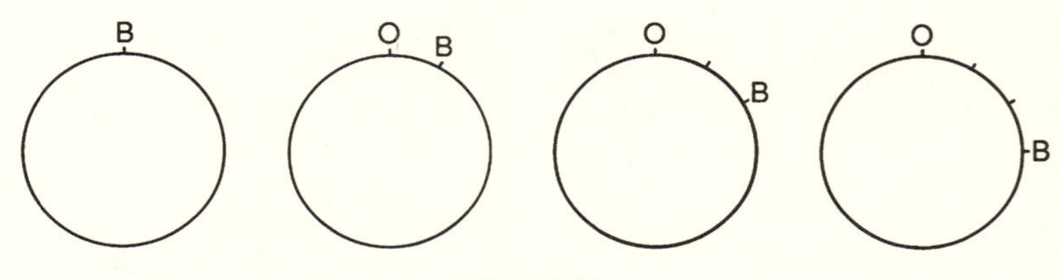

Figure 2.34

The wheel now seems to turn in the opposite sense. What has become of the ball released from B in the first instant? Here are its positions at successive instants:

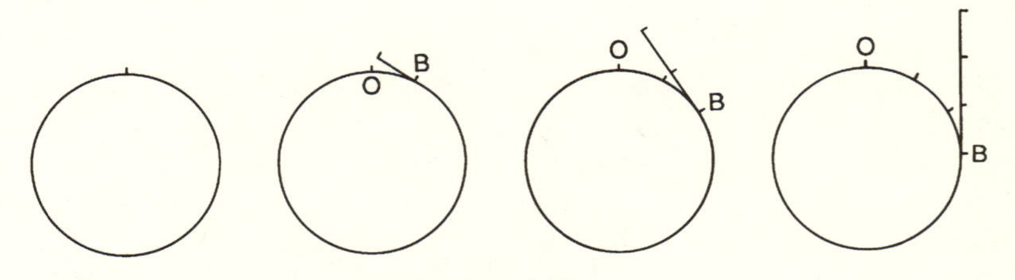

Figure 2.35

The distance traversed by the ball on the tangent is each time equal to the portion of the arc that separates B from the summit. It is as if the tangent were unrolled all around the wheel. By connecting the successive positions of the end of the tangent, the evolute of which Huygens speaks emerges (fig. 2.36).

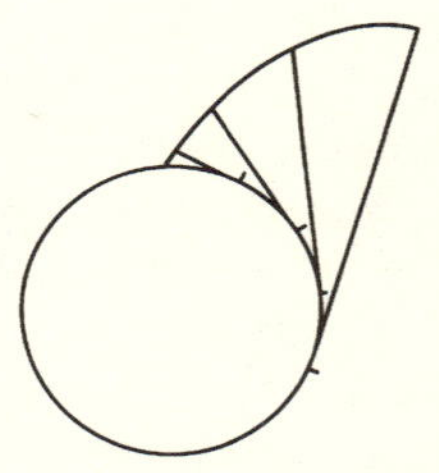

Figure 2.36

In other texts Huygens explains the properties of evolutes.[49] Here, for the argument of the *De vi centrifuga*, the principal interest of this curve arises from its being tangent to the radius at its point of departure. Huygens demonstrates this in the manner of the ancients by proving that it is impossible to draw a straight line BK that makes an angle with the direction BV of the radius and which does not cut the evolute (*HO*, 16:263–65).

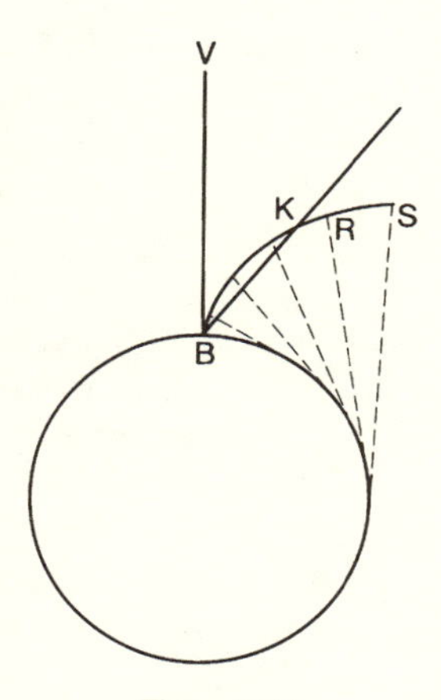

Figure 2.37

The trajectory of the ball released at B is therefore in the exact prolongation of the radius AB; it merges with the direction of the radius at the very commencement of the motion:

> Since, therefore, the globe revolving with the wheel endeavors to describe a curve in relation to the radius on which it is situated, and the curve is such that it is tangent to this radius, it appears that by this effort the thread to which the globe is attached is not otherwise stretched than if the globe endeavored to recede along the radius prolonged. (*HO*, 16:265)

Descartes was therefore right—in agreement with common sense—to claim that the centrifugal effort is radial, although the moving body has a tendency to escape along the tangent.

But all this reasoning presupposes a frame of reference in rotation. Nonetheless, Huygens returns to the fixed framework that he had adopted previously

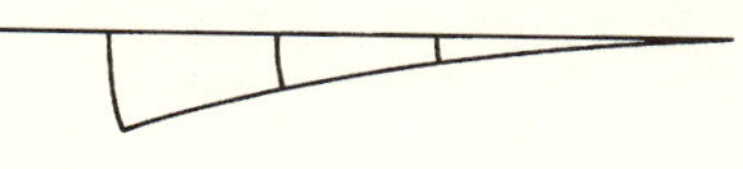
Figure 2.38

and substitutes arcs of evolutes for the successive rectilinear divergences. He shows that these arcs can be replaced by rectilinear segments "in the commencement of the separation of the globe and the wheel" (*HO*, 16:265). He thus justifies after the fact the approximation that he had employed at the beginning of the text.

The recourse to the evolute is very ingenious and very new. Newton is much less cautious in the *De motu*: he discusses neither the form nor the direction of the successive divergences. On a decisive point, however, Newton is more rigorous: his force is *centripetal* and not *centrifugal*. Huygens, it might be said, calculates the effort in the wrong direction. Circular motion is for him a primary datum, in relation to which he evaluates the effort toward deviation. Yet Huygens is well aware that the choice of frame of reference can modify the nature of the effect being evaluated:

> The motion of a body can be at the same time truly uniform and truly accelerated, according as one relates its motion to different other bodies. (*HO*, 16:197)

The formulations of the *De vi centrifuga* are lacking in consistency on this subject. Sometimes Huygens asserts that the released ball recedes with a uniform motion (for example in the demonstration of Proposition 16, cited above on pp. 133–34); elsewhere he declares that the ball recedes from the man with an accelerated motion (see *HO*, 16:263, cited above on p. 134). The two assertions are true, but in relation to different frameworks. It is only in a rotating framework that the recession of the ball appears to be an accelerated motion.

In relation to what or whom is the recession produced? From the moment that Huygens chooses to evaluate "the effort to recede from the man" who is placed on the rim of the rotating wheel, he causes a centrifugal acceleration to appear.

His contemporaries, even the most well-informed on dynamical questions, will follow Huygens—and Descartes—on this point, and will not pay sufficient attention to the consequences of choosing a rotating frame of reference. Thus Johann Bernoulli:

> I understand by centrifugal force the same thing as Monsieur Huygens, as he explains at the end of his treatise on pendulum clocks; for example, if one swings round the hand a sling charged with a stone, *the hand will always feel a force* due to the stone's effort to recede from the center; it is this force that is called centrifugal. (Johann Bernoulli to l'Hôpital, 5 March 1695, *Briefwechsel*, 1:270)

Huygens makes objective and measurable the tension in the thread, but he does not concern himself with the direction of the effort. The correct evaluation of the force or effort presupposes a norm, a virtual frame of reference, in relation to which the divergence due to the force is calculated. If this norm is poorly chosen, what is not force is counted as force, or the force is evaluated in the wrong sense.

These ambiguities in the work of Huygens lead to a better appreciation of the Newtonian thesis of an absolute space and time. Newton judged it necessary to assume an absolute framework as the foundation for his conception of force and the cause of motions.

The equivocation that weighs upon the *De vi centrifuga* is linked to a dynamic difficulty, or more precisely, to Huygens' lack of a dynamic analysis. Can a tension result from a tendency to uniform motion? Is effort correlative to an increase in velocity? The linking of effort and accelerated motion is not at all justified in his account. This link could be analyzed in another perspective that is more clearly dynamical: a constant push, that is, a push reproducing itself at each instant, would generate an increase of velocity proportional to the time, and if impeded, it would maintain a permanent tension against the obstacle.[50]

But Huygens avoids as much as he can the consideration of motive forces. Richard Westfall, in his book on force in the seventeenth century, very justly entitled his chapter on Huygens, "Christiaan Huygens' Kinematics" and shows that the characteristic of this work is "the constant effort to eliminate dynamic concepts" (Westfall *FNP*, 177).

Torricelli showed how momento generates an acceleration because it is repeated instant after instant. The static pressure of a ball on a table—or the tension due to a globe suspended on a thread—must, according to Torricelli, be analyzed as an ever-renewed struggle between the instantaneous momento of the weight and the ever-reborn reaction of the table or thread. If the table is withdrawn, or if the ball is liberated from the thread, an accelerated motion will be produced, because the momento will act continually to increase the velocity; the momenti accumulate if they are not "extinguished" (see above, p. 91).

It is true that Huygens did not exactly accept the Galilean thesis of a reproduction in time of the momento of weight and that he refused to admit as a primary datum what Galileo had taken as his point of departure, namely that in equal times the increments in velocity are equal. He even believed, as has been shown, that he could deduce this thesis from other principles in which the relation of force and time was less apparent.

NEWTON AND CIRCULAR MOTION BEFORE THE *PRINCIPIA*

THE COMPARISON BETWEEN GRAVITY AND THE EFFORT TO RECEDE FROM THE CENTER, ACCORDING TO NEWTON

At about the same time as Huygens, the young Newton also studied circular motion and discovered independently the formula for evaluating "centrifugal force." He gave it the same name Descartes had given it in his *Principia philosophiae*: "the effort to recede from the center" [*conatus recedendi a centro*]; and his reasoning like that of the Cartesians took circular motion as a primary datum. Yet the manuscript that contains these calculations and the formula of evaluation also contains some very new thinking on the equilibrium between centrifugal effort and gravity, as extended to the case of the celestial bodies:

> If a body A turns on a circle AD towards D, its effort to separate from the center will be such that it would carry the body, in a time AD (which I suppose very small), from the circumference to the distance DB; for it would traverse this same distance in the same time if by an unimpeded endeavor it moved freely along the tangent.
>
> Now since this endeavor, provided it were to act in a straight line in the manner of gravity, would impel bodies through distances which are as the squares of the times: to know through what space they would be impelled in the time of a single revolution ADEA, I ask for a line which may be to BD as the square of the

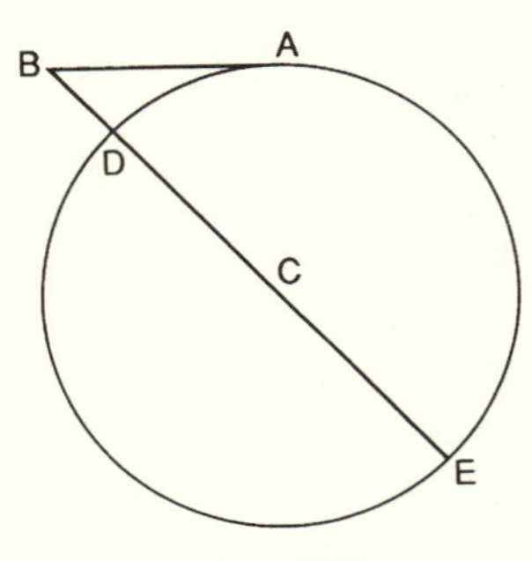

Figure 2.39

circumference ADEA to the square of AD. Now BE/BA = BA/BD (by [Book] 3 of [Euclid's] Elements). But since the difference between BE and DE, and also between BA and DA is supposed infinitesimally small, I substitute one for the other in each case and it follows that DE/DA = DA/DB. And then by making DA^2 (or DE × DB) to $ADEA^2$ as DB to $ADEA^2/DE$ I obtain the required line (namely the third proportional of the circumference to the diameter) through which its endeavor of receding from the center would impel the body in the time of a complete revolution when applied constantly in a straight line.

For example, since the third proportional equals 19.7392 semidiameters, if the endeavor of approaching to the center [of the earth] in virtue of gravity were exactly equal to the endeavor of receding from the center at the equator due to the diurnal motion of the Earth: then in a periodic day it would impel a heavy body through 19¾ terrestrial semidiameters, that is through 69087 miles: and in an hour through 120 miles. And in a first minute through $^1/_{30}$ mile or through $^{100}/_3$ paces, that is $^{500}/_3$ feet; and in a second minute through $^5/_{108}$ feet or $^5/_9$ inches. But actually the force of gravity is of such a magnitude that it moves heavy bodies down about 16 feet in one second, that is about 350 times further in the same time than the endeavor from the center [would move them], and thus the force of gravity is many times greater than what would prevent the rotation of the Earth from causing bodies to recede from it and rise into the air.

Corollary: Hence the endeavors from the centers of divers circles are as the diameters divided by the squares of the periodic times, or as the diameters multiplied by the [squares of the] numbers of revolutions made in a given time. So that since the Moon revolves in 27 days 7 hours and 43 minutes or in 27.3216 days (whose square is 746½) and is distant 59 or 60 terrestrial semidiameters from the Earth, I multiply the distance of the Moon 60 by the square of the lunar revolution, 1, and the distance of the surface of the Earth from the center, 1, by the square of the revolutions, 746½, and so I have the ratio 60 to 746½, which is that between the endeavor of the Moon and the surface of the Earth of receding from the center of the Earth. And so the endeavor of the surface of the Earth at the equator is about 12½ times greater than the endeavor of the Moon to recede from the center of the Earth. And so the force of gravity [at the surface of the earth] is 4000 and more times greater than the endeavor of the Moon to recede from the center of the Earth.

And if the Moon's endeavor from the Earth is the cause of her always presenting the same face to the Earth, the endeavor of the lunar and terrestrial system to recede from the Sun ought to be less than the endeavor of the Moon to recede from the Earth, otherwise the Moon would look to the Sun rather than to the Earth. . . .

> And so the force of gravity will be 5000 times greater than the endeavor of the Earth to recede from the Sun. . . .
>
> Finally since in the primary planets the cubes of their distances from the Sun are reciprocally as the squares of the numbers of revolutions in a given time: the endeavors of receding from the Sun will be reciprocally as the squares of the distances from the Sun. (Herivel, 195–97)

The primary objective of the text is to compare gravity and centrifugal force. It is a matter of replying to the objections of the anti-Copernicans by showing that gravity is sufficiently strong to counterbalance the centrifugal force caused by the earth's rotation, even at the equator where the effort to recede would be most intense. Newton's result in this manuscript is that gravity is 350 times as strong as the force required to retain a body at the surface of the earth. This discussion is a prolongation of the argumentation of Galileo's *Dialogo* on the two principal systems of the world (see below, p. 141ff.).

The essential conceptual instrument comes from Descartes' *Principia philosophiae*: it is the notion of "effort to recede from the center." But in contrast to Huygens, Newton admits in advance that this effort is of the same nature as gravity, that it acts in the same mode (*ad modum gravitatis*). And so the distance traversed must be, a priori, proportional to the square of the time. Newton deduces from this that, if the body deviates by the distance DE during the time AD, it will deviate by the distance $ADEA^2/DE$ during an entire period. He calls this distance "the third proportional of the circumference to the diameter" because it satisfies the proportion x : ADEA :: ADEA : DE. Expressing ADEA and DE in terms of the radius results in $4\pi^2r^2/2r$, which gives very nearly the number that Newton calculates—19.7392 times the radius.

It is still unknown what this distance traversed might signify. The absurdity is in a sense more obvious than in Huygens' work, where the lengths of the deviations remained very small in relation to the circular trajectory. What rectilinear distance would the force cause to be traversed during the time of an entire period? The moving body has had time to occupy successively all directions and to return: it is not a deviation with respect to a direction that is being calculated. And what is this rectilinear path that is traversed in accelerated motion, when the body has left the wheel and there is no longer a cause of acceleration?

In reality, the only thing that matters, as with Huygens, is the very first commencement of motion. The initial deviation is to be compared with a fall during a very brief time. Thus, the centrifugal force can be compared with gravity, and it can be proved that even at the equator, the weight of bodies suffices to keep them on the earth.

The corollary goes well beyond this discussion of the rotation of the earth and the anti-Copernican arguments. This passage of the manuscript is historically very important because it is probably the oldest document from Newton's hand that attests to the intuition of a universal gravitation, though in a form still implicit and shrouded. The evaluation of centrifugal effort is first extended to the case of the moon. Newton compares the effort of bodies to recede at the surface of the earth with the moon's effort to recede owing to its rotation around the earth. He concludes that terrestrial gravity is 4,000 times stronger than the centrifugal effort of the moon but says nothing of a possible "gravity" that would retain it in its orbit. The reasoning about a compensatory force, or a comparison between gravity and centrifugal force, remains limited to the case of terrestrial bodies.

The manuscript next treats of the comparison between the centrifugal effort of the moon in relation to the earth and the effort of the earth-moon system to recede from the sun. Newton hypothetically puts forward a rather questionable argument: if the moon always turns the same face toward the earth, it is an indication that its effort to recede from the earth is greater than the corresponding effort of the earth-moon system to recede from the sun. (Newton is presupposing here that the moon is not a perfect sphere.)

Finally, the idea of centrifugal effort is extended to the other bodies of the solar system. Kepler's "third law," which asserts the constant proportionality between the cube of the orbital radius and the square of the period, makes it possible to compare the efforts to recede of different planets; these efforts are inversely proportional to the square of the distance from the sun.

What difference is there here from the *De motu* of 1684? Here it is a question of centrifugal effort, not of centripetal effort. More specifically, in the cases of terrestrial bodies, the moon, and the planets, it is a question of a variation of centrifugal effort as a function of distance. The idea of a gravity that retains these bodies is only implicit. That this idea underlies the discussion is evident if one extends to these other cases the reasoning on equilibrium that is applied to heavy bodies at the equator: the effort to recede is counterbalanced by gravity.[51] Newton does not say what counterbalances the centrifugal effort of the moon and planets, but an analogy is at least suggested by the text. Once it is known that the centrifugal efforts of the planets diminish as the square of the distance, and that in the case of the earth it is gravity that keeps the centrifugal effort in check, the question that arises is whether the force that retains the planets does not also diminish as the square of the distance.

Such is the way the young Newton reasoned some fifteen years before the *Principia*. The approximate date of the text can be inferred from evidence given by D. Gregory (see Herivel, 194) and above all from a parallel contained in a letter that Newton sent to Huygens on receiving from him his *Horologium oscillatorium*:

> I am glad that we are to expect another discourse of ye Vis centrifuga, wch speculation may prove of good use in natural Philosophy & Astronomy as well as Mechanicks. Thus for instance, if ye reason why ye same side of ye moon is ever

> towards ye earth be ye greater conatus of ye other side to recede from it,[52] it will follow (upon supposition of ye earths motion about ye Sun) that ye greatest distance of ye sun from ye earth is to ye greatest distance of ye Moon from ye earth, not greater then 10000 to 56. (*Corresp.*, 1:190, 2:446; see also *Corresp.*, 2:436)

The question of date is important in order to relate to these different stages the discovery of universal gravitation. For the study of the mathematization of force, this manuscript text is invaluable. Newton here presents in a brief and still rudimentary form the linkage between dynamics and geometry that he develops and enriches in the *De motu* and *Principia*. It is a matter of knowing how much the body would be "deviated" in relation to the circumference (*quantus . . . deferret a circumferentia*). With Newton, as with Huygens and Descartes, it is the deflection or deviation that represents and measures the effort. The new and fertile point of view consists in envisaging this deflection under a double aspect—at once as a dynamical and as a geometrical reality. The segment DB must be examined in two ways:

> as a path traversed in accelerated motion (It is then proportional to the square of the time because the centrifugal effort acts "in the mode of gravity.")
>
> as a divergence between the tangent and the circle (It is then proportional to the square of the arc, for very small arcs, and it is therefore proportional to the square of the time because the arcs are themselves proportional to the elapsed time.)

In Newton's text, these two aspects correspond to two stages of the reasoning: the dynamical point of view permits the statement that the distance x traversed during an entire period must agree with the proportion $x/\mathrm{BD} = \mathrm{ADEA}^2/\mathrm{AD}^2$. The geometrical point of view then permits the assertion that AD^2 can be replaced by $\mathrm{DE} \cdot \mathrm{DB}$.

Deviation in Galileo's *Dialogo*

Up to this point, an essential element has been lacking in this account. Before seeking to calculate in different ways the measure of the deflection caused by the force, thinkers had to understand that the object on which attention had to focus was precisely this deflection or deviation. The Cartesian theory of "impeded motion" could lead to this idea, but another celebrated text, Galileo's *Dialogo*, contains a much more explicit discussion in the passage that treats of the anti-Copernican arguments relative to projectile motion. If the earth rotates, must not its enormous motion of rotation throw bodies up toward the sky? It is this discussion that Newton takes up and continues in the manuscript that we have studied.

The discussion between Salviati and Sagredo is very long and interrupted with detours and repetitions, so only two very important excerpts will be cited. In the first, Galileo dismisses a confusion concerning the notion of velocity:

> Salviati: Up to this point we have conceded to Ptolemy, as an unquestionable effect, that, since the casting off of the stone is caused by the speed of the wheel moving about its center, the cause of this casting off increases as the speed of rotation [*vertigine*] augments. Whence it is inferred that, the terrestrial rotation being immensely greater than that of any machine we could make to rotate artificially, the resulting extrusion of stones, animals, etc., must be extremely violent.
>
> But I now remark in this discourse a very great fallacy, when we compare indifferently and absolutely such speeds with one another. It is true that if I made the comparison between velocities of the same wheel, or of equal wheels, the one turning most rapidly would throw stones with greater impetus [*impeto*], and as the velocity increased, the cause of the projection would increase in the same proportion.
>
> But suppose the velocity were increased not by increasing the velocity of the same wheel, that is, by giving it a greater number of turns in equal times, but by increasing its diameter and making the wheel greater, so as to keep the time of revolution the same for the small and the large wheel. The velocity on the large wheel becomes greater as the circumference is enlarged. In this case no one will believe that on the large wheel the cause of extrusion would increase in the same proportion as the velocity of its circumference, as compared to the velocity of the circumference of the small wheel.
>
> For that is quite false, as can be shown in a gross way with a ready experiment: we can throw a stone with a stick a yard long that we could not throw with one six yards long, even though the motion of the end of the long stick, that is, the motion of the stone that is held there, were twice as rapid as the motion of the point of the small stick, which would be the case if the velocities were such that in the time of one turn of the large stick, the small one made three turns. (Second Day, EN, 7:237–38)

To evaluate the projection due to a motion of rotation, it is necessary first to indicate in what sense this motion is understood, and what, in this case, the velocity means. It should be said in passing that here Galileo draws very explicitly the consequences of the principle of inertia—which he did not always do (see Koyré *EG*, Troisième Partie).

Having established the distinction between angular velocity and velocity with which the arc is traversed, the interlocutors turn to the quantitative evaluation of the force of projection:

> Let us turn to the analysis of our problem. And for a better understanding let us give a figure for it. Let there be two unequal wheels with center A, BG being the circumference of the small one and CEH that of the large one, and ABC a radius perpendicular to the horizon. Through the points B and C let us draw the rectilinear tangents BF, CD, and on the arcs BG, CE let us take two equal parts BG, CE, and let us suppose that the two wheels are turned about their centers with equal

velocities, so that two moving bodies, which could be, for example, two stones placed in the points B and C, are carried through the circumferences BG, CE with equal velocities. Thus in the same time that the stone B traversed the arc BG, the stone C would pass over the arc CE.

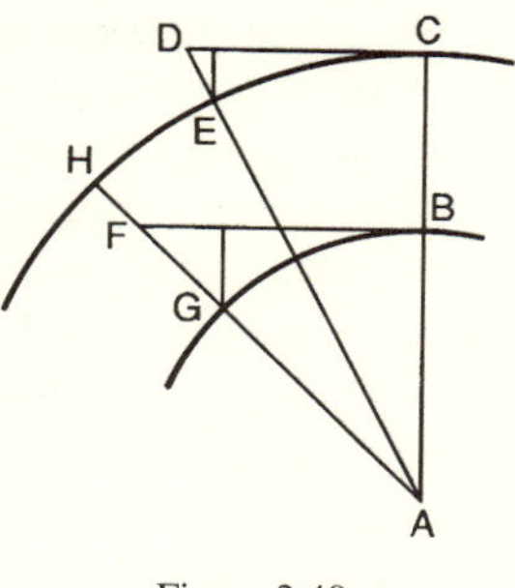

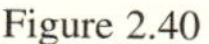
Figure 2.40

I say now that the rotation [*vertigine*] of the small wheel is much more powerful [*potente*] to produce the projection of the stone B, than the rotation of the large wheel is to produce that of the stone C. Since, as already explained, the projection must be made along the tangent, when the stones B and C are to separate from their [respective] wheels, and begin the motion of projection from the points B, C, the impetus generated by the rotation will throw them along the tangents BF, CD. Therefore along the tangents BF, CD the two stones have equal impetuses to move forward, and it is thus that they would advance if they were not deviated by some other force [*forza*]. Is it not so, Signore Sagredo?

Sagredo: It seems to me that the business indeed happens in this way.

Salviati: But what force, do you think, can make the stones deviate from their motion along the tangents, to which motions they were in fact impelled by the impetuses due to rotation?

Sagredo: Either it is their own gravity, or some adhesive that keeps them placed on or attached to the wheels.

Salviati: But to cause a moving body to deviate from the motion to which it is entrained by its impetus, is there not necessary more or less force, according as the deviation is to be greater or less? That is to say, according as it must, in the deviation, pass over a greater or smaller distance?

Sagredo: Yes, because we have already concluded that, in causing a body to be moved, the greater the velocity with which it is to be moved, the greater the moving virtue must be.

Salviati: Now consider how, to cause the stone on the small wheel to deviate from the motion it would make on the tangent BF, and to keep it attached to the wheel, it is necessary that its own gravity should pull it back through the length of the secant FG [*la ritiri per quanto e lunga la segante FG*], or through the length of the perpendicular dropped from G onto the line BF, while on the larger wheel the drawing back [*ritiramento*] need not be greater than the secant DE or the perpendicular dropped from the point E onto the tangent DC . . . and thus a much greater force will be required to keep the stone B conjoined to its small wheel than to keep the stone C conjoined to its greater one. (Second Day, EN, 7:240–42)

Salviati's reasoning is supported by the representation of two concentric circles along which two bodies move with equal velocities, velocity being here understood as linear velocity: the bodies traverse the arcs of equal length BG and CE in the same interval of time. The impetuses that tend to make them follow the tangents are therefore of the same intensity and cause them to traverse in the same time the equal rectilinear segments BF and CD.

Nonetheless the rotation on the small circle is "more powerful" for extruding the body towards the exterior of the circle. Why? The body tends to traverse the tangent, it is "the motion to which it is entrained by its impetus"; if it is deflected, there must be "some other force." And the intensity of this force can be evaluated by considering the effect it must have, that is, by finding out how much it must deflect the body from its normal trajectory.[53] Consequently the gravity—or the adhesive, whatever it may be—that keeps together the elements of the wheel must be greater in the case of the small wheel because it has to pull back the body through the length FG, which is longer (given that the time is the same) than the deviation DE on the larger wheel.

What is the precise direction of the deviations? Galileo leaves this question undetermined and does not choose between oblique deviations, such as GF, and perpendicular deviations. But the very fact that Salviati hesitates and leaves the question undecided already shows a certain subtlety in the analysis several decades before the refinements introduced by Huygens and the invention of the evolute.

The essential ideas of this text were taken up again in the course of the seventeenth century, as has been seen in the writings of Newton.[54] The idea of an equilibrium or compensation between centrifugal effort and gravity (or the cohesion of the body in rotation) serves as a mainspring in the young Newton's manuscript text on circular motion and the centrifugal efforts of the celestial bodies. The measuring of force by the deviation is a common theme in the different reasonings that have been examined in the work of Huygens and Newton. Finally, like Newton in the *De motu*, Galileo interprets the divergence FG as a distance traversed (the body "in the deviation" must "pass over a greater or smaller distance" in the same time).

Hooke's Entreaty

Can an analysis of the planetary motions in terms of centrifugal effort and gravity, with the one compensating the other, lead far enough to give birth to a veritable dynamical theory with a detailed development?[55] Several obstacles present themselves. In the first place, there are always the risks of mistaking the direction of the force and taking as accelerated a uniform rectilinear motion. Also, how can the theory of centrifugal force be applied to noncircular trajectories traversed with variable velocities? Whatever the answer to this question, it is a totally different analysis that Newton comes to propose in the *De motu* and the *Principia*, in which the very term "centrifugal force" is nearly absent, except in a few particular situations. The fundamental elements of the analysis of curvilinear motion are henceforth inertial motion and the exterior force that causes the body to deviate.

It is very probable that Hooke's intervention was decisive in Newton's intellectual evolution, leading him to a new conception of curvilinear motion. An exchange of letters took place between the two in 1679. Later Hooke drew from it an argument for claiming priority over Newton. The latter rejected Hooke's claim (*Corresp.*, 2:437–40) but acknowledged that these letters had been a stimulus to him.[56]

These exchanges were a renewal of contact, the two men having already in 1672 had an occasion to confront one another and indeed to collide over the theories of light. This time the motive was quite official. Hooke had been a secretary of the Royal Society since 1677, and his office required that he encourage exchanges of ideas, discussions, commentaries on all sorts of hypotheses and experiments. In the name of the Royal Society, therefore, he entreated Newton to participate in the common scientific activity, and then he added some lines in his own name:

> For my part I shall take it as a great favor if you shall please to communicate by Letter your objections against any hypothesis or opinion of mine, And particularly if you will let me know your thoughts of that of compounding the celestiall motions of the planetts of a direct motion by the tangent & an attractive motion towards the centrall body. (Hooke to Newton, 24 November 1679, *Corresp.*, 2:297)

Among the theories that Newton is invited to discuss, Hooke has chosen in particular a hypothesis relative to the motion of the planets, which he had propounded to the Royal Society, beginning in 1666, in the following terms:

> I have often wondered why the planets should move about the sun according to Copernicus's supposition, being not included in any solid orbs . . . nor tied to it, as their center, by any visible strings; and neither depart from it beyond such a degree, nor yet move in a strait line, as all bodies that have but one single impulse, ought to do. . . . But all celestial bodies, being regular solid bodies, and moved in a fluid, and yet moved in a circular or elliptical line, and not strait, must have some other cause, besides the first impressed impulse, that must bend their motion into that curve. (Birch, *History Royal Society*, 2:90; *Corresp.*, 2:299 n.7)

It is once again the question that Kepler posed around 1600 (see above p. 68ff.), but with the principle of inertia now taken as point of departure. In his exposition of 1666, Hooke proposed two possible explanations:

> that this fluid, the aether, has a density whose degree increases with distance from the Sun
>
> that there is a continually attractive property of the body situated at the center, the Sun. (Birch, *History Royal Society*, 2:90; *Corresp.*, 2:299 n. 7)

The second hypothesis was developed in a work published later, five years before the exchange with Newton. Here is how Hooke set forth there the three propositions essential to his theory:

> 1. That all Coelestial Bodies whatsoever, have an attraction or gravitating power towards their own Centers, whereby they attract not only their own parts, and keep them from flying from them, as we may observe the earth to do, but that they do also attract all other Coelestial Bodies that are within the sphere of their activity. . . .
>
> 2. The second supposition is this, That all bodies whatsoever that are put into a direct and simple motion, will so continue to move forward in a streight line, till they are by some other effectual powers deflected and bent into a Motion, describing a Circle, Ellipsis, or some other more compounded Curve Line.
>
> 3. The third supposition is, That these attractive powers are so much the more powerful in operating, by how much the nearer the body wrought upon is to their own Centers.
>
> Now what these several degrees are I have not yet experimentally verified. (Hooke, "An Attempt to Prove," 27–28; cf. Westfall *NR*, 382, and Koyré *EN*, 180–84)

Here is the "deflection" of the *De motu* and of Descartes' *Principia philosophiae* but in a context completely foreign to Descartes, that of the "Magnetic Philosophy" which admits "attractive powers."[57]

It was concerning this theory that Hooke was soliciting the objections of his correspondent. Was Newton acquainted with it? He claimed that he was not, having for some years past been "endeavoring to bend [him]self from Philosophy" (*Corresp.*, 2:300). Nevertheless, to prove his good will, Newton proposed an experiment that should make it possible to discover the diurnal rotation of the earth:

> I shall communicate to you a fansy of my own about discovering the earth's diurnal motion. In order thereto I will consider ye Earth's diurnal motion alone without ye annual, that having little influence on ye experimt I shall here propound.
>
> Suppose then BDG represents the Globe of ye Earth carried round once a day about its center C from west to east according to ye order of ye letters BDG; & let A be a heavy body suspended in the Air & moving round with the earth so as perpetually to hang over ye same point thereof B.
>
> Then imagin this body A let fall & it's gravity will give it a new motion towards y^e center of y^e Earth without diminishing y^e old one from west to east. Whence the motion of this body from west to east, by reason that before it fell it was more distant from y^e center of y^e earth

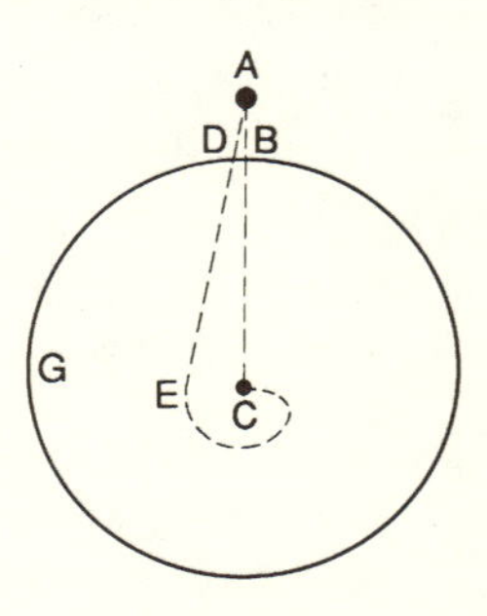

Figure 2.41

then the parts of y^{e} earth at wch it arrives in its fall, will be greater then the motion from west to east of y^{e} parts of y^{e} earth at wch y^{e} body arrives in it's fall: & therefore it will not descend in ye perpendicular AC, but outrunning ye parts of ye earth will shoot forward to ye east side of the perpendicular describing in it's fall a spiral line ADEC, quite contrary to ye opinion of ye vulgar who think that if ye earth moved, heavy bodies in falling would be outrun by its parts & fall on the west side of ye perpendicular. The advance of ye body from ye perpendicular eastward will in a descent of but 20 or 30 yards be very small & yet I am apt to think it may be enough to determin the matter of fact. (Newton to Hooke, 28 November 1679, *Corresp.*, 2:301–2)

The untrained, who have not assimilated the principle of inertia, believe that, if the earth rotates, a falling body will fall westward of the vertical because the ground will slip away eastward underneath it during its fall. On the contrary, Newton claims, if the body participates in the earth's motion of rotation and keeps this motion during its descent, then it must fall a little in advance of the vertical.

The essential condition of the problem is that the body A keep its motion "from west to east," even when this motion is combined with a motion of descent: "it's gravity will give it a new motion towards ye center of ye Earth without diminishing ye old one from west to east." What does this condition signify? Has it a physical meaning? Is it compatible with what the *Principia* will establish for the trajectory of a body under the action of a central force? It is the law of areas that establishes the standard for such a path, and when Newton takes up this problem again in more confident terms in section 9 of the *Principia*, he will use the law of areas as the fundamental tool in the study of rotating orbits.[58]

Some days later Hooke replied to Newton and corrected him on a decisive point: would the body reach the center of attraction? The curve drawn by Newton seems to end at the center C. Is that possible? Even if one supposes the earth entirely penetrable (and nonetheless attractive), will the body subject to a gyratory motion combined with attraction arrive finally at the very center of the earth, as Newton seems to admit?

But as to the curve Line which you seem to suppose it to Desend by, . . . Vizt a kind of spirall which after sume few revolutions Leave it in the Center of the Earth my theory of circular motion makes me suppose it would be very differing and nothing att all akin to a spirall but rather a kind Elliptueid.

At least if the falling Body were supposed in the plaine of the equinoxciale supposing then ye earth were cast into two half globes in the plaine of the equinox and those sides separated at a yard Distance or the like to make Vacuity for the Desending Body and that the gravitation to the former Center remained as before and that the globe of the earth were supposed to move with a Diurnall motion on its axis and that the falling body had the motion of the superficiall parts of the

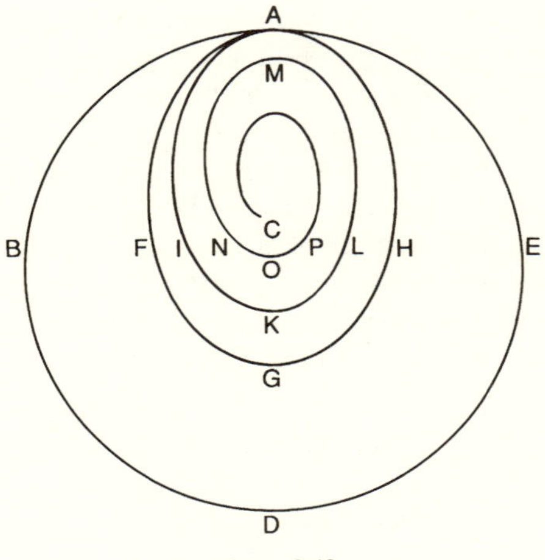

Figure 2.42

> earth from whence it was Let fall Impressed on it, I conceive the line in which this body would move would resemble An Elleipse: for Instance Let ABDE represent the plaine of the equinox limited by the superficies of the earth: C the Center therof to which the lines of Gravitation doe all tend. Let A represent the heavy Body let fall at A and attracted towards C but Moved also by the Diurnall Revolution of the earth from A towards BDE &c. I conceive the curve that will be described by this desending body A will be AFGH and that the body A would never approach neerer the Center C then G were it not for the Impediment of the medium as Air or the like but would continually proceed to move round in the Line AFGHAFG &c. . . .
>
> But w[h]ere the Medium through which it moves has a power of impeding and destroying its motion the curve in wch it would move would be some what like the Line AIKLMNOP &c and after many resolutions would terminate in the Center C. (Hooke to Newton, 9 December 1679, *Corresp.*, 2:305–6)

Hooke's idea is ingenious: imagining the earth divided into two hemispheres separated by a very narrow space retains an essential condition of the problem, namely the existence of a center of forces, but allows the attracted body to approach the center as much as it tends to do. By virtue of this fiction, bodies falling to the surface of the earth become similar to satellites or to planets, unconstrained in their motions in the cosmic milieu. The falling motion is assimilated to planetary motion.[59]

Under these conditions, will bodies fall to the center, as one could believe in extrapolating the ordinary falls of terrestrial bodies, or will they circulate around the attracting center, precisely in the manner of satellites and planets? Hooke claims, without giving justification (except for a new and vague invocation of "his theory"), that these bodies will approach the attracting center during a portion of their travel then recede from it; there will be a point of nearest approach, here marked G (the "perigee" in the terminology of astronomy).

THE LETTER TO HOOKE OF DECEMBER 1679 AND THE SUM OF MOTIONS GENERATED

By what right does Hooke assert that the body must recede after having approached? Newton's response is primarily devoted to this question. Newton accepts the idea and develops it by proposing a plausible reasoning that would justify Hooke's conjecture. To show that the distance of the moving body from the center of force must pass through a minimum before increasing again, Newton employs a completely new mode of reasoning: he evaluates—or compares—sums of "motions generated" by the action of gravity. Here is the text of this response:

> I agree wth you y^{t} . . . if [the body's] gravity be supposed uniform it will not descend in a spiral to ye very center but circulate wth an alternate ascent & descent made by it's vis centrifuga & gravity alternately overballancing one another.
>
> Yet I imagin ye body will not describe an Ellipsoeid but rather such a figure as is represented by AFOGHIKL &c. Suppose A ye body, C y^{e} center of y^{e} earth, ABDE quartered wth perpendicular diameters AD, BE, wch cut y^{e} said curve in F & G; AM y^{e} tangent in wch y^{e} body moved before it began to fall & GN a line drawn parallel to y^{t} tangent.

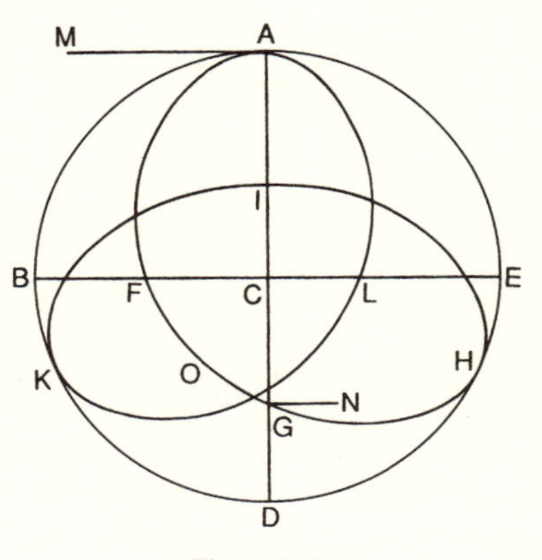

Figure 2.43

> When y^{e} body descending through y^{e} earth (supposed pervious) arrives at G, the determination of its motion shall not be towards N but towards y^{e} coast between N & D.
>
> For ye motion of ye body at G is compounded of ye motion it had at A towards M & of all ye innumerable converging motions successively generated by ye im-

presses of gravity in every moment of it's passage from A to G: The motion from A to M being in a parallel to GN inclines not ye body to verge from ye line GN. The innumerable & infinitly little motions (for I here consider motion according to ye method of indivisibles) continually generated by gravity in its passage from A to F incline it to verge from GN towards D, & ye like motions generated in its passage from F to G incline it to verge from GN towards C.

But these motions are proportional to ye time they are generated in, & the time of passing from A to F (by reason of ye longer journey and slower motion) is greater then ye time of passing from F to G. And therefore ye motions generated in AF shall exceed those generated in FG & so make ye body verge from GN to some coast between N & D.

The nearest approach therefore of ye body to ye center is not at G but somewhere between G & F as at O.

And indeed the point O, according to ye various proportions of gravity to the impetus of ye body at A towards M, may fall any where in ye angle BCD in a certain curve wch touches ye line BC at C & passes thence to D.

Thus I conceive it would be if gravity were ye same at all distances from ye center. But if it be supposed greater nearer ye center ye point O may fall in ye line CD or in ye angle DCE or in other angles y^t^ follow, or even no where. For the increase of gravity in ye descent may be supposed such y^t^ ye body shall by an infinite number of spiral revolutions descend continually till it cross ye center by motion transcendently swift.

Your acute Letter having put me upon considering thus far ye species of this curve, I might add something about its description by points *quam proximé*. But the thing being of no great moment I rather beg your pardon for having troubled you thus far wth this second scribble wherin if you meet wth any thing inept or erroneous I hope you will pardon ye former & ye latter I submit & leave to your correction remaining Sr

Your very humble Servant
Is. Newton

(*Corresp.*, 2:307–8)

This letter is extremely valuable evidence concerning Newton's evolution in the study of forces. At the beginning of the text Newton refers to the conceptual framework he had utilized in his first letter: the couple made up of attraction and centrifugal force, which balance each other and are alternately victorious. In the argumentation that follows, centrifugal force no longer plays any role. Hooke had proposed a different analysis, in which the *vis centrifuga* is absent: nothing else is necessary in order to render an account of curvilinear motion than the principle of inertia and the action of gravity. This is in agreement with Hooke's supposition in 1674: the body tends to continue its uniform motion in a straight line, until it is "deflected" or "incurved" by an exterior power. This is what Hooke characterizes as his idea of "compounding the celestiall motions of the planetts of a direct motion by the tangent & an attractive motion towards the centrall body" (see above, p. 147). Has Newton perceived that this analysis differs from his own? Whatever the situation, it is this new framework he accepts in his reasoning: in the initial instant, the body tends to traverse the tangent AM, then it receives "impresses of gravity" that modify its motion.[60] Each impress of gravity generates a new motion which is then compounded with the initial motion and the motions already received. The impress and compounding are repeated in each instant.

The scansion of the impresses of gravity is given by the unfolding of time, as in Galileo's theory. The body receives "all ye innumerable converging motions [towards the center of force] successively generated by ye impresses of gravity in every moment." Also, the motions received along an arc are proportional to the time in which this arc is traversed.

This way of considering gravity is precisely the one that Descartes judged to be too abstract, fictive, and geometrical, destitute of foundation. Without knowing anything of the mode of action, without saying anything about the central body (here scarcely "physical") or of the milieu, Newton, like Galileo, supposes that gravity acts in a constant and regular way in the course of time: at each instant the impress that a body receives is identical with the impress it received the moment before. He further assumes, implicitly, that the body conserves all the impulsions received in the form of motions that are compounded with the preceding motion.

The study of these successive impresses should make it possible to answer the question disputed between Hooke and Newton: what is the figure of the trajectory, or more precisely, is there a point of maximum approach, followed by a moving away? Newton, piqued by having been caught in an error and corrected, wants to regain the ascendency; not only does he know very well that the body would not in all cases reach the center, but he can even prove that there exists a point of maximum approach which is not situated in G, opposite the apogee A, as in the "ellipsoeid" of Hooke. This point, Newton claims, is somewhere on the trajectory before point G, because in point G the body tends in a direction that departs from the center; the "determination of its motion" (this is the Cartesian formula) is "towards ye coast between N & D."

The aim of the reasoning is therefore to show that the initial motion of the body, directed along AM, has been progressively altered so as finally, at point G, to be directed at an angle between GN and GD. To obtain this result, it would be necessary in principle to evaluate the sum of all the motions successively generated by gravity, all through the instants of the motion AFG, the received impresses being always directed towards the center C and being considered as of equal intensity. In fact Newton does not form the sum, properly speaking, of all these "innumerable & infinitly little motions"; he limits himself to comparing the totality of motions received between A and F with the totality of motions received between F and G. He separates the motions generated by gravity during the journey from A to G into two masses and weighs against each other what the body has received along AF and what it has received along FG.

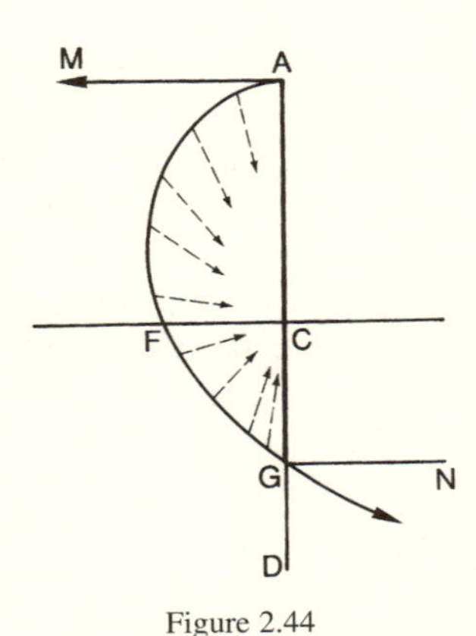

Figure 2.44

The initial motion, directed along AM, matters little since it is exactly aligned with GN and in the opposite direction. It contributes nothing toward inclining the moving body to go above or below the direction GN. The only significant contributions come from the two groups of impresses received after the initial instant. Those that are received in the first quadrant, between A and F, have a downward component in the figure, since gravity tends always toward C; those that are received in the second quadrant, between F and G, have, on the contrary, an upward component.

What is the final resultant of these opposing contributions? Do all the impresses received combine to direct the body upward, or downward, when it has arrived at the point G?[61] The answer comes from comparing the times required for the two parts of the journey. In the first quadrant the path is longer and the motion slower; the body therefore takes more time and hence receives more impresses of gravity than during its passage through the second quadrant. The resultant motion is therefore, on the whole, directed downward rather than upward. One can thus say that at G the direction of motion is such that the body is receding from the center.

The decisive passage of the reasoning, and the one most subject to caution, is in the comparison of the times of traversal of the two different portions of the trajectory. The moving body receives more impresses of gravity between A and F because it remains there longer. There are two reasons for this: the arc AF is longer than FG; and the body has received fewer impulsions accelerating its motion toward C. The comparison remains rough and qualitative and depends on the supposed form of the trajectory.

But, inversely, how to determine the trajectory without knowing the new motions received after the initial instant? The time of passage and the form of the trajectory seem to depend inextricably on each other. Here one realizes the importance of the law of areas for the problem at hand, and how Newton lacks it at just this point.[62] If it were known that the sector swept out is proportional to the time, it would no longer be necessary to know the form of the trajectory and the velocity with which it is traversed to evaluate the impresses of force received. It is probable that Newton at this date had not yet demonstrated, nor even definitely accepted, the proposition that Kepler had advanced but which was not universally accepted among the astronomers.

It is difficult to pinpoint exactly what Newton knew at the moment of his exchange with Hooke. May he have preferred to keep for his own use certain results of the highest importance and leave Hooke in ignorance of the first foundations that might lead him to a more sure-footed examination of the celestial motions? The "old lion," as Johann Bernoulli later called him, kept so many things to himself without publishing them . . .

Yet a later manuscript, perhaps intended to serve as preface for a second edition of the *Principia*, implies very clearly that Newton did not recognize the truth of the law of areas until after the exchange with Hooke and that it is just this law that opened for him the way to a demonstration of the ellipticity of the orbits:

> And [Dr. Hooke] added that they would not fall down to the center of the earth but rise up again & describe an Oval as the Planets do in their orbs. Whereupon I computed what would be the Orb described by the Planets. For I had found before by the sesquialterate proportion of the tempora periodica of the Planets with respect to their distances from the Sun, that the forces wch kept them in their Orbs about the Sun were as the squares of their mean distances from the Sun reciprocally; & I found now that whatsoever was the law of the forces wch kept the Planets in their Orbs, the areas described by a Radius drawn from them to the Sun would be proportional to the times in wch they were described. And by the help of these two Propositions I found that their Orbs would be such Ellipses as Kepler had described. (Add. 3968, fol. 101, in Cohen, 293)

The law of areas was the key to the dynamics of the *De motu* and the *Principia*, and Newton did not yet use it when he corresponded with Hooke.

The Trajectory of Comets (Flamsteed and Newton)

Entreated by Hooke, Newton showed himself capable of estimating very cleverly the shape of the trajectory of a body subject to an attraction. Soon afterward, another occasion arose that offered a further opportunity for deploying his skill in dealing with orbital motion. It was a discussion with Flamsteed, the astronomer royal, concerning the motion of comets. This time it was a matter of celestial motions, and the attracting body was no longer the earth, but the sun.

In November 1680, a comet began appearing before sunrise. Then, after an interval in which nothing celestial was visible, another comet began appearing in mid-December, this time after sunset. Flamsteed claimed that it was the same comet returning after having approached the sun.[63] Newton opposed this idea on the basis of observations made in various parts of Europe, especially in Rome (*Corresp.*, 2:342), but Flamsteed convinced him that the disagreement resulted from a misunderstanding in the determination of dates (the Gregorian calendar had not yet been accepted in England) (*Corresp.*, 2:349).

The astronomer royal imagined a very complicated mechanism to account for the motion of the comet, including an attraction of a magnetic nature and a vortex in rotation about the sun. Here is how Flamsteed explained his conjectures:

> I conceave therefore that the Sun attracts all the planets and all like bodys that come within our Vortex, more or lesse according to the different substance of theire bodyes and nearenesse or remotenesse from him;
>
> [and] that it drew the Comet by its northerne pole, the line of whose motion was at first really inclined from the North into the South part of the heavens but was by this attraction as it drew nearer to him bent the Contrary way, as if *ee* were the plane of the Ecliptick, *p* y^e North pole of y^e Sun, *ad* y^e line of its first inclination which by y^e attraction of y^e pole *p* is bent into y^e curve *abc* in which Gallet observed it to move. . . .

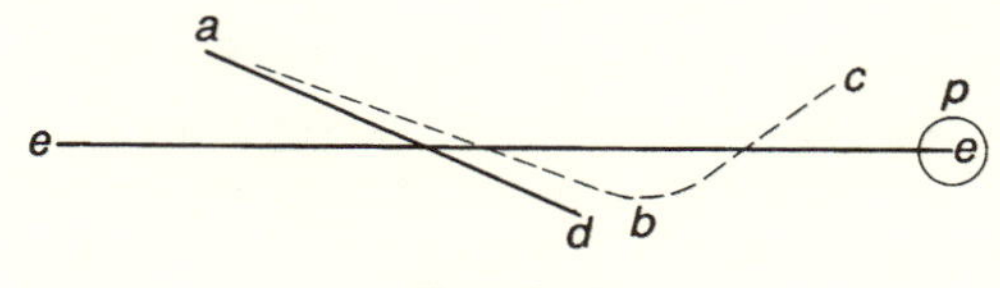

Figure 2.45

> When it came within y^e Compasse of y^e orbis annuus this attraction of y^e Sun would have drawne it neare him in a streight line, had not the laterall resistance of y^e Matter of y^e Vortex moved against it bent it into a Curve;
>
> as if in the figure of y^e [this] Page $\alpha\gamma$ were y^e line in which y^e motion of y^e Comet were directed; this by y^e attraction of y^e Sun is bent into β,*B*, & when y^e Comet comes to *B* it would be carried on to y^e Sun in y^e streight line *BC*, were it not for y^e motion of the Vortex beareing it out of y^t line from *e* to *g*.
>
> the body of y^e Comet I conceave to have a'lwayes the same part carried foremost in y^e line of its motion so y^t when it comes to C it moves contrary & crosse

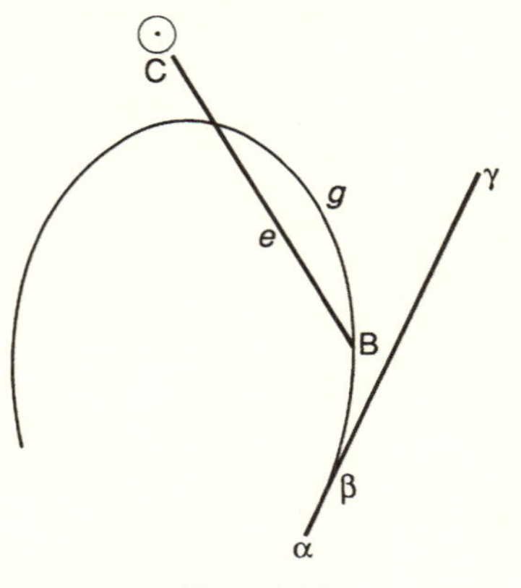

Figure 2.46

ye motion of ye Vortex till haveing ye contrary End opposite to ye Sun hee repells it as ye North pole of ye loadstone attracts ye one end of ye Magnetick needle but repells ye other.

This act of repulsion would carry ye Comet from ye Sun in a streight line, were it not that the crosse motion of ye Vortex bendes it back. (Flamsteed to Halley, 17 February 1680, *Corresp.*, 2:337–38)

Flamsteed presents a good sample of the Magnetic Philosophy of the English disciples of Gilbert and Kepler. The solar attraction, quasi-magnetic in nature, depends on the nature of the attracted bodies and their orientation with respect to the sun since like magnets, the celestial bodies have a friendly side and an unfriendly side. The behavior of the comet and the sun is reminiscent of Kepler's assumptions, except for certain details (for example, according to Flamsteed the sun has poles, contrary to what Kepler says; see above, p. 80).

The study of the trajectory is crude and naive: What does it mean to "bend" a motion? And if a body is attracted by a center of force can it, after deflection, be directed finally straight at this center? It is this latter question that Newton comments on. He corrects Flamsteed's suggestion and shows him that it would be necessary to imagine a continuous deviation that would make the comet pass behind the sun and not in front of it:

I can easily allow that ye attractive power of ye Sun as ye Comet approaches ye Sun passing from *B* to *f* & then to *g*, will make ye comet verge more & more from its former line of direction towards ye Sun, so that its line of direction wch at *B* was *Bp*, at *f* shall become *fq*.

But I do not understand how it can make ye Comet ever move directly towards ye (as at *g* where ye line of direction *gS* passes through ye center of ye Sun *S*, much less can it make ye line of direction verge to ye other side ye sun as at *h* where ye line of direction is *ht*. For if ye Comet at *g* moved directly towards ye (& the (also attracted it directly towards himself it would continue to go towards ye line

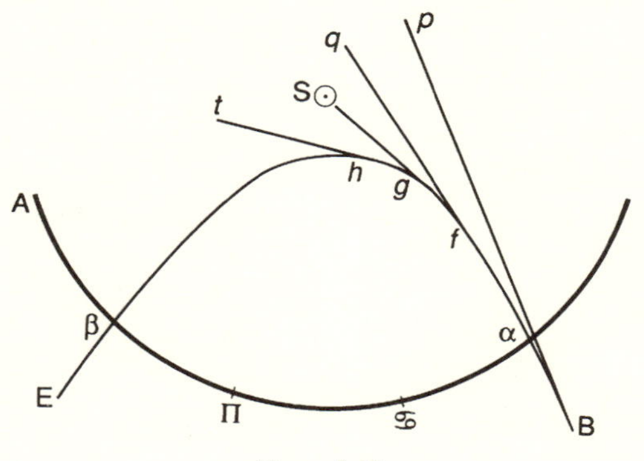

Figure 2.47

> *gS* till it fell upon y^e (, there being no cause to turn it out of y^e line of its direction *gS* towards *h*.
>
> The case is as if a bullet were shot from west to east. The attraction of ye earth by its gravity will make y^e bullet tend more & more downwards, but it can never make it tend directly downwards much less verge from east to west.
>
> Nor will y^e motion of y^e Vortex releive y^e difficulty but rather increase it. For that being according to y^e order of y^e letters & marks $A\beta\pi\alpha$ would make the Comet verge from y^e line *gS* rather towards y^e line *fq* then towards *h*.
>
> The only way to releive this difficulty in my judgmt is to suppose y^e Comet to have gone not between y^e (& Earth but to have fetched a compass about y^e (as in this figure. (Newton to Crompton for Flamsteed, 28 February 1680/81, *Corresp.*, 2:341)

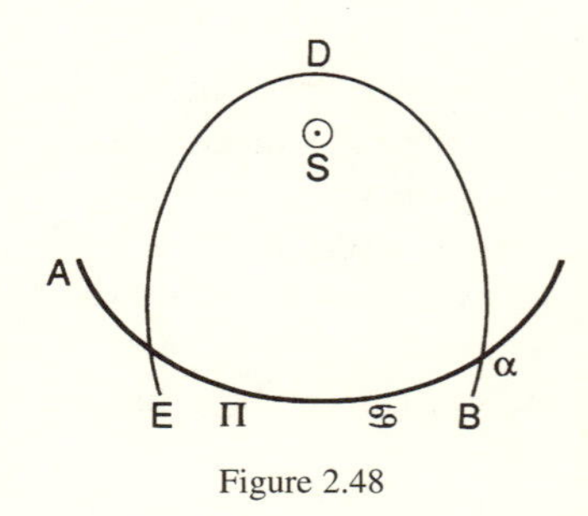

Figure 2.48

In the remainder of the letter, Newton discusses the idea of a magnetic influence of the sun on the planets and comets. Will not a magnetic body lose its virtue when it becomes very hot, as is the sun? And when near a large magnet is placed another small one floating freely in a walnut shell on water—a case very like the situation of the comet and the sun—will not the large magnet make it turn nimbly about into an agreeable position and then attract it? It is impossible that afterward its unfriendly side would again be found turned toward the large magnet (*Corresp.*, 2:341–42).

It is, above all, the discussion of the trajectory that is of interest in this correspondence. The comet has penetrated into a region so close to the sun that it cannot be followed observationally; only a physico-geometrical study can make possible the determination of its motion as it approaches the center that attracts it.

This time Newton plays, with respect to Flamsteed, the role that Hooke had played with respect to Newton himself some months before. Flamsteed speaks to him of a body moving toward direct collision with the center of force, and Newton retorts that it is impossible in the general case. Just as the cannonball launched obliquely can have a trajectory increasingly incurved without ever being able to turn so as to head directly for the center that attracts it, so the comet is deviated by the sun but never directed towards it.

CHAPTER III

THE MATHEMATICAL METHODS

INDIVISIBLES OR ULTIMATE RATIOS?

"I Here Consider Motion according to ye Method of Indivisibles"

In his letters to Hooke of 1679, Newton gave evidence, as we have seen, of a new capacity—that of analyzing trajectories under the influence of a force. The context was that of a fiction which generalized terrestrial gravity. The following year, in his discussion with Flamsteed, Newton exploited the same resource for the study of actual celestial trajectories. He compared heavy bodies projected round the earth with comets in their passage near the sun and implied that planetary motion could be treated similarly. But he did not go beyond rough qualitative determinations concerning the respective situations of the orbit and the center of force (could the attracted body come to be directed straight at the center? and must the orbit pass in front of the center of force or "fetch a compass" round it?) and concerning also the phases of approach and recession (is there a perihelion or perigee? in what quadrant of the associated circle must it be situated?).

Hooke had urged Newton: develop the preliminary ideas, progress beyond simple qualitative determinations, construct the detailed orbit:

> It now remaines to know the proprietys of a curve Line (not circular nor concentricall) made by a centrall attractive power which makes the velocitys of Descent from the the tangent Line or equall straight motion at all Distances in a Duplicate proportion to the Distances Reciprocally taken. I doubt not but that by your excellent method you will easily find out what that Curve must be, and its proprietys, and suggest a physicall Reason of this proportion. If you have had any time to consider of this matter, a word or two of your Thoughts of it will be very gratefull to the Society (where it has been debated) And more particularly to, Sr, your very humble Servant
>
> Robert Hooke.
>
> (Hooke to Newton, 17 January 1679/80, *Corresp.*, 2:313)

It is, in sum, the question that Halley will pose: how to pass from the law of force to the trajectory that the attracted body must follow? If the action of the force at different distances is known (Hooke expresses this action in terms of "velocities of descent"), how to trace the curved line that the body subject to the force describes? Hooke speaks in this connection of Newton's "excellent method" as undoubtedly permitting the determination of the trajectory. He seems to believe that Newton possesses certain tools appropriate to the study of these dynamical situations. What can these tools be?

The most plausible response is to search in the letters exchanged by the two men during the preceding weeks. There Newton had deployed a certain know-how, particularly in fixing the general shape of the trajectory, by considering the impresses of force successively received. He compared sums of infinitely small motions generated by gravity. In the very middle of this reasoning, Newton appears to give a very precise characterization of his method when he enunciates parenthetically what could appear to be a profession of faith:

> The innumerable & infinitly little motions (for I here consider motion according to y[e] method of indivisibles) continually generated by gravity . . . (*Corresp.*, 2:308; see above, pp. 151–52)

Coming from Newton's pen, this declaration is somewhat surprising. He does not speak of fluxions nor of infinite series, the new tools which he had himself created and which must render obsolete the "method of indivisibles." Still, it would be necessary to know what this formula signified around 1680. As will be seen later, the term "method of indivisibles" did not have a univocal meaning; at this time, it could designate nearly anything. According to the most orthodox interpretation, that of the adepts of Cavalieri's *Geometria indivisibilibus continuorum nova quadam ratione promota* (1653), the method consists in employing, as intermediary in the reasoning, geometrical elements that have one dimension less than the object studied (see below, p. 176ff.): plane figures are compared by means of lines that they contain, and volumes by means of plane surfaces that they contain.

Is it reasoning of this sort that Newton is invoking? The context should be able to clarify the importance of this reference to indivisibles. Newton is in the process of evaluating "motions generated" on different segments of the trajectory, and his reasoning has only a vague and distant similarity to Cavalieri's procedures: he evaluates the little motions generated by the force at each instant and by means of them compares two total deviations along two portions of the path. The comparison is a summary one and ends only in a simple judgment of inequality. The "method of indivisibles," if this is rightly called such, is here very poor and rudimentary.

But the reference to indivisibles appears in another light if one pays attention to the words that precede the term. Newton is considering infinitely small motions, and he proposes to compare sums of similar elements (which sums must be finite deviations). It is a crude way of speaking and is contrary to the classical tradition. In invoking the method of indivisibles, Newton seeks, rather, a guarantee. Others before him had believed it legitimate to consider finite magnitudes as composed of an infinity of infinitely small elements. He is therefore well founded in composing a motion of "the innumerable & infinitly little motions . . . continually generated by gravity" in the passage of the moving body from one point to another.

This boldness does not conform to the very strict requirements of Cavalieri, but it is here very useful to Newton. The invocation of indivisibles enables him to sidestep a difficult question: what signification to give to the infinitely small elementary motions that make up a finite motion. Is there between these elementary motions and the finite motion the same sort of dimensional difference as between lines and the surface that contains them? The notion of indivisibles presents a very useful ambiguity: one can leave undecided the dimension of the element and not seek to know whether the infinitesimal motions are in relation to the total, finite motion as points in relation to the line or as finite segments (this ambiguity, explicitly accepted, has already been encountered in Barrow's text; see above, pp. 110–11).

This is probably the direction in which Hooke was urging Newton to go when speaking to him of his "excellent method" (see above, p. 159). It is a matter of resolving the "inverse problem," or of responding to "Halley's question," that is, of constructing the curve traversed from the knowledge of the attractive force. It is very natural to imagine first an initial direction of projection, then to seek to evaluate "ye innumerable converging motions successively generated" by the central force (see p. 151). Kepler himself had reasoned in this way in the Epitome (see above, p. 82): he knew the value of the "libration" at each point, and he deduced from it the total libration accumulated along a given arc. It is true that the libration according to Kepler was very different from the centripetal impresses since it represented a departure in relation to the circular orbit and not in relation to the rectilinear inertial motion. Could the trajectory be constructed by following this reasoning? It would be necessary to suppose an initial direction of the moving body, then to modify it little by little in incorporating the successive inflections caused by the force.

The Generalization of the Law of Fall and Ultimate Ratios

It is not in this way, however, that the *De motu* proceeds. In it there is apparently no summation of the impresses of force and no construction of the trajectory starting from an initial direction of propulsion. The manner of reasoning is very different: the curved trajectory is given in advance, and the force is evaluated at a point of this trajectory by measuring the "deflection." Instead of considering the accumulation of impresses of force and deducing the deviation from it step-by-step, Newton studies locally a trajectory given in advance. The known deviation in the neighborhood of a point serves to measure the force at this point; then the variation of this measure along the known orbit (circle, conic, etc.) makes it eventually possible to assign the law of the attractive force.

If there is such a disparity between the reasoning of 1679 and that of 1684, that is, between the argumentation sketched in the letters to Hooke—which the latter urged carrying further—and the mode of treatment of force in the *De motu*, this is because of fundamental mathematical difficulties: How to sum the variable impresses of force? How to obtain a trajectory from these successive impresses?

In 1684, Newton chooses another route, which enables him to reply to different and simpler questions. The scope of these new procedures is more limited than what Hooke or Kepler had hoped for. The punctiform or local study of the force by means of deviations does not lead to a resolution of the "inverse problem." Other implements would have been required, primarily a global evaluation of the force received.

These implements are absent from the *De motu*, but in the *Principia*, Newton will advance beyond the study of the force at a point. In the middle of Book I, there is a theorem of summation that permits a beginning of the solution: Proposition 39 makes possible a global evaluation of the force received along a given path and opens the way to a determination—still imperfect—of the trajectory as a function of the law of force (see below).

Nevertheless, there is in the *De motu* a summation of impressions of force, but it is hidden in one of the fundamental postulates laid down at the start: Newton supposes that the deflection is proportional to the square of the time (Hypothesis 4; see above, p. 18). This assertion is rendered plausible by the analogy between centripetal force and gravity. What Galileo asserted with respect to falling bodies is also valid for every motion carried out under the action of a centripetal force.

It is a matter here, too, of summing impresses of force, but, in the simplest case, a succession of identical impresses. Galileo had shown that accumulation of the actions of gravity, repeated identically at each instant, gives a traversed distance proportional to the square of the time (see above, p. 107). Newton extends this enunciation to a much wider class of phenomena: the action of every centripetal force yields the same effect as gravity, at least if only the beginning of the motion is considered. At this stage of his work, therefore, Newton does not seek to sum the variable impresses of force; he is satisfied to assert—at first without demonstration—that the summation of constant impresses, as carried out by Galileo, remains true at each point of a trajectory traversed under the action of a centripetal force, provided that the domain studied is sufficiently restricted.

Then, in the manuscripts that follow the first version of 1684, he gives to this enunciation a mathematical demonstration (Lemma 2 in Manuscripts C and D of Hall). The requirement of a proof does not bear on the summation of actions as such—Galileo's result has already been established—but only on its extension to variable forces: How to demonstrate the law of fall in its "nascent" form? How to study mathematically a "commencement of motion"?

This text belongs to the period separating the first redaction of *De motu* from that of the *Principia*. Between 1684 and 1686, Newton progressively deepens and enriches the first version of the *De motu*, giving it more rigorous dynamical and mathematical foundations. The generalization of the law of fall (Hypothesis 4) is developed considerably, in the form of Lemma 2 (Hall, 268), which is of great interest because it contains the first use of "ultimate ratios," the subject of section 1 of the *Principia*. The demonstration of this law is the kernel of the entire section on ultimate ratios.

In its initial version, the *De motu* does not lack mathematically debatable steps that are difficult to justify by the canons of traditional geometry (see pp. 55–56): Newton deforms figures while claiming that certain relations are preserved; he substitutes one infinitely small magnitude for another; he announces certain properties of beginning motions and nascent magnitudes; he compares parts of magnitudes generated at the same moment.

Initially Newton does not seek to justify the totality of these procedures; his effort toward rigor is concentrated on the notion of the beginning of motion and the nascent magnitude, as required to extend Galileo's law.

The demonstration returns to the diagram and the essential elements of Galileo's theorem (or rather of the two theorems, since Newton treats at once both Theorems 1 and 2 of the Third Day of the *Discorsi*). Time is represented by the vertical axis ABD; the increasing velocities (or from another point of view, the momenta that are accumulating) are represented by the horizontal segments associated with the points of ABD, that is, with the instants of time. Galileo deduced that the spaces traversed are proportional to the squares of the time. He did not draw his argument from the proportionality between the areas but employed a sort of equivalence: the accelerated motion can be replaced by a mean uniform motion (Theorem 1), hence the distances traversed in accelerated motion can be reduced (in Theorem 2) to the formula:

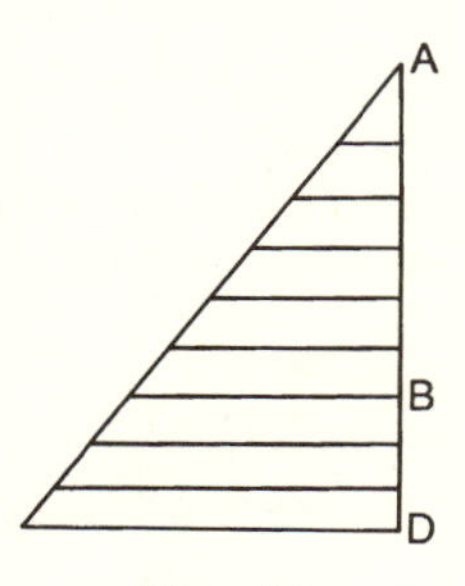

Figure 3.1

$$S : S' = (V : V') \cdot (T : T'),$$

or since V is proportional to T:

$$S : S' = (T : T') \cdot (T : T').$$

Newton is more direct: he considers that the result follows from the proportionality between the areas. If AB and AD are the times, the spaces traversed are as the areas ABC and ADE, that is, as the square of AB to the square of AD.

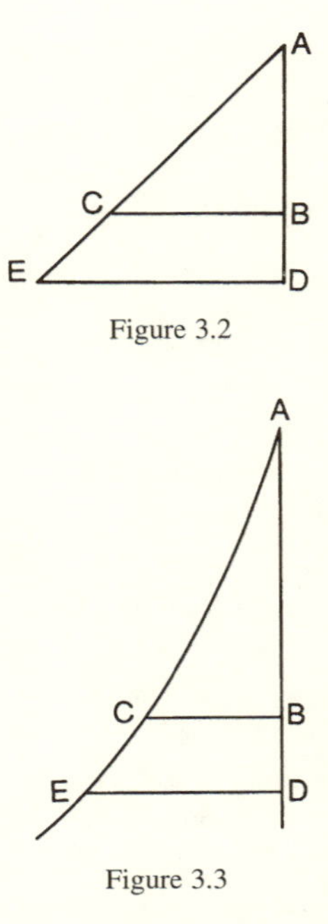

Figure 3.2

Figure 3.3

But centripetal force, unlike gravity, does not produce constant impresses. From the moment the body is displaced, the intensity of the force changes. (In all these texts, Newton supposes implicitly that the force depends only on the distance.)

If the force is variable, Galileo's proportion is no longer exactly verified. The line that joins the ends of the horizontal segments is no longer rectilinear but a curve (if the increase in the velocity is continuous), as for instance ACE in figure 3.3. Nonetheless, if the two triangles are very small, if they are nascent or evanescent, the straight line can be identified with its tangent, and the sides of the two triangles will be rectilinear and will enclose the same angles. Consequently the triangles ABC and ADE in their nascent state are to one another as the square of AB to the square of AD, and hence the distances traversed are proportional to the squares of the elapsed times. When the times AB and AD tend toward A, each in its own way, and become smaller, the ratio of areas approaches the ratio of the squares of the segments.

The difficulty lies in the fact that the triangles are in agreement with this relation only at the very instant of their common disappearance, when they are engulfed in the point A. Newton employs an artifice to preserve a finite situation and to enunciate relations with respect to finite, well-determined magnitudes: while the segments AB and AD, which represent the times, diminish and vanish, they remain proportional to two fixed finite segments, Ab and Ad. The vanishing configuration is thus represented in the finite by another configuration from which can be read the relations between the vanishing segments:

> *Lemma 2*. The space described by a body urged by any centripetal force at the beginning of its motion, is as the square of the time.
>
> Let the times be represented by the lines AB, AD proportional to the given lines Ab, Ad, and the spaces described under a uniform centripetal force are represented by the areas ABF, ADH bounded by the perpendiculars BF, DH and any straight line AFH, as Galileo explained.
>
> But let the spaces described under a nonuniform centripetal force be represented by the areas ABC, ADE bounded by any curve ACE that touches the straight line AFH in A. Draw a straight line AE meeting the parallels BF, bf, dh in G, g, e, and let AFH extended meet bf, dh in f and h.
>
> Since the area ABC is greater than the area ABF, and less than the area ABG, and the curvilinear area ADEC is greater than the area ADH and smaller than the area ADEG,

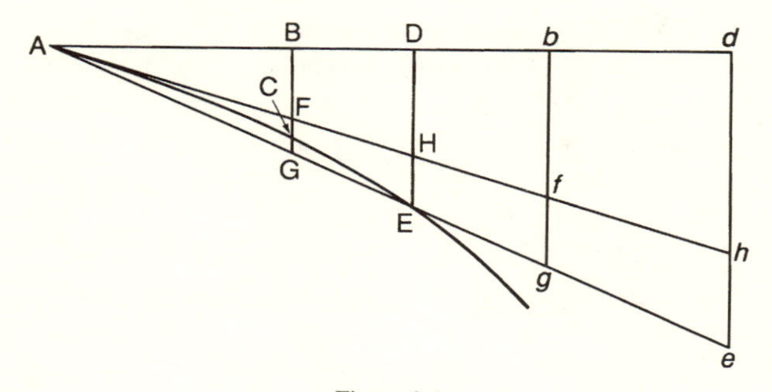

Figure 3.4

[the ratio of] the area ABC to the area ADEC[1] will be greater than [the ratio of] the area ABF to the area ADEG, and smaller than [the ratio of] the area ABG to the area ADH.

that is, greater than [the ratio of] the area Abf to the area Ade, and smaller than [the ratio of] the area Abg to the area Adh.

Now let the lines AB, AD decrease in their given ratio until the points A, B, D meet and the line Ae coincides with the tangent Ah, and so the ultimate ratios [*ultimae rationes*] of Abf to Ade and of Abg to Adh become the same as the ratio of Abf to Adh.

But this ratio is the duplicate of that of Ab to Ad or AB to AD; therefore the ratio of ABC to ADEC, intermediate between these two ultimate ratios, will become the duplicate of the ratio of AB to AD, that is, the ultimate ratio of the evanescent spaces, or the first ratio of the nascent spaces, is the duplicate of the ratio of the times. (Hall, 244–45; Herivel, 295–96; *NMP*, 6:76–77)

The segments BC and DE are the equivalent of the horizontal lines of Galileo's diagram. They represent velocities that increase, uniformly in the case of gravity, but here at a variable rate. The line joining the extremities of the velocities is no longer a straight line but a certain curve ACE. Newton proposes to show that, nonetheless, Galileo's proposition remains true provided that the triangles ABC and ADE are taken in their nascent (or evanescent) state. AB and AD represent the time; they are variables and diminish until B and D meet in the point A. Yet in diminishing they retain to each other a constant ratio, namely the ratio of Ab to Ad, two fixed finite segments on the straight line ABD. These two segments Ab and Ad, which do not vary, are a kind of finite and unmoving image of the vanishing segments. The straight line AFHfh is also fixed, since it is tangent to the curve of the velocities in A. The rest of the figure shifts while following the motion of the points D and B. As D and B approach A, the rectilinear secant AE approaches the tangent Ah, finally coinciding with it, and the points g and e end by coinciding with f and h.

The reasoning deals with three sorts of triangles, all standing on the horizontal straight line ABD. Their lower side can be on the tangent (the rectilinear triangle ABF), on the curve (the curvilinear triangle ABC), or on the secant (the rectilinear triangle ABG). The areas of these triangles are such as to satisfy the following inequalities:

$$\text{ABF} < \text{ABC} < \text{ABG},$$
$$\text{ADH} < \text{ADEC} < \text{ADEG}.$$

From these inequalities Newton draws an inequality between ratios of areas:

$$\text{ABF} : \text{ADEG} < \text{ABC} : \text{ADEC} < \text{ABG} : \text{ADH}.$$

The ratio of the two vanishing curvilinear triangles is thus caught between two ratios of rectilinear triangles, and these triangles are also vanishing.

The next step consists in replacing the two outside ratios by ratios between finite triangles (in lower-case letters). These are indeed the same ratios, since AB and AD in the course of their variation maintain to each other the constant ratio of Ab to Ad, therefore:

$$\text{Abf} : \text{Ade} < \text{ABC} : \text{ADEC} < \text{Abg} : \text{Adh}.$$

Finally, when D and B together approach A, the rectilinear secant AGE approaches the tangent AFH. The outside ratios become very close to the same ratio between two finite rectilinear triangles: the ratio of Abf to Adh. The ratio sought is found between two finite ratios that tend toward the same value when D and B approach A; it can therefore be asserted that "ultimately" the ratio of the areas ABC and ADEC must be equal to the ratio of the rectilinear areas contained by the tangent.

Galileo's proposition is thus true "*ultimo*," at the last instant of the variation. The "beginning of its motion" is here read rather as an ending, since it is an evanescence or vanishing. Time is made to go backward rather than forward because it is more convenient, and indeed indispensable, to start from the finite situation in order then to shrink it to the infinitesimal. The complexity, the richness of configuration, of the law of free fall is found associated with each point of space and enclosed in the first instants. Everything becomes engulfed in the initial point, and yet relations can be discerned there: the areas of the triangles have a determinate ratio at the moment of their disappearance.

It is not sufficient, in fact, to be able to manipulate infinitesimals; it is also necessary to succeed in establishing relations between them and to discern multiplicity and ratios in the midst of a vanishing configuration.

In this lemma, the areas do not have a determinate ratio except at the very instant in which they disappear. In the *De motu*, there are ultimate relations of a different nature in which the ratios or properties attained "ultimately" are preserved from what was valid in the finite case. Thus the law of areas (Proposition 1) holds for any central orbit of finite width, and the areal measure of time on an elliptical orbit extends to the degenerate ellipse that is the straight line (Problem 5, see above p. 53ff.).

But in the present situation, it is not the same: the initial or final instant is decisive and alters the relations radically. Cases of this kind are particularly resistant to a classical treatment. It no longer suffices to base the reasoning on a principle of continuity that would guarantee the passage from the finite to the infinitesimal. So it is that Newton has recourse to this astonishing procedure of similitude in the finite: the relations between vanishing magnitudes are faithfully imitated and enlarged by a finite configuration (here the triangles Abg, Ade) in which can be read the ratios of vanishing entities.[2]

The rudiments of the entire method of ultimate ratios are present in Lemma 2 of the *De motu*. Lemma 2 serves as point of departure and initial nucleus for the whole of the *Principia*'s section 1, where essentially the same proposition is taken up in a strictly geometrical form, in Lemma 9, which treats of triangles; then the result is applied, in Lemma 10, to the distance traversed under the action of a variable force. In this context, the previous demonstration receives new foundations:

> Application of the generalized similitude to curvilinear figures (Lemma 5, deriving implicitly from Lemmas 2–4)
>
> The properties of "continuous curvature" (Lemma 6)
>
> Finally, a general principle that makes it possible to assert equality between two magnitudes or ratios that tend toward each other (Lemma 1)

Thus Newton, in giving a broader basis to his extension of the theorem of fall, in moving back step-by-step towards always earlier foundations, ends by formulating an original method of mathematics, which is not strictly speaking infinitesimal but is destined to reach the same enunciations and results as the infinitesimal procedures.

The method of ultimate ratios (or of "first" ratios, or of "limits of sums and ratios" of "nascent and evanescent quantities" [*Princ.*, 37]) is in fact proposed at the beginning of the *Principia* as an alternative to indivisibles (see the final scholium of section 1). The method of indivisibles is "too harsh" (*durior*, *Princ.*, 37), it is better to use ultimate ratios, by means of which one proceeds "more safely" (*tutius*, *Princ.*, 37). Certain abuses of language will henceforth be permitted and should not abuse the reader:

> If hereafter I should ever consider quantities as made up of constant particles, or should use little curved lines in place of straight ones, I would not be understood to mean indivisibles, but evanescent divisible quantities. (*Princ.*, 37)

Yet the method that Newton is proposing is not weaker or less fruitful:

> For by these [first or last ratios, or limits] one accomplishes the same things as by the method of indivisibles. [*His enim idem praestatur quod per methodum indivisibilium.*] (*Princ.*, 37)

THE METHODS OF INDIVISIBLES

What is "the method of indivisibles"? Have indivisibles the same dimensions as the magnitude that contains them? Is it necessary to assume a postulate of continuity in the passage from the finite to the infinitesimal?

The authors of the seventeenth century associated the method of indivisibles primarily with Cavalieri, even though it is evident that most did not open his books (for example, in 1671, in his *Theoria motus abstracti*, the young Leibniz claimed to give "the foundation of the method of Cavalieri" by means of points of inassignable extent (*LMS*, 6:68); but he conceived the indivisibles as infinitesimals of which the continuum would be composed, while Cavalieri had explicitly and repeatedly claimed the contrary). There were, at the time, proponents of many other theories of indivisibles. A balanced presentation should include especially Kepler, whom Cavalieri read and criticized and whose infinitesimal methods were briefly mentioned in connection with the demonstration of the law of areas (see above, p. 75ff.). It would also be necessary to indicate the role of Roberval, whose *Traité des indivisibles* was published in 1693 but which probably harks back to fragments of a teaching presented fifty years earlier. The foundations of his doctrine are much vaguer than those of the Italians who will be discussed, and he often presented the indivisibles as simple fictions.

This discussion will be restricted to the theories of two principal Italian authors—Cavalieri and his popularizer and unfaithful disciple, Torricelli—while placing them in relation to those of their mutual teacher, Galileo.

Within this limited area are already two divergent conceptions of indivisibles, two methods based on radically different foundations. Cavalieri refuses to compose the continuum out of the infinitely small; he attempts to found his theory on relations of proportionality between a "mass" of lines or surfaces. Torricelli, on the contrary, describes veritable infinitesimals that are homogeneous with the figures they make up, that is, of the same dimensions as the magnitudes considered. Previously (p. 110 above) it was shown that in a similar vein Barrow refused to distinguish between points and "indefinitely small lines [*lineolis*]" or between instants and "very small times [*tempusculis*]." Both points of view are fruitful and can generate coherent mathematical theories.

The importance of the postulate of continuity is not the same in the two doctrines. Cavalieri does not need to claim that the relations are preserved in the infinitely small, while this principle is essential to Torricelli's construction, which relies on Galileo's conceptions to justify the idea that the ratios between magnitudes still hold for the "ultimate vestiges" of these magnitudes.

These theories are of interest for their content and for the light they throw on opposing notions of the indivisible. But they are important also because of the adherents they would acquire in England through the works of Gregory and Barrow.

The Physico-Geometrical Continuum of Galileo

Any analysis of these theories must begin with Galileo, the Master or the Old One (*il Vecchio*), who dispensed advice and patronage—his arguments and conjectures furnished nourishment, stimulus, or warning with regard to diverse Italian theories of indivisibles. Nevertheless Galileo did not work as a mathematician in the full sense of the word. He composed beautiful works of pure mathematics, but he was more interested in natural philosophy, and the rigor of the Greek mathematicians was not his principal concern.

He provided no exposition of methods of mathematical demonstration with infinitesimals, but in several texts written towards the end of his career, he spoke of the composition of the continuum. His last work, the *Discorsi* of 1638, contains very long discussions of indivisibles. The First Day proposes a hypothesis on the cohesion of matter as preparation for the Second Day's discussion on the resistance of materials. This cohesion, he suggested, was due to infinitely many and infinitely small vacua, which would link the atoms together. Galileo supported and illustrated this physical conjecture with several mathematical analogies, and he asserted in passing his conviction that every continuous magnitude is constituted of an infinity of elements infinitely small.

It is thus in a text of physics, and uniquely in order to corroborate and make plausible a conjecture about matter, that Galileo expounds his conception of indivisibles. The initial question is this: what is the cause that makes the parts of solid bodies hold together? Galileo distinguishes several possibilities. In the first place, certain bodies owe their cohesion to the interlacing of the fibers that compose them. For instance, a cord resists traction because of the mutual disposition of the threads of hemp; it can even be supposed that a piece of wood owes its cohesion to the interlacing of its fibers. In the second place, it may be that two parts of the same material, for example two slabs of very flat and well-polished marble, are held together by the force of the vacuum alone, as seen when one attempts to separate them. For Galileo, the abhorrence of nature for a vacuum is therefore a cause of cohesion, and he indicates how its intensity could be measured.

But these two reasons do not suffice to explain the cohesion of matter. How does each fiber of hemp keep itself whole? To what does each of the slabs of marble owe its solidity? Galileo proposes a conjecture (*fantasia*): in matter there could be very small and very numerous vacua. It would be the repugnance for a vacuum that would hold together the pieces of matter. The explanation is therefore the same as for the slabs of marble, but this time on the scale of "atoms" (the term is Galileo's).

A first justification for this hypothesis is found by observing how fire acts. When fire liquifies metal, that is, when it liberates the molecules of the metal from their cohesion, it is because the "very subtle particles" of fire penetrate into all the empty interstices between the particles of metal. As soon as the particles of fire have taken the place of the interstitial vacua, the molecules of metal move freely with respect to one another. It can therefore be asserted that it is the pores of the metal that maintain its cohesion: the tiny vacua "attract" the particles toward each other and keep them linked (EN, 8:67).

But how many of such vacua are required? If an infinite number is necessary, how to conceive that this infinity is contained in a finite portion of matter? At this point a thesis of more general import makes its appearance: in a finite extent there can be an infinity of vacua.

To show this, Galileo has recourse to a geometrical paradox drawn from the *Mechanical Questions* attributed to Aristotle. It is the problem called "Aristotle's wheel," of which the essential feature is the following: if a large circle rolls without slipping along a horizontal base, point B, which rests on the base at the first instant, will return to it after a complete turn. If a smaller circle is

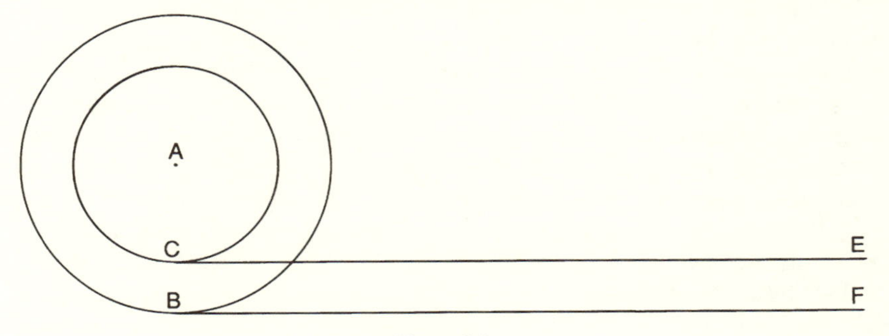

Figure 3.5

traced within this wheel with the same center A and turning with the same angular velocity, point C of this circle, which was the lowest point at the first instant of the rotation, will also return to its initial position after a complete turn. Yet the length of the two circles is different: the large circle, which has always remained supported on the base BF, has traversed a length exactly equal to its circumference. What then is to be said of the motion of the small circle? How is it possible that it should have also traversed a length CE equal to BF?

To resolve this paradox, Galileo takes a detour and first considers hexagons instead of circles. The large hexagon does not actually roll; by a series of topplings it comes successively to rest each of its sides on the base. During this

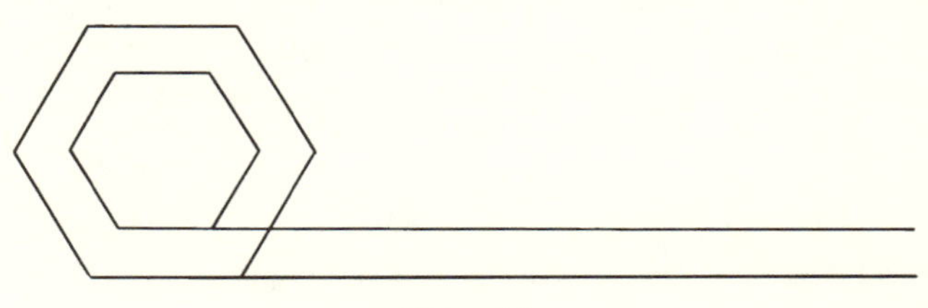

Figure 3.6

time the small interior hexagon comes also successively to rest each of its sides on a horizontal straight line parallel to the base, but it leaves unoccupied segments at regular intervals.

What happens in the case of the hexagons indicates the way to understand what happens in the case of circles. Let the number of sides be multiplied, and consider, not hexagons, but polygons of twenty or a thousand sides. Then, in the latter case, the small interior polygon will rest a thousand times on the base, and each time it will leave a small horizontal segment untouched. The straight line will be made up of a thousand small segments separated by a thousand small vacua.

Yet the circle can be considered as a polygon of an infinite number of sides (EN, 8:71). One can therefore consider that the small circle traverses the line CE by making an infinite number of infinitely small jumps. The horizontal line CE will be made up of an infinite number of points, partly full and partly empty (*infiniti punti parte pieni e parte vacui*, EN, 8:71).

This rather surprising solution to the paradox of Aristotle's wheel furnished Galileo with an analogy for explaining a number of physical phenomena: first, the presence of the innumerable small vacua that cause cohesion, and second, the cases of the unlimited expansion or condensation of matter.

Expansion is possible in virtue of the interposition of the vacua. If the line is decomposed into an infinity of its elements, it can be put together again in such a way as to obtain a greater length; it suffices to introduce enough infinitesimal vacua (EN, 8:72). And in a symmetrical fashion it is conceivable that the line, once decomposed into its points, could be contracted into a smaller line without any interpenetration of one part into another (EN, 8:96). It suffices to reason in reverse on the same polygons as before: if one takes for norm the rotation of the small polygon, whose perimeter is then exactly equal to the horizontal line, the large polygon will traverse a line of the same length, although its perimeter is greater. This is brought about by a contraction (*condensazione*, EN, 8:93) of the length of the larger circle.

Parts "without Magnitude"

These transformations would be inconceivable if the vacua referred to had a certain size. It is necessary to assume that the elements are parts "without magnitude." The Italian term is *parti non quante*, and an exact translation is impossible because Galileo uses it in two different senses—different at least to modern eyes. In some places, *non quanti* signifies "escaping all numeration," "infinite," for instance, in *i lati non son quanti, ma bene infiniti* (EN, 8:71). On the other hand, in most cases *non quanti* means "without size," "without assignable magnitude," "indivisible," for example, in *l'esser le parti infinite si tire in consequenza l'esser non quante . . . mi diciate se le parti quante nel continuo nel vostro credere son finite o infinite* (EN, 8:80). For this last question of Salviati to have meaning, it must be possible to conceive without contradiction of *parti quante* (parts of measurable magnitude), which are also infinite in number; but it then follows that the magnitude that contains this infinity of *parti quante* must itself be of infinite size. To contain an infinity of parts or elements within a magnitude of finite size, it is necessary that these elements be *parti non quante*, without assignable magnitude.

This last argument is essential to the more general reasoning that Galileo proposes in the same text apropos of the composition of the continuum. If one admits the resolution of the line into all its points, and its recomposition—with more or fewer vacua—in the way Salviati explains, it is necessary also to admit a general thesis regarding the continuum: the line is composed of points, the divisible is composed of indivisibles (EN, 8:72). The old question of the composition of the continuum is thus explicitly posed and receives an answer which is not that of the Aristotelian tradition.

Galileo demonstrates that if the continuum is made of parts that are always again divisible, it must be made of parts without magnitude and infinite in number (EN, 8:80). In brief, if the continuum is always divisible, it must be made of indivisibles! It is Aristotle who provides the definition of the continuum that serves as the point of departure: the continuum is that of which the parts are always again divisible (*Physics*, 6.1). And here is the reasoning by which Galileo draws from this definition consequences opposed to those of Aristotle:

> Since the line, and every continuum, is divisible into parts always divisible, I do not see how one can avoid [concluding] that these continua are composed of an infinity of indivisibles; for a division and a subdivision that can be carried on perpetually suppose that the parts are infinite, because otherwise the subdivision would be terminable; and from the parts being infinite it follows that they are *non quante*, because infinitely many parts endowed with magnitude make an infinite extension; and thus we have a continuum composed of infinite indivisibles. (EN, 8:80)

A manuscript text by Galileo contains the same argument in a clearer form:

> I say it is most true and necessary that the line be composed of points, and the continuum of indivisibles. . . . Open your eyes, for goodness' sake, to this light that has remained hidden perhaps until now, and recognize clearly that the continuum is divisible into parts always divisible only because it is constituted of indivisibles. For if the division and subdivision must be able to go on forever, it must necessarily be that the multitude of the parts is such that one can never go beyond it, and therefore the parts are infinite [in number], otherwise the subdivision would come to an end, and if they are infinite, they must be without magnitude [*non sieno quante*], because an infinity of parts endowed with magnitude compose an infinite magnitude [*infiniti quanti compongono un quanto infinito*],

and we are speaking of terminated magnitudes [*quanti terminati*]; therefore the highest and ultimate components of the continuum are indivisibles infinite in number. (EN, 7:745)

The requirement that the parts be "without magnitude" appears here in its proper light: it is necessary that there always be available parts, indefinitely, and yet this infinity of parts must remain within limits of a finite magnitude. The component parts must therefore be of another sort than the parts reached at any one stage of the subdivision. While the successive divisions always arrive at parts of a certain size, parts endowed with magnitude (*quante*), the ultimate elements must be without size, without magnitude (*non quanti*). The argumentation can be reduced to two essential theses:

1. There is necessary an infinity of elements in order that the division can always continue.
2. This infinity must be able to be contained within boundaries of finite magnitude.[3]

Once the necessity of such elements without magnitude has been proved, it remains to maintain the doctrine against several traditional objections. If entities without magnitude are added, how can a magnitude be generated? If a line of a certain length is composed of a determinate number of points, it must be possible to divide and redivide this line, until one of the indivisible points must be divided. Galileo avoids these difficulties by claiming that an infinity of indivisibles is necessary to compose a continuous magnitude (EN, 8:78).

But if there is an infinity in every magnitude, then one infinite is greater than another; the infinity of points of a long line will be "larger" than the infinity of points of a short line. The solution is radical and consists in declaring that the attributes "larger" and "smaller" are meaningless when it is a matter of infinite quantities (EN, 8:78). Galileo illustrates this idea by the paradox that the whole numbers are just as numerous as the squares of these same numbers.

The Paradox of the "Bowl" and the Postulate of Continuity

Galileo insisted more than once on the powerlessness of human intellect to comprehend the infinite and the infinitely small.[4] The strange and incomprehensible character of infinity is illustrated by various examples of "metamorphoses," relations between the elements of a figure or of a physical object. Galileo therefore seemed very far from admitting what has been called here the postulate of continuity, that is to say, the supposition that relations are preserved in the passage from the finite to the infinitely small. Yet the First Day of the *Discorsi* contains an important passage that argues in the other direction and in which Torricelli found a basis on which to support his own theory of indivisibles.

The reasoning aims to prove that a line can be equal to a point. If the solution that Galileo proposed for the wheel of Aristotle is accepted, it is necessary to admit that a line can be stretched out or contracted at will, and this admission makes it possible to explain the differences in the distances traversed by the concentric circles. It is even necessary to arrive at an account of the extreme case, that of the central point on the wheel: how can this point traverse the whole horizontal line, and be in a certain way equal to it?

Galileo attempted to render the paradox acceptable by showing in another example that the idea of an equality between a point and a line can impose itself on the mind. To this end he examined a geometrical figure that had served several authors of the time for calculating the volumes of spheres or spheroidal solids. A half-sphere with center L, resting on its summit A, is inscribed within a cylinder having the same diameter GH, and within the same boundaries is constructed a right cone (fig. 3.7).

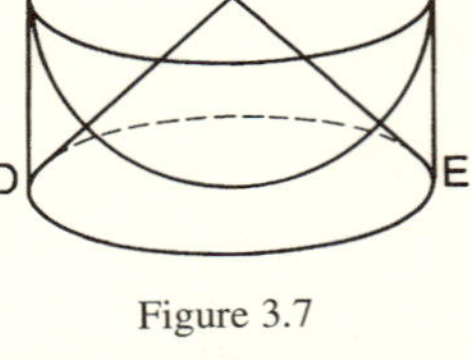

Figure 3.7

The mathematician Luca Valerio had proved (*De centro*, pt. 2, Prop. 12) that the cone is equal to what remains of the cylinder when the half-sphere is taken away—Galileo calls this residual figure a "bowl" (*scodella*). The equality between the bowl and the cone makes it possible to calculate the volume of the sphere, if the volume of the cylinder and that of the cone are known.

The equality of the cone and the bowl can be demonstrated by means of horizontal planes that cut them at the same height: if several horizontal planes pass through the half-sphere, the cone, and the bowl, then the disc cut from the cone by a plane is equal in area to the bracelet (*armilla*) cut from the bowl by the same plane (beginning with the identical disc DE when the plane is at the bottom, passing through the common base of the cone and cylinder).

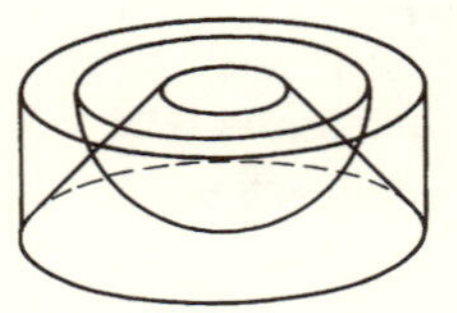
Figure 3.8

Cavalieri and Torricelli knew and made use of this figure while generalizing it (cf. Cavalieri, *Geometria indivisibilibus* [1653], bk. 3, Cor. 1, 34). Torricelli even wrote an entire treatise, which remained unpublished, on a somewhat more extended class of "bowls" (*De solidis vasiformis*, *TO*, vol. 1, pt. 2). He presents the reasoning on the bowl and the crown as a canonical example of the use of indivisibles:

> Let us suppose now that when I have two solids (for example in the well-known hemispherical bowl, on the base of which one erects a cone of the same height), let us suppose, I say, that if we find that having drawn planes always perpendicular to the axis and therefore parallel to one another, the bracelet arising in the cutting

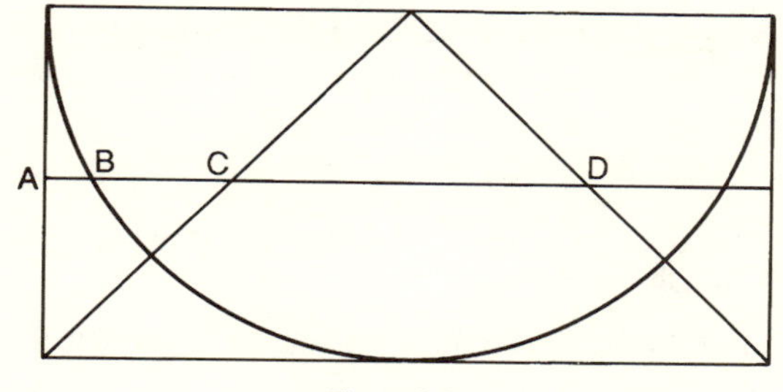

Figure 3.9

of the bowl is ever equal to the circle CD arising from the cutting of the cone; the bowl will then be equal to the cone.

This is only for an example, since this has already been proved, and can be proved in many ways without the help of indivisibles. But this will serve to persuade us that in proceeding thus the argumentation is correct. Each bracelet being equal to each circle corresponding to it, and all the cutting planes being parallel to one another, it follows that all the bracelets taken together, that is to say, the bowl, are equal to all the circles, that is to say the cone. (*De solidis vasiformis*, *TO*, vol. 1, pt. 2, 106)

Similar processes of passing from surfaces to volumes will be discussed later at greater length. What is of interest to Galileo in the *Discorsi* is another aspect of the geometrical situation. He quickly proves that each disc (generated by a radius such as HP) is equal to the corresponding bracelet (generated by a segment such as GI). $IC^2 = IP^2 + PC^2$, but $IC = AC = GP$; therefore: $GP^2 = IP^2 + PC^2$. But $PC = PH$, and hence, $GP^2 = IP^2 + PH^2$, or, $GP^2 - IP^2 = PH^2$.

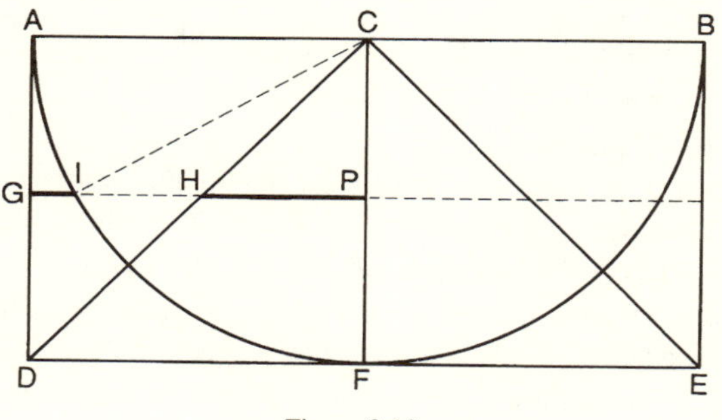

Figure 3.10

The crucial question for Galileo is what happens when one arrives at the apex of the cone. The width of both disc and bracelet diminishes, and the limiting position finally puts in correspondence a point and a circle. The bracelet and the disc have disappeared; there remain only a line and a single point. Nevertheless,

> must one not declare these to be equal, since these are the ultimate remainders and vestiges left by equal magnitudes? [*Li quali, perche non si devon chiamare eguali, se sono le ultime reliquie e vestigie lasciate da grandezze eguali?*] (EN, 8:75)

It is a "marvel" (EN, 8:76). It could be possible that there is an infinite difference between a line and a point, but

> since in the diminution of the two solids the equality between them is maintained to the end [*sino all'ultimo*], it seems very appropriate to say that the last and ultimate results of the diminution remain equal to one another, and that the one is not infinitely larger than the other. (EN, 8:75)

The two magnitudes must still be equal "at the moment of their ultimate diminution" (EN, 8:75).

Torricelli, in the great work in which he proposes his novel conception of indivisibles, will recall this precedent which corroborates his own theory:

> What Galileo says of a point being equal to a line is true. (*TO*, vol. 1, pt. 2, 321)

The relations between the magnitudes, if they remain constant during the variation of the latter, must still be valid for the "ultimate remainders and vestiges" of these magnitudes.

Cavalieri's Rigorous Theory

It is only with Cavalieri that the doctrine of indivisibles becomes a coherent, elaborated theory based on explicit and carefully chosen principles.[5] These are far from the "fantasies" of Galileo. Yet Cavalieri recognized a sort of debt, since after receiving the *Discorsi*, he congratulated and thanked the old teacher:

> One can say of Your Excellency that with the guide of good geometry and the polestar of your high genius you have been able to navigate successfully across the immense ocean of the indivisibles, of vacua, of the infinite, of light, and of a thousand other foreign and difficult things, each of which was enough to shipwreck as great a genius as there is. Oh how much the world will be indebted to you for having smoothed the road to things so new and delicate! . . . And I too must be obligated to you not a little, since the indivisibles of my *Geometria* come to be indivisibly illustrated by the nobility and clarity of your indivisibles. (Cavalieri to Galileo, 21 June 1639, EN, 18:67)

It is necessary to recognize the part of courtesy and respect in these expressions of praise and thanks. Certainly Galileo's discussion in the *Discorsi* could give luster (*illustrari*) to the indivisibles of Cavalieri and prepare minds (*ispianare la strada*) for the doctrine expounded in the *Geometria indivisibilibus*.

But the indivisibles of Cavalieri are very different, and the mathematician of Bologna takes care immediately to make things precise. The lines that follow, after these words of praise, focus very explicitly and clearly on the differences between the two conceptions:

> As for myself, I have not ventured to say that the continuum is composed of indivisibles, but I have shown that between the continuous [magnitudes] there is no other ratio than there is between the mass of indivisibles (on condition of taking them parallel, when we speak of lines and plane surfaces, the particular indivisibles I have considered). (Cavalieri to Galileo, EN, 18:67)

Facing Galileo, Cavalieri points to the originality of his own theory:

> He does not take a stand as to the composition of the continuum.
> Between the continuum and the atoms of magnitude, he is content to utilize an indirect link, namely sameness of ratio: the ratio between the mass of indivisibles could be transmitted to the continuous magnitudes containing these indivisibles.

Cavalieri's strategy therefore consists in evading the philosophical question and in leaving undetermined the link between indivisibles and magnitudes. Repeatedly Cavalieri explains that the theory is valid "whether the continuum is composed of indivisibles, or whether there is in the continuum something other than indivisibles" (Cavalieri, *Geometria indivisibilibus* [1653], bk. 2, Prop. 1, scholium, 111). The indirect link will be sufficient for the needs of geometers. The demonstrations require only this sort of transfer of relations: the continuous magnitudes are related to one another in the same way as the aggregates of indivisibles that can be cut from them.

The transmission of proportionality is the mainspring of his method. A Latin formula that Cavalieri employs summarizes this manner of proceeding very well:

> *continua sequi indivisibilium proportionem* (Ibid., bk. 7, 483),

that is, "continuous magnitudes have the same ratio as their indivisibles." It is therefore necessary to begin by establishing the ratio between the masses of indivisibles, since then it can be transferred to the magnitudes themselves.

What is an indivisible according to Cavalieri? It is known already, from the letter quoted above, that the indivisibles with which Cavalieri is concerned are only lines and planes (and not points, of which there had been a question for Galileo and which will be discussed again in connection with Torricelli). It is known also that the indivisibles will always be taken mutually parallel (*equidistanti*). Cavalieri says that he had first attempted to compare solids of revolution by comparing the surfaces that generate them, hence by comparing indivisibles that were not parallel but radiating from a center; but the absurd results he thus arrived at dissuaded him from this path. He reproaches Kepler for having relied on such indivisibles in the *Astronomia nova* (Cavalieri, *Geometria indivisibilibus*, preface).

The use of indivisibles is very strictly controlled in Cavalieri's theory. First, a terminological remark is necessary: there are no "indivisibles," properly speaking, in the demonstrations. Cavalieri never uses these words in his demonstrative reasonings but only in the commentaries and notes exterior to the demonstrations proper. The term "indivisible" belongs, so to speak, to the metalanguage of the theory. In the demonstrations, instead, is the terminology: "all the lines of the surface" or "all the planes of the solid."

How are "all the lines of a surface" determined? For simplicity, only the case of plane figures will be considered. The process involves a slicing of the given figure in accordance with very precise conditions. The figure must first be enclosed between two parallel and opposed tangents EO and BC. (The notion of tangent here does not imply any idea of "continued curvature" or of differentiability: it is simply a straight line in relation to which the curve is entirely on one side.)

Through these two tangents there pass two parallel planes that cut the plane of the figure. One of the planes is made to slide toward the other, which remains stationary. As the mobile plane moves to successive positions, at each instant it cuts a certain straight line on the plane of the figure, and its trace on the figure is a certain segment LH that varies with position (the mobile plane can even trace on the figure several disjointed segments if the figure is irregular).

The aggregate (*congeries*, *aggregatum*) of all the lines such as LH forms

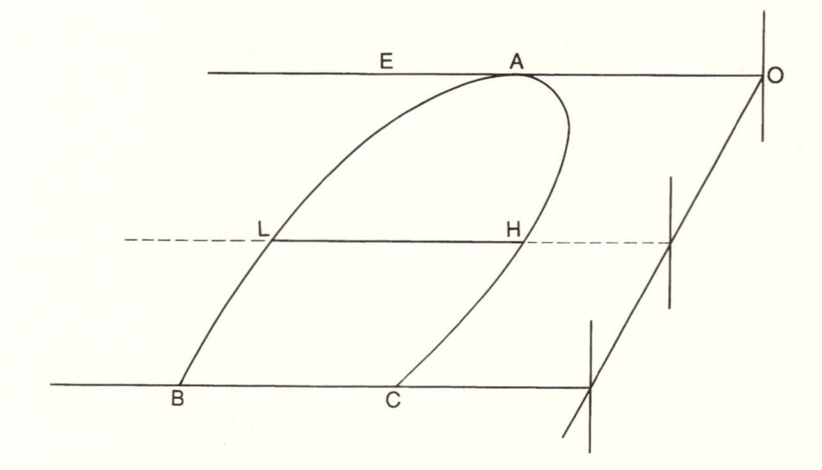

Figure 3.11

what Cavalieri calls "all the lines of the figure ABC, taken according to the rule BC."

Definition 1 of book 2 of the *Geometria indivisibilibus* indicates what should be understood by "all the lines":

> If through opposed tangents of any given plane figure are drawn two parallel planes, perpendicular or oblique in relation to the plane of the given figure, and prolonged indefinitely in either direction, and one of these planes moves towards the other while remaining always parallel to it until it merges with it (by Postulate 2 of Book 1), the individual straight lines that are generated during the whole motion as intersections of the moving plane and the given figure, we shall call, when they are taken all together, "All the lines of such a figure, taken with one among them for a rule." (Cavalieri, *Geometria indivisibilibus* [1653], bk. 2, Def. 1, 99)

The final specification is very important: All the lines *taken according to the rule BC*. The aggregate of the lines is determined each time by the direction chosen. A theorem will be necessary to ensure an invariance: the aggregates of lines will be equivalent for a given figure if the direction of slicing is changed; these aggregates will still be equivalent if the areas are equal although the figures are different. Exactly what this equivalence between aggregates signifies (Cavalieri speaks of an "equality"; see ibid., Prop. 2, 112) is not easy to specify.

The indivisible is not at all an ultimate or evanescent state of finite magnitude, but rather the product of a slicing—an assembly of entities of one dimension lower than that of the magnitude itself. Reasonings like those of Galileo about the bowl lack rigor in Cavalieri's eyes and cannot serve to found a true theory. The Bolognese mathematician had shared with Galileo his criticisms apropos of the "bowl," first in basing himself on arguments of dimensionality (EN, 16:136), then in specifying that it was not necessary to count the extreme terms when one compared two aggregates of lines:

> In the conception of all the lines of a plane figure or of all the planes of a body one must not, according to my definitions, include the extremes, although they appear to be of the same kind. I call all the lines of a plane figure the common sections of the plane cutting the figure as it moves from one extreme to the other or from one tangent to the opposite tangent: now, since the beginning and end of the motion are not motion, one must not, therefore, account the extreme tangents as among the aggregate of the lines; and since I intend the same thing for planes and solids, it is no marvel that these extremes remain unequal as in the example of the bowl. (Cavalieri to Galileo, 16 December 1634, EN, 16:175)

The indivisible is not an evanescent or ultimate entity. Nonetheless, time plays an essential role in the foundations of the theory. With each instant is associated a position of the plane, and hence also a certain segment (or several segments) in the interior of the figure. It is the continuity of the motion and of the time that ensures that "all" the lines of the figure have indeed been taken (as A. Koyré has justly remarked[6]). Cavalieri explains this in his *Exercitationes* in response to criticisms by Paul Guldin:

> Suppose a square, and through two of its sides let there pass two indefinitely extended planes erected perpendicularly . . . and let one of these planes move continuously towards the other. . . . At any moment this moving plane will cut the plane of the square. . . . Since in the square of which I speak one cannot assign any of these lines through which the plane does not pass at some time, that is, at a certain moment (this is why I have said that these lines are described by this plane), thus all these lines, assembled in mind so that none is supposed excluded, I have called "all the lines.". . . And if I have said that these lines are produced in the totality of the motion, I have understood this to mean: in all the instants of the time during which the total motion is made. (Cavalieri, *Exercitationes*, 198–99; *Geometria degli indivisibili*, 809)

Because the motion of the cutting plane is "continuous" (in the present-day sense), it omits none of the lines of the figure. The idea of motion is presupposed from the first postulates of book 1 of the *Geometria*; these authorize making a plane to glide parallel to a fixed direction (Cavalieri, *Geometria indivisibilibus* [1653], bk. 1, 14; *Geometria degli indivisibili*, 80).

Proportions between Infinite Aggregates

Once the meaning of "all the lines of a figure" has been clarified, the next step consists in making these aggregates of lines enter into relations of proportionality. This is a particularly delicate element of the theory, and Cavalieri asks Galileo for advice on it beginning in 1621:

> I would like to know whether all the lines of a plane have a proportion to all the lines of another plane, because as one can always draw more and more of them, it seems that all the lines of a given figure are infinite, and are thus excluded from the definition of the magnitudes that have proportions among them; but since, if one enlarges the figure, the lines also become larger, as there are then those of the first figure and in addition those that are in the excess of the enlarged figure over the given figure, it seems that these lines are not excluded from the definition [of the magnitudes that have proportions between them]; I wish that you might deliver me from this uncertainty. (Cavalieri to Galileo, 15 December 1621, EN, 13:81)

Galileo's response is not known, but it is possible that the much later text of the *Discorsi* gives his opinion on this point: in the domain of the infinites, the words "greater" and "smaller" lose their meaning.

Cavalieri's response is different and in its essence already contained in the question he directs to Galileo: since the lines cut off in a figure are enclosed in the outline of the figure, and since there is a proportion between the figures themselves, there can be a proportion between all the lines of one figure and all the lines of another. Cavalieri seeks to demonstrate this in Theorem 1 of book 2 of his *Geometria*:

> All the lines of any plane figures . . . are magnitudes that have ratios with one another. (Cavalieri, *Geometria indivisibilibus* [1653], bk. 2, 108–9)

How to demonstrate a proposition so new? How to establish a proportion between two infinite aggregates? Cavalieri reasons on two figures in which "all the lines" are cut according to a common rule EQ. Consider first the case in which the figures have identical heights (if not, the excess of the highest over the smallest is cut off and shifted further on the same line returning thus to the case of figures of the same height).

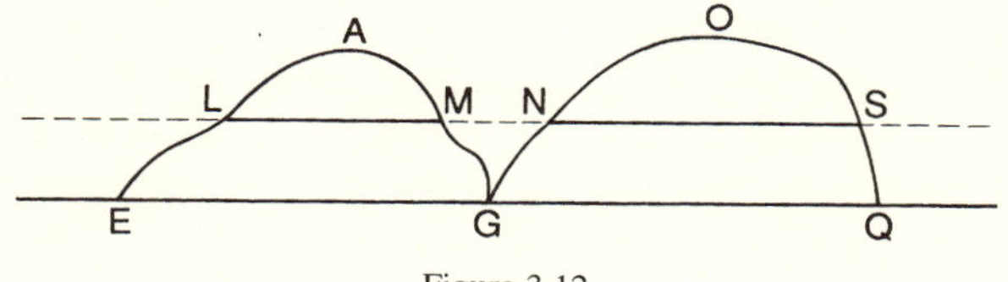

Figure 3.12

The lines called "all the lines of EAG" and "all the lines of GOQ" are the segments such as LM and NS cut off by the successive positions of the lines such as LS always parallel to the base or "rule" EQ. Here is the essential passage of the reasoning:

> If the straight line [= segment] NS is shorter than the straight line LM, it can, being prolonged indefinitely, become at some moment longer than the latter. If we suppose this to be done for all the other lines that are at the same distance from the rules EG and GQ, it is clear that each of the lines that are in the figure GOQ, being prolonged, becomes longer than [the corresponding line] in the figure AEG. . . .
>
> It is therefore clear that all the lines of the figure AEG will be a part of all the lines of the figure GOQ thus prolonged; the latter will therefore be the whole, since the first are included in the last. . . .
>
> Yet the whole is greater than the part; therefore all the lines of the figure GOQ have been prolonged so as to be made longer than all the lines of the figure EAG . . . [and inversely if we exchange the roles of the figures GOQ and EAG].
>
> Yet two magnitudes are said to have a ratio between them if, being multiplied, either of them can exceed the other; therefore it is clear that all the lines of the figures EAG and GOQ, in the case in which the heights are equal, have a ratio to each other. (Cavalieri, *Geometria indivisibilibus* [1653], 108–9; *Geometria degli indivisibili*, 201–3)

Thus is satisfied the criterion imposed by Euclid: two magnitudes have a ratio or a proportion if they are capable, being multiplied, of exceeding one another (Euclid, bk. 5, Def. 4; restated differently by Archimedes, *On the Sphere and Cylinder*, Post. 5, in *Opera*, 2:10). The ancients knew, for instance, that an angle of contact between a curve and a straight line and a rectilinear angle between two straight lines were not comparable magnitudes, being unable to exceed each other mutually, and therefore unable to have a ratio with one another (cf. Proclus, *In Primum Euclidis*, 234).

What is not Euclidean here is the supposition that the operation of prolongation has been carried out an infinity of times ("If we suppose this to be done for all the other lines. . ."). One can also see the danger there is in employing the axiom: "the whole is greater than the part" (Axiom 8 of the first book of Euclid, accepted by the great majority of traditions) in the case of infinite aggregates.

The key to the demonstration is the idea previously encountered in the letter to Galileo of 1621: the infinity of lines is contained within the limits of the figures. Cavalieri expounds the same thesis in response to Guldin: the aggregate of the lines of a figure is not absolutely infinite, but only infinite "under a certain aspect" [*secundum quid*] (Cavalieri, *Exercitationes*, bk. 3, chap. 8, 202). In fact this infinity is contained within the contours of a limited figure; since the figures themselves are comparable, the aggregates of the lines therein contained must also be comparable:

> If for example we set side-by-side with the square considered above another square that is equal to it, will it not be manifest that all the lines of the first square are obtained again in the second square, and consequently that both taken together are double the first one alone, although we do not know the number of either aggregate? It is here as with the algebraists, who do not know what it is they call root, side or cossa, or what inexpressible roots can be, but who in multiplying and dividing them, etc., . . . are finally led to the discovery of what they seek, as by obscure and winding ways. (Ibid., 202–3; *Geometria degli indivisibili*, 814)

The number of lines matters little; they can be compared and manipulated as the algebraist manipulates his unknowns and his radicals, without knowing, to begin with, what they are.

After having shown how he gives a sense to the proportionality between indivisibles, Cavalieri completes the foundations of his method with two theorems:

> *Theorem 2*. All the lines of two figures are equal if the figures are equal.
>
> *Theorem 3*. There is the same proportion between the figures as between the aggregates of lines. (Cavalieri, *Geometria indivisibilibus* [1653], 112)

The demonstrations of these propositions, which raise some questions, will not be examined here. It is important only to know how Cavalieri has defined his indivisibles and how he makes use of them. Theorem 3 is the "supreme foundation" of this new geometry:

> We thus see that to find what proportion holds between two plane or solid figures, it is sufficient to find, in the plane figures, what proportion all the lines of these figures have to each other, and in the solid figures, what proportion all the planes of these solid figures have to each other, as determined according to any rule. Such is the grand principle on which I found my new geometry. (Ibid., Th. 3, Cor. 115)

Examples of Application

A very simple example will show how this method is used. To demonstrate that every parallelogram is divided by its diameter into two triangles of equal area, it is sufficient to compare all the lines, such as BM and HE, taken at equal distances respectively from the side of C and from the side of F. These lines are equal each to each. Therefore all the lines of one triangle are equal to all the lines of the other, taking CD as the common rule. Hence, in virtue of the fundamental theorem (Theorem 3), the figures themselves are equal.

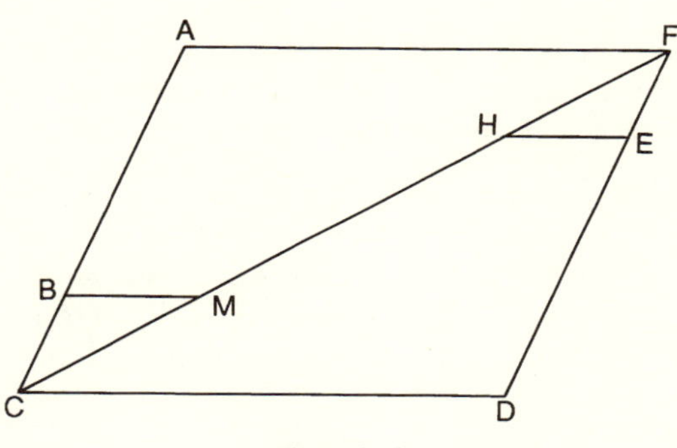

Figure 3.13

Most of the demonstrations are more indirect and very difficult to follow, and in their very bold utilization of indivisibles, some of them contain surprises. For example, Cavalieri rearranges one-by-one all the lines of a figure by transporting them according to a given direction to a vertical or oblique straight line that serves as a barrier or curb (ibid., bk. 2, Prop. 15, 128).

Infinite aggregates of a more complex sort are introduced for the sake of demonstrations of quadrature. Thus Cavalieri defines "all the squares of a figure," that is, all the squares erected parallel to one another on all the lines of a given figure. Then can be compared, for example, "all the squares" of two parallelograms of the same height:

Proposition 9. If parallelograms are of identical height, all their squares, with the base from which the height has been determined taken for rule, will be between them as the squares of the bases. . . .

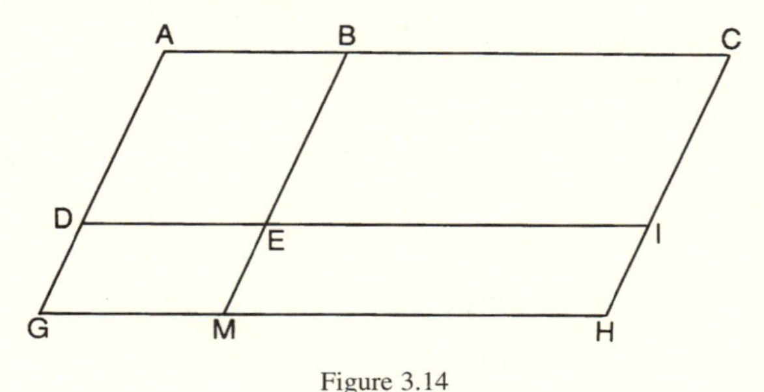

Figure 3.14

Let us draw in the interior of the parallelograms AM and MC any straight line DI parallel to GH. . . . Since DE is equal to GM and the similar plane figures described by homologous sides or lines are equal, it results that the square of DE will be equal to the square of GM, and the square of EI to the square of MH. Therefore the square of GM will be to the square of MH as the square of DE to the square of EI. And since DI is any parallel to GH, it follows that as one square is to one, so are all to all [*ut unum ad unum, ita omnia ad omnia*], that is, GH being taken for rule, all the squares of the parallelogram AM will be to all the squares of the parallelogram MC, as the square of GM to the square of MH. Q.E.D. (Ibid., 121; *Geometria degli indivisibili*, 219)

Other results of quadrature, of a more valuable kind, are obtained by the same route. Thus it can be shown that the ratio between all the squares of a given square ABCD and all the squares of the triangle ABD constructed on the diagonal is the ratio of 3 to 1 (Cavalieri, *Geometria indivisibilibus* [1653], bk. 2, Prop. 24).

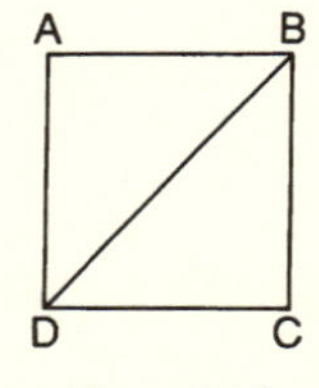

Figure 3.15

Cavalieri applies this method to the area of the segment of a parabola. He puts in correspondence each of the ordinates KH of the parabola AHC and each of the squares constructed on IK. All the squares of the triangle AED then represent all the ordinates of the curvilinear triangle ADC. Here is the text of this demonstration as given in the *Exercitationes*, where it is briefer and more direct:

Let there be any parabola OAC, with vertex A, base OC, and diameter AB. . . . I say that the rectangle TC is one and a half times the parabola in size.

Let BA be prolonged to F so that AF is equal to AD, and complete the rectangle FADE; in this rectangle draw the diagonal AE; between CE and BF, and parallel to CE and BF, draw any straight line GR to serve as rule. . . .

The square of CB is to the square of HS as BA to AS (by the first book of the

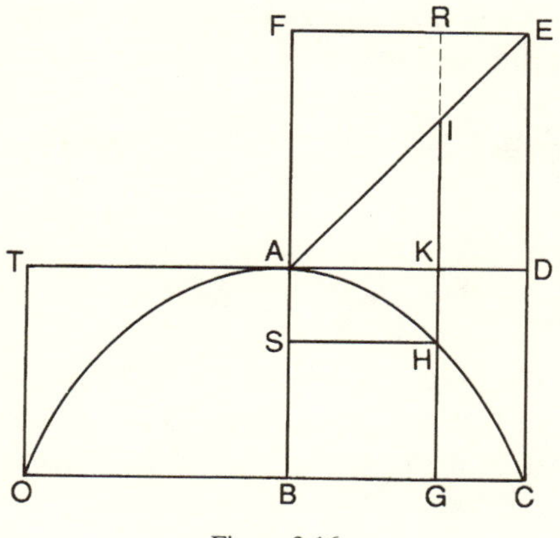

Figure 3.16

Conics of Apollonius, Prop. 20). . . . As BC is equal to AD, that is, to AF or again to RK, and SH is equal to AK, that is, to KI . . . , hence the square of RK is to the square of KI as GK to KH.

Thus all the squares of the rectangle FD will be to all the squares of the triangle ADE, while taking CE for rule, as all the lines of the rectangle BD to all the lines of the curvilinear triangle ADC, while taking CE again for rule. But in Prop. 24 we have shown . . . that all the squares of FD are triple of all the squares of triangle ADE. Therefore all the lines of rectangle DB will be triple of all the lines of the curvilinear triangle ADC. But, by Prop. 3, all the lines of one figure are to all the lines of another as the figures themselves. Therefore the rectangle BD is triple of the curvilinear triangle ADC, so that it is one and a half times the half of the parabola. Q.E.D. (Cavalieri, *Exercitationes*, bk. 1, 81–82)

An Innovative Disciple: Torricelli

Unhappily, the *Geometria indivisibilibus* of Cavalieri is a difficult book and not very inviting. The first hundred pages (of book 1) is a long preamble devoted above all to plane and solid similitudes and to generalized cones and cylinders. Cavalieri will suppress this part in the simplified presentation that he will later give in book 1 of the *Exercitationes*. When one finally reaches the discussion of the method of indivisibles itself, the ponderousness of the procedures is very off-putting, and the results obtained are nearly always well-known. The respect for Euclidean ritual—at least in appearance—brings with it a discouraging prolixity. Also, Cavalieri's Latin is not very flowing. . .

By happenstance, the new theory found a spokesman of genius in the person of Torricelli, who presented the method of indivisibles in a direct and easily accessible form and rendered it more attractive with new and surprising results. This popularized presentation of indivisibles is found in the *Opera geo-*

metrica of Torricelli (1644) in two texts of remarkable originality. In the first (*De dimensione parabolae*), the author puts together twenty-one different demonstrations of the same proposition—the quadrature of the parabola. The first ten are in the style of the ancients; that is, the demonstration proceeds indirectly by a double reduction to the absurd according to the method that has since been called the method of exhaustion. The last eleven utilize indivisibles, and the reader cannot fail to be aware of the contrast between the two methods, though both obtain the same result. The method using indivisibles appears much more direct and natural.

The second text (*De solido hyperbolico acuto*) demonstrates that the volume of a certain hyperbolic solid of infinite height is equal to that of a finite cylinder. The novelty lies in the determination of the volume of a body with an infinite dimension, and also in the use of curved indivisibles.

These pages will be much more frequently read than the laborious demonstrations of Cavalieri, yet has not Torricelli's success generated a grave misunderstanding? He presents his work as an illustration of the new geometry of Cavalieri, but the simplification is so great that it raises the question of infidelity to the latter's work. The essential divergence consists in this: for Torricelli, the indivisibles constitute continuous magnitude. Each demonstration ends in a pronouncement of the sort: "all the lines, or if one prefers, the figure itself" [*omnes lineae sive figura ipsa*] (for an example see above, pp. 174–75).

On the other hand, the work contains no fundamental propositions on the very strict conditions for slicing figures, on proportions between infinite aggregates, or on the passage from infinite aggregates to plane and solid figures. Torricelli, with humor, is content to advise his reader to navigate across the ocean of indivisibles in the pages of Cavalieri's *Geometria*, but announces that he himself will be content to go modestly along the coast.

Yet Torricelli is much more than a popularizer and a simplifier. He reflects penetratingly on the foundations of the method, he discovers a whole series of paradoxes that nourish his meditation, and finally, he constructs a theory of indivisibles very different from and much bolder and more fruitful than Cavalieri's. These steps are attested to uniquely in manuscripts that remained unpublished until the Faenza edition in 1919, but it is very probable that these new ideas were disseminated in the English milieu of the 1660s.

The fundamental paradox on which Torricelli reflected is the following. The two parts of the accompanying rectangle on each side of the diagonal are equal in area. Yet considering all the lines such as FE and all the lines such as GE will necessarily produce this absurd result: that the triangle ABD is to the triangle DBC as all the lines of the one are to all the lines of the other, that is, as FE is to EG, or as AB is to BC.

Cavalieri avoided this sort of absurdity by imposing a direction of slicing (Cavalieri's correct demonstration of the equality of the two halves of the parallelogram is on pp. 183–84 above). But Torricelli, rather than imposing

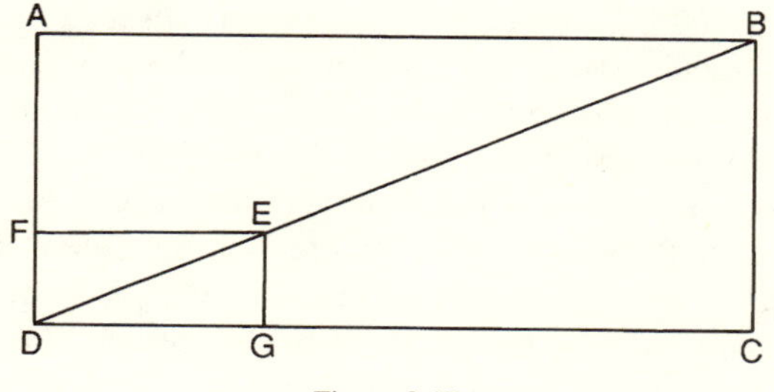

Figure 3.17

restrictions on the use of indivisibles, attempts to penetrate more deeply into the mystery. He collects paradoxes of this sort, for the most part variations on this fundamental paradox, and draws up several lists of them (*TO*, vol. 1, pt. 2, 20–23, 47–48, 417–26).

Here for example is what happens in the case of the surface of the sphere and that of the circumscribing cylinder: if one considers "all the circles of the cylinder" and "all the [horizontal] circles" of the sphere, it is clear that those of the cylinder have a constant size, while those of the sphere vary between a unique maximum in the middle (where the cylinder touches the sphere) and two circles reduced to points, above and below. It would be necessary therefore to admit that the surface of the sphere is smaller than that of the cylinder, which is false since they are equal (Archimedes, *Opera*, 1:149–57).

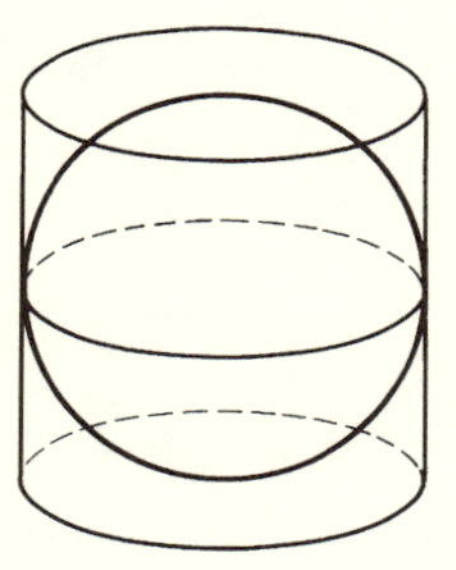

Figure 3.18

Other examples are more refined and richer in surprises. Torricelli takes pleasure in accumulating absurd reasonings in which a perverse slicing leads to a manifestly false conclusion. One of these examples provides a glimpse of the audacious solution that Torricelli suggests. Parallelograms that have the same height and equal bases are equal in area. That is true, for instance, for AEBH and AECI, included between the two horizontal parallels and having AE as common base. That still remains true if the base is diminished by half: DEBJ is still equal to DECK. But what happens if the base is diminished

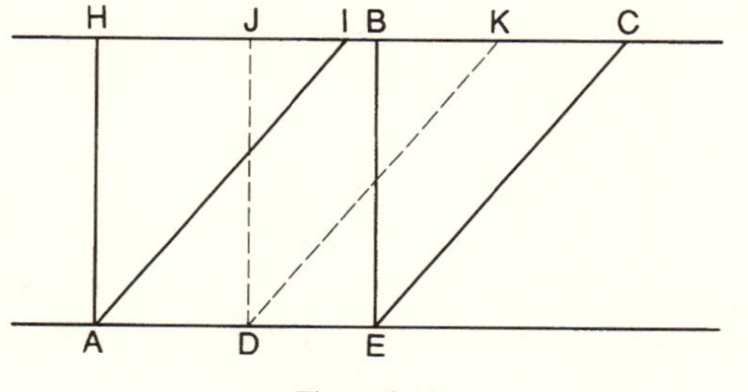

Figure 3.19

indefinitely? The "ultimate residues," as Torricelli calls them, are straight lines and no longer parallelograms, the segment EB on the one hand, and the segment EC on the other. It is obvious that these segments are unequal. Yet they are the vestiges of equal surfaces. How is this possible?

Torricelli's solution consists in assigning a thickness to each of these segments, but different thicknesses according to their respective situations. Here BE will be thicker than CE. Both are supposed to "pass adequately" through point E, or "occupy [it] adequately," in Torricelli's terms. With the breadth of point E on the horizontal base serving as common reference and with BE and CE inclined on this horizontal base, since BE meets the base at a right angle and CE is very inclined, if they are both to "occupy" point E adequately, it is necessary that CE be thinner than BE.

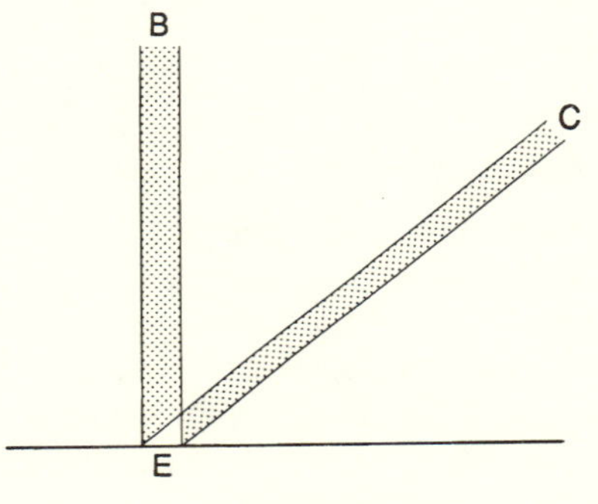

Figure 3.20

One is thus truly in the presence of the infinitely small: infinitesimals having as many dimensions as the figure in which they are contained. The segment of a straight line is an infinitely thin rectangle, the disc is a cylinder of infinitely small height. The indivisible is therefore no longer, as it is for Cavalieri, the result of a slicing of a continuous figure; it is reached as a "vestige," or "ultimate residue" of the finite figure.

The crucial feature of this theory is that there are differences, relationships of greater and lesser, in the realm of the infinitely small. The text in which Torricelli expounds this conception most clearly (entitled *Delle tangenti delle parabole infinite*) opens with this affirmation that indivisibles are not all equal to one another; one point can be larger than another point, one line larger than another line, one surface higher or thicker than another:

> That all indivisibles are equal to one another, that is, points equal to points, lines equal in thickness to lines, and surfaces equal in depth to surfaces, is an opinion that in my judgment is not only difficult to prove, but even false. Let there be two concentric circles, and from the center let there be drawn all the lines toward all the points of the larger circumference. There is no doubt that the same

number of points will be generated by the transits of these lines through the smaller periphery, and each of these will be as much smaller than each of those [the points of the larger circumference], as the diameter is less than the diameter. (*TO*, vol. 1, pt. 2, 320; *Opere* [1975], 505–6)

Torricelli added a note in the margin of this page opposite the example of the concentric circles:

> because the lines are tapered [*accuminate*], but if one draws a single line, the points will be equal. (*TO*, vol. 1, pt. 2, 320; *Opere* [1975], 505–6)

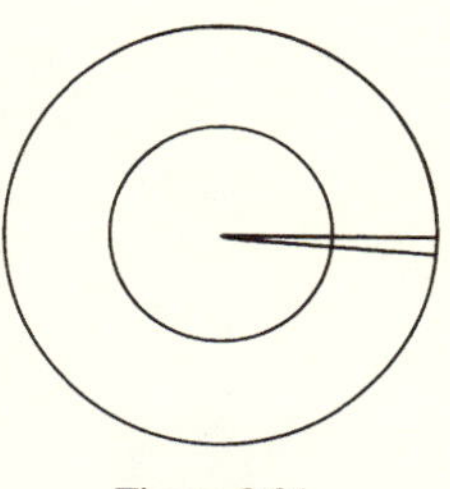

Figure 3.21

Each radius is to be considered as a very thin sector of a circle; the points must become thinner and thinner nearer the center, and the lines converging toward the center are more and more closely crowded together.

Subsequent to this text, Torricelli takes up again the fundamental example of the parallelogram mentioned earlier:

> If there are two parallelograms on the same base AB, and from all the points of AB there are drawn the infinity of lines parallel to the sides, both in parallelogram AC and in parallelogram AD, all the AC taken together will be equal to all the AD taken together. But they are also equal in number (because in either case there are as many lines as points on AB); hence one [of the lines of one set] is equal to one [of the lines of the other set]. But they are of unequal length; therefore, although indivisibles, they are of unequal widths, reciprocally proportional to their lengths. (*TO*, vol. 1, pt. 2, 321)

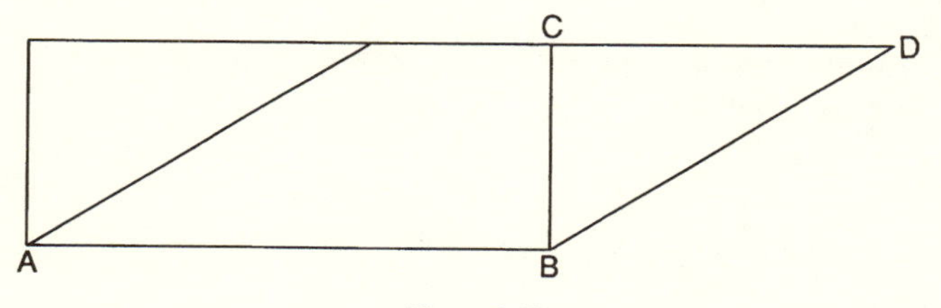

Figure 3.22

Among the paradoxes Torricelli collected, the simplest concerned a rectangle cut in two by the diagonal: one half is "filled" by lines proportional to one side of the triangle, and the other half is likewise "filled" by lines equal in number but of very different size. In his great text on the new indivisibles, Torricelli returns to this paradox again, using a different means of analysis:

> In the preceding parallelogram ABCD, if on the diameter BD we take any point E, the semi-gnomon EBC will be equal to the semi-gnomon EBA. But if we divide BE in half at the point I, the divided or narrowed semi-gnomon IBC will

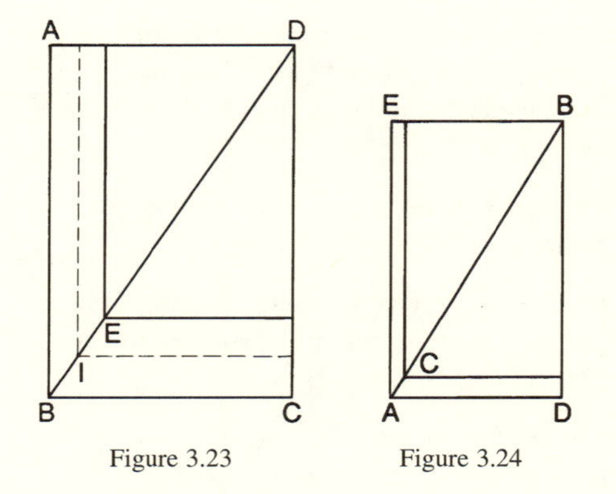

Figure 3.23 Figure 3.24

> be equal to the narrowed semi-gnomon IBA. And if this division is made, or supposed made, an infinity of times, we shall come to have, in place of semi-gnomons, a line BC equal to a line BA. I say equal in quantity, not in length, because although they are both indivisible, BC will be, in relation to BA, as much wider as the latter is longer. And in truth if both must occupy adequately the diametral point B, it is necessary that CB, which is nearer to being perpendicular, be also wider than the line AB, which is more inclined.
>
> That is seen clearly in the next figure, in which AB is the diameter, and on this is the point AC through which pass adequately the two lines AD, AE; it is clear that in order that AE occupy quite adequately this point, it suffices that it be of little breadth, because it is very inclined. But AD, because it is close to the perpendicular, must be much wider, and that in a proportion inverse to the lengths. (*TO*, vol. 1, pt. 2, 321–22)

Each segment thus has its "quantity," which differs from its length: this quantity is its infinitesimal area, the product of its length by its larger or smaller infinitesimal width.

Torricelli extends this reasoning from gnomons constructed on the diagonal of a rectangle to other, more complex "gnomons," constructed on a curve: in the case of the diagonal the two semi-gnomons are equal; in the case of a curve of the form $Y^m = AX^n$, Torricelli proves that the curvilinear semi-gnomons are to each other as exponent to exponent.

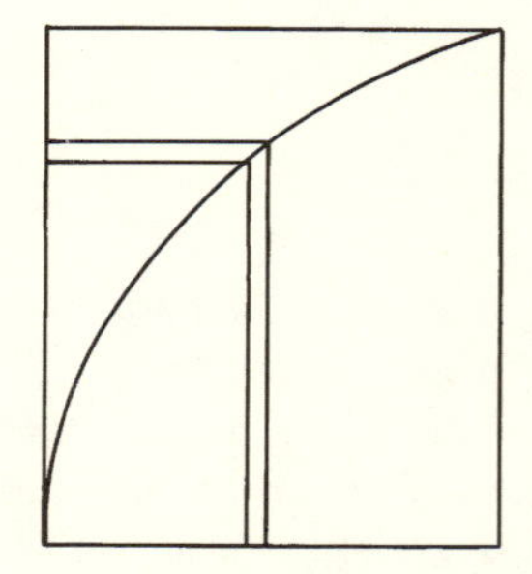

Figure 3.25

Torricelli extracts from this a very original method for drawing tangents. It must be emphasized that this would be impossible with the tools of the *Geometria indivisibilibus* of Cavalieri, where the only indivisibles are lines and planes, while here it is necessary to use infinitesimals of length, by considering in curves the very small segments that "compose" them.

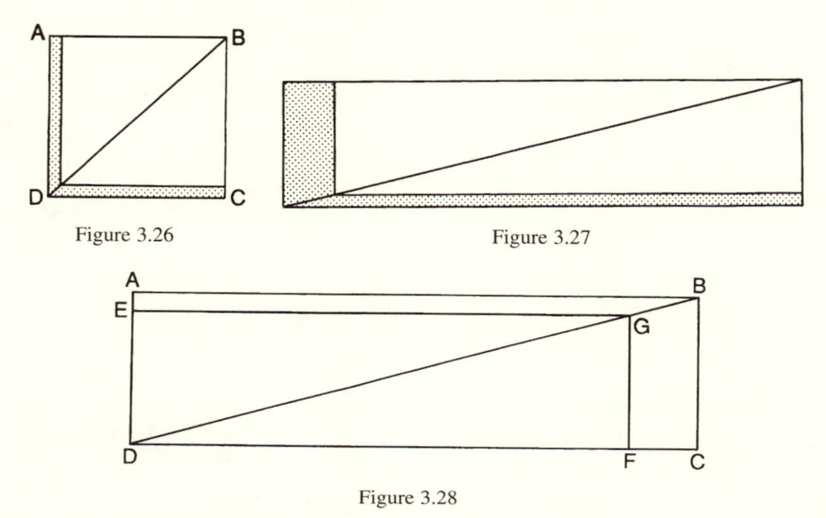

Figure 3.26

Figure 3.27

Figure 3.28

To enter into this reasoning, it is necessary first to admit that the thickness of lines is variable according to their inclination in relation to a line of reference: if a point B on a line BD is "occupied adequately" by two lines AB and AC, and if the inclination of the latter in relation to a line of reference BD is the same, their infinitesimal thickness will be the same. For example, the sides of the square have the same thickness since they have the same inclination to the diagonal and both occupy point D adequately (fig. 3.26). If the inclination is not the same, the thickness of the lines will be different (fig. 3.27). This thickness is such that the total area (length multiplied by infinitesimal breadth) must remain the same on the two sides, since the segments EG and GF, for example, are the vestiges of the semi-gnomons ABGE and FCBG which are equal in area.

In the case of a curve $Y^m = AX^n$ (which Torricelli calls an "infinite parabola"), the semi-gnomons cb and ab are to each other as the exponents *m* and *n*, and Torricelli assumes that this relation remains true when these gnomons have become segments of indivisible thickness.

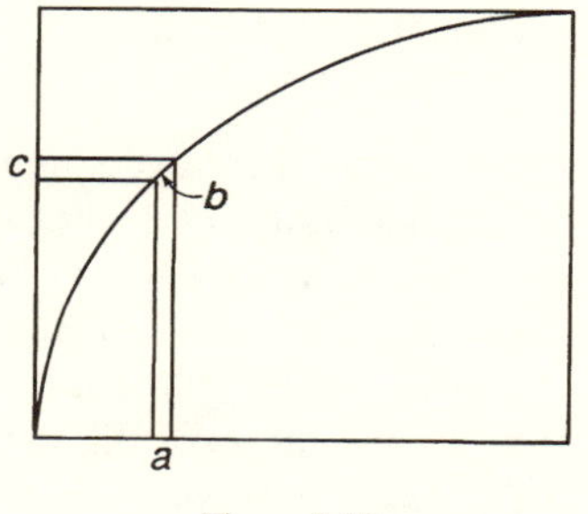

Figure 3.29

To trace the tangent to one of these curves at a point B, it will only be necessary to superpose the two situations, rectilinear and curvilinear, by considering simultaneously the semi-gnomons relative to the diagonal and the semi-gnomons relative to the curve. Here is the schema of the reasoning (see fig. 3.30). The location of point E is not known, that is, the inclination of BE is unknown since it is precisely the direction of the tangent in B that is being sought.

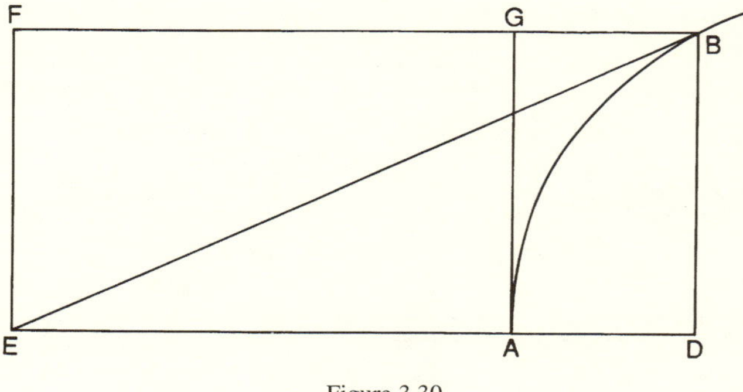

Figure 3.30

DB and BG have an infinitesimal thickness given by the ratio m/n of their areas and the ratio of their lengths. On the other hand, the tangent EB must be the prolongation of the point B "occupied adequately" by the lines GB and DB. This tangent will be the diagonal of a certain rectangle, and in this rectangle the lines FB and BD are "supplementary" (they are equal in area, because they are semi-gnomons).

The infinitesimal thickness of GB is known: it suffices to find the line FB which, endowed with the same infinitesimal width as GB, will have an area equal to that of the line DB.

Here is Torricelli's very brief text:

> Let ABC be one of the infinite parabolas, and we are to give the tangent for the point B. Let BE be the tangent, BD the *applicata* [ordinate], and let the figure DF be completed. Let us conceive that through the same point B through which passes the tangent also pass adequately the lines BD and BF. BD and BF will therefore be supplementary lines; and the length ED will be to the length DA as the length FB to the length BG, or again as the quantity FB to the quantity BG (because FB and BD are supplementary), that is, as the exponent to the exponent. (*TO*, vol. 1, pt. 2, 322–23)

It is important here to distinguish clearly between the "quantity" of a segment (of a "line") and its length: the quantity is the infinitesimal area, which is determined here by supposing that GB and BD are vestiges of curvilinear semi-gnomons and that FB and BD are vestiges of rectilinear semi-gnomons.

It is possible to present this reasoning in shortened form. If e and e' designate the infinitesimal thicknesses of the segments BD and BG, then: $\mathrm{BD} \cdot e / \mathrm{BG} \cdot e' = m/n$ (because the semi-gnomons relative to the curve ABC are in the constant ratio of the exponents). From this can be deduced the ratio of the thicknesses: $e/e' = m \cdot \mathrm{BG}/n \cdot \mathrm{BD}$. A length of BF such as to satisfy $\mathrm{BF} \cdot e' = \mathrm{BD} \cdot e$ is needed (because the semi-gnomons relative to the diagonal are equal). The result is $\mathrm{BF} = (e/e') \cdot \mathrm{BD}$. That is,

$$\mathrm{BF} = (m/n)(\mathrm{BG}/\mathrm{BD})\mathrm{BD} = (m/n)\mathrm{BG}.$$

Consequently the subtangent ED must be to the abscissa BG or AD in the ratio of m to n. Thus it is shown how the infinitesimal thicknesses serve only as intermediaries of calculation and disappear naturally at the final step.

Spiral and Parabola: An Example for Comparing the Two Theories

Previously (pp. 112–13 above) it was shown how the attribution of a size to points permits an in-depth analysis of accelerated motion and a veritable demonstration of Galileo's law of fall.

With these same tools, Torricelli is able to rectify curves or compare the lengths of different curves, which is unquestionably not possible in the perspective of Cavalieri, at least directly. The most striking and original result, which is also a splendid example of the application of the new indivisibles, is the comparison in length between an arc of a spiral and an arc of a parabola. Nonetheless the indivisibles of Cavalieri here assist the infinitesimals of Torricelli, since it was in the *Geometria indivisibilibus* that Torricelli found the idea of a comparison between the spiral and the parabola; but it was a comparison dealing only with areas enclosed by the curves and carried out in accordance with Cavalieri's canons. Thus, to the same objects will be applied these two very divergent doctrines.

Cavalieri demonstrated, in book 6 of his *Geometria*, the equality of two curvilinear areas: one which is enclosed between an arc of a spiral and the associated circle and another which is contained between a certain parabola and the sides of the associated triangle. The utility of this result, in the book, is to permit an evaluation of the area under the spiral. Yet the comparison between the two curves was interesting in itself, and Cavalieri saw emerge in it "the admirable connection and, so to speak, the affinity between parabolic and spiral spaces" (Cavalieri, *Geometria indivisibilibus* [1653], bk. 6, Prop. 20, scholium; compare Cavalieri to Torricelli, 7 May 1644, *TO*, 3:180). Here is the text of Cavalieri's demonstration:

> *Proposition 9*. The space comprised between the spiral generated by the first revolution, and the initial line of the revolution, is the third [of the surface] of the first circle.
>
> Let AIE be the spiral generated by the first revolution, and AE the initial position of the turning line. Let the first circle ESM be traced with A as center and AE as radius. I say that the space AIE is one-third of the circle EMS.
>
> Having taken a point V on AE, let us describe the circle VIT with center A and the interval AV, and join AI, prolonged to S. Let us then describe the right triangle OQR [see fig. 3.32], whose side OQ adjacent to the right angle is equal to AE, and whose side QR is equal to the [entire] circumference SME, and let us complete the rectangle QZ. Cut off OX equal to AV, and through X draw a straight line XY

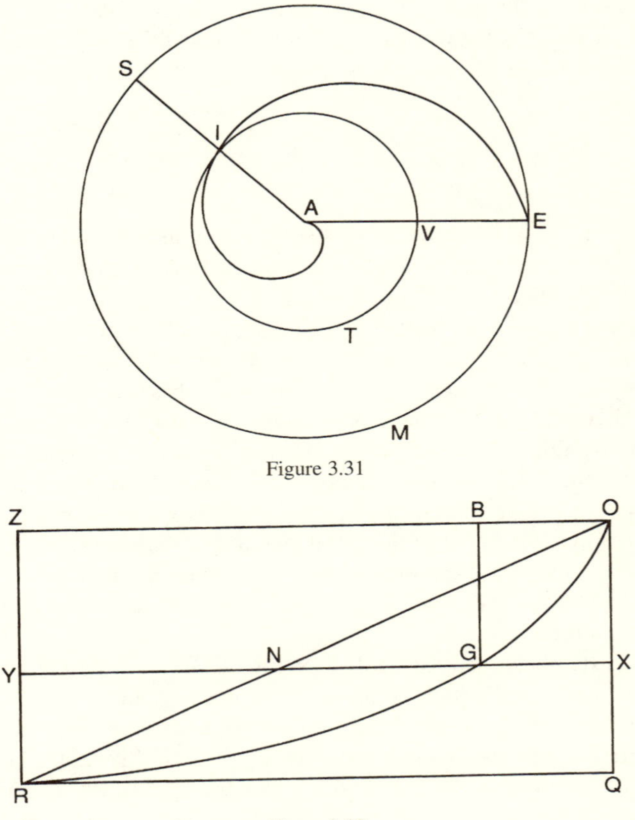

Figure 3.31

Figure 3.32

parallel to RQ, cutting ZR in Y and OR in N. Let us draw with OZ as axis the semi-parabola RGO with vertex O and passing through point R (see Book 4, Prop. 20); this cuts YX in G. Through G draw GB parallel to OQ, falling on OZ in B.

Since (Book 1, Prop. 38 and 40, Scholium) the square of ZR is to the square of BG as ZO to OB, therefore RQ is to GX as the square of QO to the square of OX, that is, as the square of EA to the square of AV. But this ratio is also that of the circumference ESM to the circumference [= the arc] ITV; in fact they have between them the ratio compounded

of the ratio of the circumference ESM to the [whole] circumference IVT, that is (Book 6, Prop. 3, Cor. 2), of the ratio of EA to AV,

and of the ratio of the [whole] circumference IVT to the circumference [= the arc] ITV, that is, of the [whole] circumference MSE to the circumference [= the arc] SME, that is (Book 6, Prop. 7), of the ratio of EA to AI or AV.

Yet these two ratios of EA to AV when compounded form the ratio of the square of EA to the square of AV (*Elements*, VI, 23); therefore the [whole] circumference MSE is to the circumference [= the arc] ITV as the square of EA to the square of AV; that is, as RQ to XG.

But RQ is equal to the circumference MSE; therefore also GX will be equal to the circumference [= the arc] ITV, and we shall thus show that any circumference [= arc] concentric to A and contained between the spiral AIE and the straight line AE, but outside the space of the spiral AIE, is equal to a line drawn in the tri-line OGRQ parallel to RQ, namely those that cut off from the side of the points O and A respectively equal parts on OQ and AE; and because OQ and AE are supposed equal, all the lines of the tri-line OGRQ, in taking RQ as rule, will be equal to all the circumferences of the tri-line contained between the spiral AIE and the circumference MSE.

Similarly, since RQ is to NX as QO to OX, or EA to EV, or the circumference MSE to TIV, and on the other hand RQ is equal to MSE, consequently NX is equal to the [whole] circumference TIV, and we shall show likewise that all the lines of the triangle ORQ are equal to all the circumferences of the circle MSE. Hence all the lines of the triangle ORQ are to all the lines of the tri-line OGRQ, or (Book 2, Prop. 3) the triangle ORQ is to the tri-line OGRQ as all the [whole] circumferences of the circle MSE are to all the circumferences [= the arcs] of the spiral figure contained between the straight line AE and the circumference MSE. By conversion of ratio, the triangle ORQ or OZR will be to the figure OGR as all the circumferences of the circle MSE to all the circumferences of the spiral space AIE, that is, as the circle [in area] to the space AIE (because the nature of the curve AIE agrees with what is required in Prop. 6, as follows from Prop. 7 [= the property of strict growth of the curve in relation to successive concentric circles]).

Since on the other hand the semi-parabola OGRZ is equal to ⅔ of the triangle OZR, it follows by subtraction that the figure OGR is the third of the triangle OZR. Consequently the spiral space AIE is a third of the circle MSE, which was to be demonstrated. (Ibid., Prop. 9, 436–39)

The key to the demonstration consists in the proportionality between the arcs such as VTI and the square of the length AV, radius of the corresponding circle. The reasoning can be schematized as follows, by decomposing a ratio into two component ratios:

$$\frac{\text{large circle whole}}{\text{arc of small circle}} = \frac{\text{large circle whole}}{\text{small circle whole}} \times \frac{\text{small circle whole}}{\text{arc of small circle}}.$$

Yet the two ratios here compounded are the same ratio: it is the ratio between the radius of the large circle and that of the small circle. In fact, in the spiral (that of Archimedes), the rotation is proportional to the increase of the radius—Torricelli, in his more kinematic style, will say that this ratio is the ratio of the time. To employ an abbreviated and anachronistic notation: the angle β, in relation to 360°, is as AV to AE, or as t to 1 if $0 \leq t \leq 1$, so that $(2\pi\text{AE}/2\pi\text{AV}) \times (2\pi\text{AV}/2\pi\text{AV} \cdot t) = (\text{AE}/\text{AV}) \times (1/t) = (1/t) \times (1/t)$. Thus the circular arcs increase as the squares of the radii and can be put in correspon-

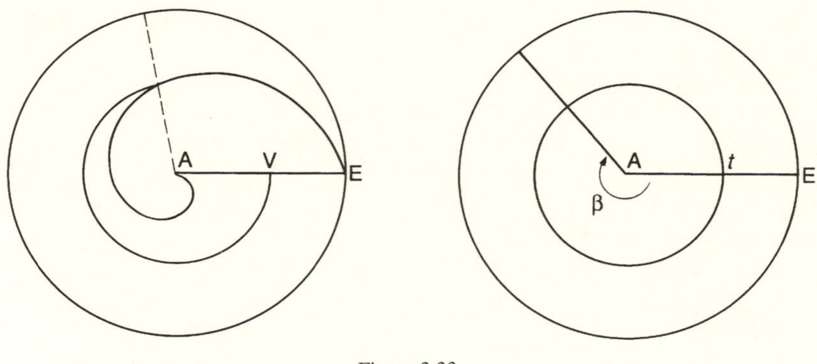

Figure 3.33

dence with the ordinates of the parabola. The fundamental theorem of indivisibles (Th. 3 of bk. 2) then permits us to identify the ratio between the areas with the ratio between "all" the ordinates of the parabola and "all" the arcs of the space within the spiral.

Here now is the manuscript text (or more exactly the two texts, since it is necessary to put two passages end-to-end), in which Torricelli demonstrates the equality between a parabolic arc and an arc of a spiral by employing his theory of points of different size and of the adequate occupation of the same point by different lines:

> Let there be a spiral of Archimedes ABC, with center A and AP its commencement, and take on this spiral any point C. The circular arc PVC having been traced, take a right angle PAF, with AF equal to half the arc PVC, and complete the rectangle PF. Trace the quadratic conic parabola AOG which has AF for diameter. I say that the spiral line ABC is equal to the parabolic line AOG.
>
> Take any point on the spiral or on the parabola, or rather on the straight line AP; let D be this point, from which one draws DO parallel to AF. At point O let the parabola have OI for tangent. Also describe the circular arc DEB starting from D, and the radius BA. And having determined the right angle BAH, draw the tangent BH to the spiral. And in addition let BT be tangent to the arc BED.
>
> The arcs CVP and LVP are to each other as the time, that is, as the radii CA and BA or PA and DA. Also, the arcs LVP and BED are to each other as PA to DA. Hence the ratio of the arcs CVP and BED is composed of the ratios of PA to DA and of PA to DA, that is, it is as the square of PA to the square of DA, or again as the straight line PG to the straight line DO, since we suppose the parabola to be quadratic.
>
> By exchange of terms, the ratio of the arc CVP to the straight line PG, that is, the ratio of 2 to 1, is also the ratio of the arc BED to the straight line DO. On the other hand, the arc BED is equal to the straight line AH [by the property of the subtangent to the spiral; see below, p. 208], by Archimedes (and we could prove it also). Therefore the straight line DO is half of the straight line AH.

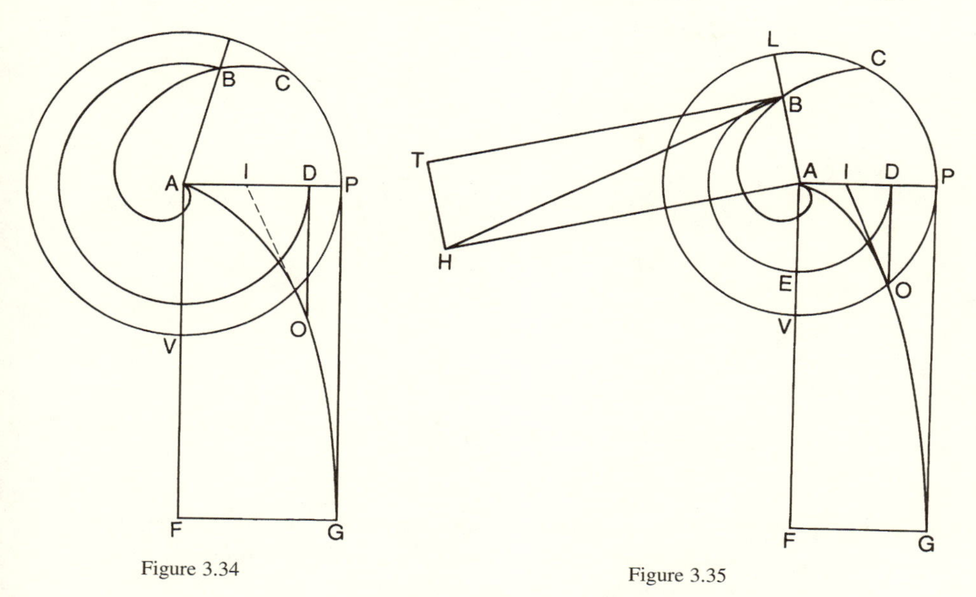

Figure 3.34

Figure 3.35

Now ID, being on a quadratic parabola, is half of AD [by the known property of the parabola's subtangent, which is equal to half of the abscissa], that is, the half of BA. Moreover, the angles at D and A are right. Therefore the triangles IDO and BAH have all their angles equal, and the angle at O is equal to the angle at H, hence equal to the angle TBH. (*TO*, vol. 1, pt. 2, 391–92; the text ends on the words: *ergo cum eadem inclinatione*, which recall the final reasoning of the other examples treated, like that concerning a more complicated spiral given by $r = kt^2$:)

Therefore the straight line DO cuts the parabola in O at the same inclination as that with which the arc DEB cuts the spiral in B. And that will always be true, whatever the point D. Therefore all and each of the points of the parabola AOG are equal in number and in magnitude, that is to say in species, to all the points of the spiral ABC. Consequently, the parabolic line AOG will be equal to the spiral line ABC. (*TO*, vol. 1, pt. 2, 388)

It is a matter of proving the equality in length between an arc of a spiral and the arc of a parabola that satisfies the condition: height AF of the parabola = ½ arc PVC associated with the arc of the spiral. This time, in contrast to what Cavalieri did, the two curves are traced on the same diagram: AP serves as radius of the spiral and as base of the parabola. We are to take a point on one and the other curve; as Torricelli explains: "or rather on the straight line AP; let D be this point"; in effect the distance traversed on AP is the fundamental variable, which could represent, if desired, time, and as a function of which will be determined a point O on the parabola and a point B on the spiral.

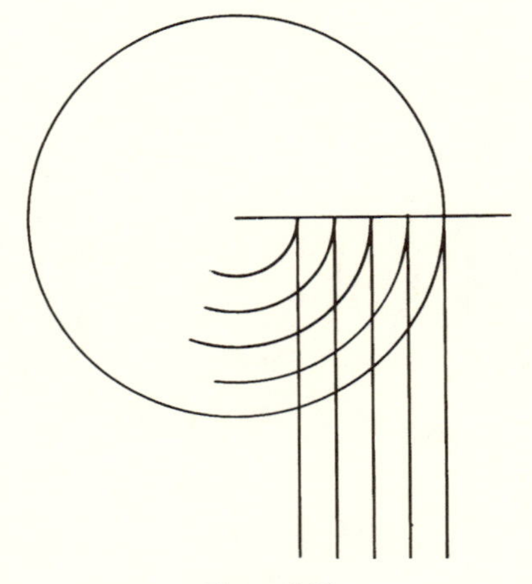

Figure 3.36

One aspect of the great originality of this reasoning consists in placing two frameworks into correspondence. The spiral is enclosed in a pattern of concentric circles, and the parabola in an orthogonal grid. The passage from the spiral in a concentric framework to the parabola in an orthogonal framework corresponds to what might be called a "setting upright" of the spiral. An observer should be able to "see" the spiral unwind to transform itself into a parabola and the circles of the circular framework becoming straight lines (this will be the point of departure for the "transmutations" of curves that James Gregory presents in his *Geometriae pars universalis*.

It is necessary to show that there are "as many" points on the two curves, and that these points, taken two-by-two, each have the same size. The placing into correspondence of the two curves point-by-point is guaranteed by the common dependence of the two frameworks in relation to the variation of the base: the horizontal displacement of the point D on AP, which determines the radius of the concentric circle and the abscissa of the parabola. Torricelli calls this displacement a "time"; it is this common time which governs the rotation of the radius and the distance of the point of the spiral from the center A, as also the distance between D and this same center. There is therefore a point of the parabola for each point of the spiral, and vice versa. But are they of the same "size"?

To demonstrate that the points are equal two-by-two, Torricelli returns to his initial intuition concerning the magnitude of points: it will be given by the variable inclination of the straight lines to the straight line of reference. Here he shows that each of the points such as O and B is the trace of lines of the same thickness that are equally inclined to the curve (or to its tangent). He starts from a common element of reference: the size of point D taken on the straight line AP. Everything else is determined in relation to it: from this point

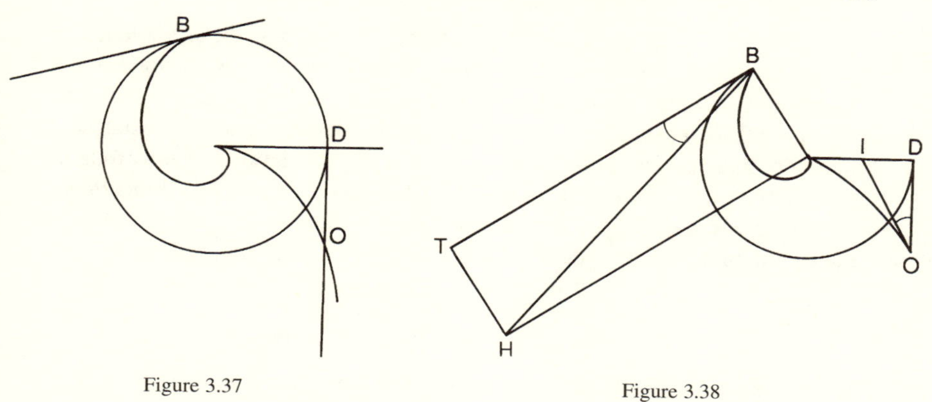

Figure 3.37

Figure 3.38

both the straight line DO and the circle DB depart at a right angle; the line OD has therefore the same thickness as the point D, as does the circle DB, or at least the tangent which replaces the circle in the neighborhood of the point B.

It will suffice to show that DO and the tangent to the circle at B meet the curves at the same angle. To this end it is necessary to replace the curves locally by their respective tangents. The final step consists in proving that the angle between the tangent to the curve and the straight line of reference is the same in both cases. Here these angles are TBH (angle between the tangent to the circle and the tangent of the spiral) and DOI (angle between the straight line of reference and the tangent to the parabola), of which Torricelli shows the equality by utilizing the properties of the tangents to the spiral and to the parabola. Thus the arc of the spiral and the arc of the parabola are composed of "points . . . equal in number and in magnitude" (*TO*, vol. 1, pt. 2, 388). The length of the two arcs is therefore the same.

A reader of the *Principia* has much to learn in these reasonings of Torricelli. To be sure, Newton was not able to read these pages themselves, for they remained unpublished until 1919. A little of their substance could nevertheless have reached him through the books and teachings of James Gregory and of Barrow, who had lived in Italy and very probably knew the Torricellian tradition, in particular through Stefano degli Angeli.

It is also true that the fundamental idea of this demonstration belongs specifically to Torricelli: only he, as far as is known, determined the breadth of a line by its inclination to a straight line of reference in which it "occupies adequately" a point.

But if one keeps in mind Theorem 4 of the *De motu*, which extends Kepler's third law from the circle to the ellipse, or if one makes comparisons with certain demonstrations of the *De methodis serierum et fluxionum* (for instance in connection with the cycloid [see below, pp. 213–14]) one can discern certain common traits in Torricelli's and Newton's reasoning: several curves or magnitudes are put into relation through a very clever system establishing relative location and successive dependence, and among these geometrical magnitudes, there is one that receives a privileged status—it is a displacement that serves as a fundamental variable and which is called time (to strike the imagination? to recall certain physical aspects of the situation?). And as a function of this basic variation, one can compare the variations of other magnitudes, such as occur if the time or the abscissa receives a very small increment (Newton calls these variations the "contemporary increases" or the "contemporary parts of the magnitudes," or their "momentaneous changes").

Indivisibles in the Public Domain

The two mathematical theories of indivisibles that have been presented, Cavalieri's and Torricelli's, each have their coherence, their fundamental presuppositions, and their procedures of calculation or demonstration. They are well defined and clearly distinct logical entities—although the second was born of the first.

But from around 1650, the methods of indivisibles were much less firmly characterized; philosophers and scientists thought themselves authorized to speak of these new objects in an extremely vague manner. Cavalieri was referred to, even invoked piously, but seldom read, and the consequences that were drawn from his doctrine and the usage made of it were often in direct contradiction with his texts.[7] Thus Wallis commences his book *De sectionibus conicis* with a very free presentation of indivisibles, which he believes to be in accord with Cavalieri:

> I suppose in beginning (agreeably to the *Geometria indivisibilibus* of Cavalieri) that any plane is so to speak composed of an infinity of parallel lines; or rather (as I for my part prefer), of an infinity of parallelograms of equal height; but of which a single one has for height $1/\infty$ of the height of the whole, that is, an infinitely small aliquot part of this height. (Wallis, *De sectionibus conicis*, 4)

If Cavalieri had in fact maintained the idea attributed to him here, certainly Wallis would have been right to prefer another conceptualization. The only ambiguous passage in the work of Cavalieri, often cited, is found near the beginning of the *Exercitationes* of 1647, where he appeals to an analogy: indivisibles contained in a magnitude are compared with threads in a fabric or pages in a book, but the difference is that the lines in the figure are indefinite in number and totally deprived of thickness (Cavalieri, *Exercitationes*, 4–5).

In truth Wallis, like many others, knew the work of Cavalieri only indirectly. He tells how he sought in vain for his *Geometria*:

> My method takes its origin from where Cavalieri's method of indivisibles concludes. Thus I thought . . . that I should name my method *Arithmetica infinitorum* as he had named his *Geometria indivisibilibus*. . . . At the beginning of the year

> 1650 I fell on the mathematical writings of Torricelli . . . where he expounds among other things the indivisibles of Cavalieri. As for Cavalieri himself, I have not had him in hand and I sought for him in vain several times at the booksellers. But his method, insofar as reported by Torricelli, was so much the more agreeable to me as . . . (Wallis, *Arithmetica infinitorum*, dedication)

Wallis thought himself to be expanding the work of Cavalieri, although he knew it only at second hand.

The indirect transmission of the theory can explain the great liberty during this era in the usage of indivisibles. The *Opera geometrica* of Torricelli, which had a large role in this diffusion, could not convey a rigorous idea of the modes of demonstration of Cavalieri. Authors as influential as Huygens and Barrow seem to have known Torricelli's book rather than Cavalieri's texts (see *HO*, 1: 132, and Barrow, *Lectiones*, 35).

Among the various degrees of indirect transmission, there is even the extreme case in which the author candidly admits that he has only heard of the theory and that it must resemble this or that:

> New Method Called the
> Geometry of Indivisibles
>
> Although the geometers agree that the line is not composed of points, nor the surface of lines, nor the solid of surfaces, nevertheless we find that in recent times there is an art of demonstrating an infinity of things, by considering surfaces as if they were composed of lines, and solids of surfaces.
>
> I have seen nothing of what has been written of this: but here is what comes to my mind regarding it (I restrict myself here to what concerns surfaces).
>
> The foundation of this new geometry is to take for the area of a surface the sum of the lines that fill it; so that two surfaces are judged equal when the one and the other are filled by an equal sum of equal lines; whether each of those of one sum is equal to each of those of the other sum; or whether there is a compensation, so that for example two of one sum which can be unequal to each other, are equal to two taken together of the other sum which are equal to each other.
>
> But in order not to give place to too much paralogism, in which one falls easily when using this method, if one is not careful, it is necessary to remark,
>
> 1. That in order that the lines be considered to fill a space, they must all be parallel to each other; whether they be straight to fill a rectilinear space, or be circular to fill circles or portions of a circle. It is easy to see the reason for this. . . .
>
> 2. In order that a sum of lines be considered equal to another sum of lines, it is not necessary to imagine that it is possible to say the number of them that each space contains (for there is no space so small that it does not contain an infinite number of them), but these sums are called equal because all the lines in the one sum and in the other cut perpendicularly two equal lines. (Antoine Arnauld, *Nouveaux éléments de géométrie,* [Paris, 1667], bk. 15, 306–7)

Cavalieri had never claimed that a planar space could be "filled by a sum of lines"; he did not compare sums of indivisibles. But the subtle neutrality of his theory was no doubt difficult to preserve once the method had succeeded.

There was thus created a sort of common doctrine of indivisibles, vague and poorly founded, of which the principal justification lay in its results. Practice is the sole efficacious norm for avoiding paralogisms and for giving a correct meaning to improper expressions, as Pascal demands:

> I shall not raise difficulties over use of this expression "the sum of the ordinates," which seems not to be geometrical to those who do not understand the doctrine of indivisibles, and who imagine that it is a sin against geometry to express a plane by an indefinite number of lines; which comes only from their lack of understanding, since one is not to understand anything else by that than a sum of an indefinite number of rectangles made of each ordinate with each of the small equal portions of the diameter, of which the sum is certainly a plane, which does not differ from the space of the semicircle than by a quantity less than any given. (Pascal to Carcavy, Pascal, *Oeuvres*, 8:352)

Those who "understand the doctrine" comprehend what one is speaking of, they know that a line is a rectangle and that by "the sum of the ordinates" it is necessary to understand a sum of products into which enter very small breadths (these are "homogeneous" and not "heterogeneous," in Pascal's terminology, probably taken from Tacquet, ibid.). This doctrine is no longer that of Cavalieri. It could be that of Torricelli, but deprived of the conceptual apparatus that gives Torricelli's its basis and solidity.

It is comprehensible that Newton could not be satisfied with a theory so fluid, and that he will attempt, in the *Principia*, to replace indivisibles by a doctrine rigorously founded on ultimate ratios. One of the elements of his construction will be precisely that to which Pascal made allusion at the end of the paragraph just cited: the procedure that consists in approaching a given magnitude by other, more and more closely approximating magnitudes, so that the difference becomes smaller than any given magnitude. The "double reduction to the absurd" of Euclid and Archimedes had proceeded in such a way.

MOTION IN GEOMETRY: THE KINEMATICS OF CURVES AND THE METHOD OF FLUXIONS

Another aspect of the mathematics of this period concerning which Newton is less explicit in the *Principia*, although it is deeply impregnated with it, is the treatment of geometrical entities in terms of motion. In the reasonings of the *De motu*, motion and time are almost constantly at work: curves are trajectories traversed by moving bodies; on these curves, points approach each other; the curves themselves are bent, and so forth. The recourse to motion is especially decisive when it is a matter of infinitesimal entities: very small segments or figures are seen as "nascent" or "evanescent."

The concern of this book is with physics, and it should therefore be expected that geometry will be animated with motion. But this explanation fails to recognize a much more profound insight: Newton inherited an entire tradition of geometry by motion—a very active and fruitful tradition during the course of the seventeenth century.

The Spiral, the Mechanical Curves, and the Direction of Tangents

The ancients were not altogether ignorant of this manner of conceptualizing. The spiral of Archimedes served as a paradigm in the seventeenth century for this treatment of curves and of geometry. Galileo often made allusion to it; he saw in the spiral the model for the composition of motions treated with mathematical rigor. Before posing his own definition of motion "naturally accelerated," he envisaged other possible motions as cases abstractly conceivable:

> For although nothing is inappropriate about imagining arbitrarily some kind of transport, and contemplating the properties that follow from it (as those who conceived helical or conchoidal lines arising from certain motions, even though nature does not use such motions, have laudably demonstrated the properties [of the lines] starting from this initial supposition). . . (EN, 8:197)

Some pages earlier Galileo had taken up nearly verbatim the first proposition of the *Spirals*, which concerns the space traversed in a uniform motion (EN, 8:192–93). The treatise of Archimedes thus constitutes one of the rare points of anchorage for the kinematics of the modern epoch.

Archimedes defined the curve as follows:

> If a straight line is drawn on a plane and if after having rotated uniformly round one of its extremities which is held fixed, it returns whence it started, and during the rotation of this line a point is carried uniformly with respect to itself through the length of the line, beginning at the extremity that is held fixed, this point will describe a spiral in the plane. (Archimedes, *Spirals*, in *Opera*, 2:50–52)

The curve is not considered as being existent from all eternity and discovered by the mental eye of the contemplative mathematician; on the contrary, it is generated by the point that describes it by being displaced in two uniform motions that are compounded together.

This manner of generating lines by the composition of motions is not peculiar to spirals. The ancients knew of analogous lines useful for the solution of hopeless problems like the quadrature of the circle, duplication of the cube, and trisection of an angle. These are the curves called "mechanical." The problems thus resolved are resolved only in an approximative and imperfect manner; there is no veritable or rigorous solution, as to *plane problems* (problems soluble with straight-edge and compass) or *solid problems* (problems soluble with the aid of conic sections).

The quadratrix, for example, was intended for squaring the circle and for dividing an angle into any arbitrary number of parts. Pappus described it as follows (*Collection*, bk. 4, chaps. 30–32): while the radius of the circle moves uniformly, sweeping out the quadrant of the circle from AB to AD, the horizontal line BC descends uniformly towards AD, until it coincides with it. The two motions must be completed in the same time, and the intersections of the radius and the horizontal line, during the course of their displacement, form the points of the curve.

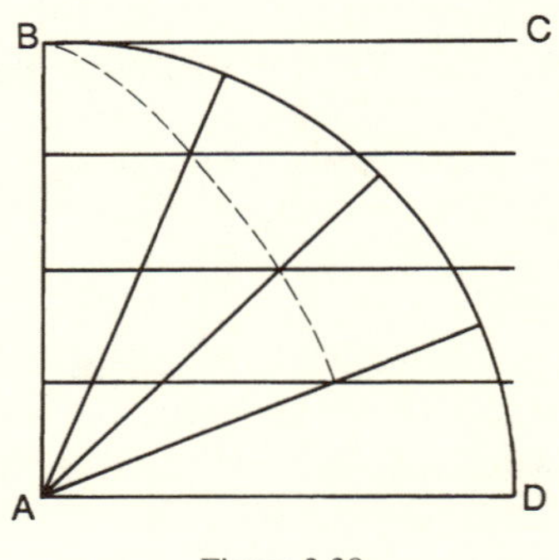

Figure 3.39

The seventeenth-century mathematicians were passionately interested in curves of this sort. (Viète had already studied the quadratrix; see *Opera*, 365–67). The stock of mechanical curves even came to be considerably enriched. Galileo, then Mersenne, invented the cycloid; Pascal imagined the "limaçon," and Roberval described the first sinusoidal (the "companion of the roulette") in the course of his study of the cycloid.

The conics themselves were studied as mechanical curves, that is, as generated by different motions. Van Schooten (*De organica*) and De Witt ("Elementa curvarum") devoted works to this procedure which they called "the organic description of curves" because it presupposes "instruments" (*organa*).

The "mechanical" description of the parabola, by composition of motions, is illustrated perfectly by Galileo's demonstration of the parabolic trajectory of projectiles (Fourth Day, EN, 8:272–73). Suppose a heavy body leaves its support with a horizontal motion; it will then be subject to gravity and therefore animated by a second motion, which will be vertical and accelerated. While the

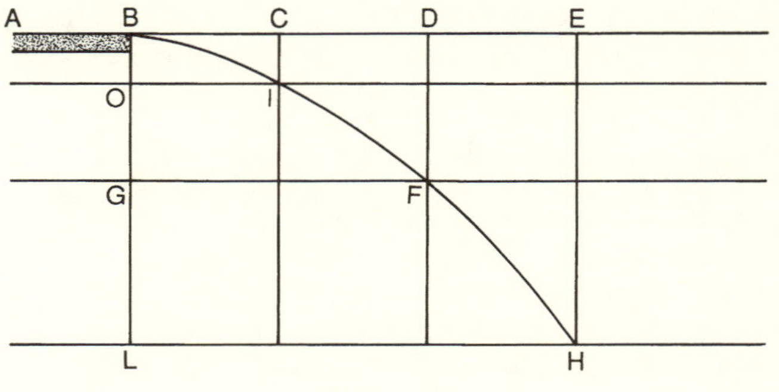

Figure 3.40

body moves uniformly from left to right and traverses a horizontal length proportional to the time, it traverses downward a distance proportional to the square of the time. The parabola is the result of the composition of these two motions because its points satisfy the two equations:

$$x = kt \quad \text{and} \quad y = Kt^2.$$

It is defined as a function of the time taken as common parameter. The uniform horizontal displacement serves as reference because it represents and measures the time: the line BE is "as the flow of time or its measure" [*tamquam temporis effluxus sive mensura*] (EN, 8:272).

This "demonstration" of the trajectory of projectiles is not a proposition of experimental physics. It is not a matter of confirming by measurements and approximations that projectiles follow a curve of such and such a shape. (Discussions continued long after Galileo as to the degree to which physical projectiles follow or depart from this trajectory.) Galileo compounded two abstract motions, perfectly defined and according to rule, and showed that the result agreed with the mathematical properties of the parabola. But neither is this simply pure mathematics: Galileo's demonstration presupposes certain physical theses, in particular the idea that different motions can be compounded in the same moving body without destroying or hindering each other. (Cf. the third hypothesis of Huygens in the *Horologium,* see above, p. 115.)

This geometry of motion occupied the ridgeline between the mathematical slope and the physical slope: Galileo borrowed from the mathematics of the ancients the theorems on the parabola to apply them to the motions of projectiles. Inversely, the consideration of trajectories was of service to Galileo's disciples, who continued the process of reciprocal fecundation. Thus Torricelli imagines new projectiles, unknown and impossible, simply in order to describe kinematically more complex curves:

> Let there be a body pushed horizontally on a plane AF, and then coming to fall, in such fashion that it possesses two impetuses [*impetus*], the one uniform and horizontal in the direction of FC, the other descending and accelerated in quadratic proportion; I say that a cubic parabola will be produced. (*TO,* vol. 1, pt. 2, 310)

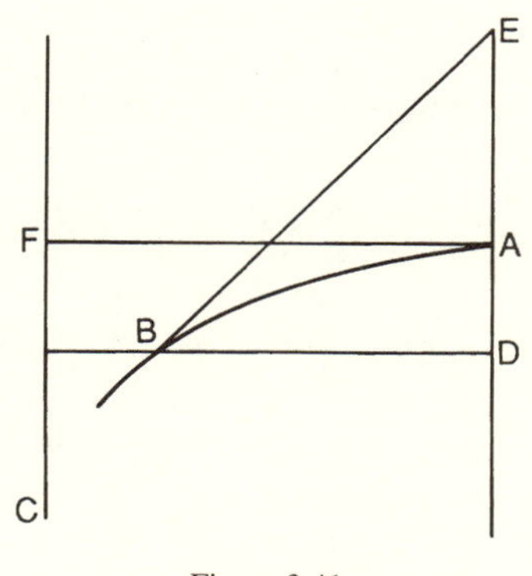

Figure 3.41

The exchange is made this time in the other direction, no longer from geometry towards physics, but oppositely. Torricelli creates new mathematical entities simply by generalizing Galileo's results on projectiles.

The consideration of motion is not simply

an aid to the imagination, a scaffolding that quickly becomes useless. It is by supposing that his curve is really described by a projectile that Torricelli finds an elegant and rapid means for determining tangents. Here is how he proceeds in the case of the "cubic parabola" defined above:

> Take ED equal to the length DA multiplied by the exponent of the parabola, hence three times DA in the present case, and the line that joins E to B will be the tangent.
>
> In fact the mobile point B which describes the parabola possesses two impetuses when it is in position B:
>
> A horizontal impetus directed along the tangent AF
> A perpendicular impetus along the diameter AD
>
> We seek the ratio of these two impetus in the following way: the horizontal impetus, during the time of the fall, has caused the space DB to be traversed, while on its side the perpendicular impetus, according to what has been said, would cause to be traversed during the duration of the fall, if it were always kept constant, a space triple of the fall AD. Consequently, the motion or the direction of the point B, which is composed of two velocities which are to one another as BD to DE, will be made along the line BE. (*TO*, vol. 1, pt. 2, 311)

In explaining this procedure for drawing tangents, Torricelli takes as admitted certain physical properties of motion: that the tangent has the instantaneous direction of motion of the moving point and that this direction can be determined by constructing the parallelogram of velocities.

He gives more detail on this question in other texts, particularly in his *Opera geometrica* of 1644 (*TO*, 2:122–23) and in a manuscript:

> We require that, if a body is moved with a certain impulse along any curved line, and is liberated from the connection that obliges it to follow this curve, it pursue its motion along a straight line, there being no new cause to incurve its path towards any side.
>
> We require also that the straight line in question be tangent to the curved line at the point in which the body will have been liberated from the preceding curvature.
>
> The truth of these requirements has already been proved in the penetrating discourses of S. Galileo in his other works. We restrict ourselves to illustrating it here.
>
> Let us imagine a narrow canal dug into a horizontal plane, the outline of it being circular or parabolic or spiral, etc. If a ball of metal that is perfectly smooth is pushed by some impulsion in this narrow canal, it will traverse and obey point by point the curvature of its borders, at least insofar as they are deep enough. But from the point where the canal ends, when the ball finds itself free on the horizontal plane, having forgotten its preceding path, it will pursue with its impetus a path

> that is no longer circular or spiral, but along a straight line. And it is precisely this straight line that will be tangent to the curve of the narrow canal, at the point where the body is liberated from its curvature. (*TO*, vol. 1, pt. 2, 377; *Opere* [1975], 489)

In the *Discorsi*, Galileo had furnished more than the elements for this evaluation of the impetus of a projectile. In the Fourth Day he showed how to determine the "impetus" of a cannonball at different points of its parabolic trajectory. Yet the construction proposed by Galileo (Prop. 4 of the Fourth Day) only makes it possible to represent the impetus on an annexed diagram drawn at the vertex of the parabola. Thus its intensity can be determined but not its direction (Galileo sought to know in particular the "force" of projection of the cannonball, or its "force" of projection on arrival). Torricelli, on the contrary, constructs the impetus as a line associated with the point where the projectile is; he thus represents directly the tangential impetus at each point.

At about the same time, during the 1640s, Roberval taught an identical method for drawing tangents. His treatise, *Observations sur la composition des mouvements et le moyen de trouver les touchantes aux lignes courbes*, was printed much later (1693). His method is based on the following principles:

> *Axiom or Principle of Invention.* The direction of motion of a point that describes a curved line is the tangent of the curved line in each position of this point. (Roberval, "Divers ouvrages," 24)

> *General Rule.* By means of the specific properties of the curved line (which you will be given), examine the several motions that the point describing it has at the place where you wish to draw the tangent: from all these motions compounded into a single one, draw the line of the compounded motion, and you will have the tangent of the curved line. (Ibid., 25)

Roberval applies this method successively to thirteen different lines (conics, various sorts of conchoids, limaçon, spiral, quadratrix, cissoid, roulette (that is, cycloid), and "companion of the roulette"). To study the parabola, he has recourse to the definition by means of focus and directrix: every point on the locus is equally distant from the focus F and the straight line (D). The point M therefore has two motions, that of the segment HM and that of FM. And when M is displaced along the parabola, HM and FM increase equally. It can there-

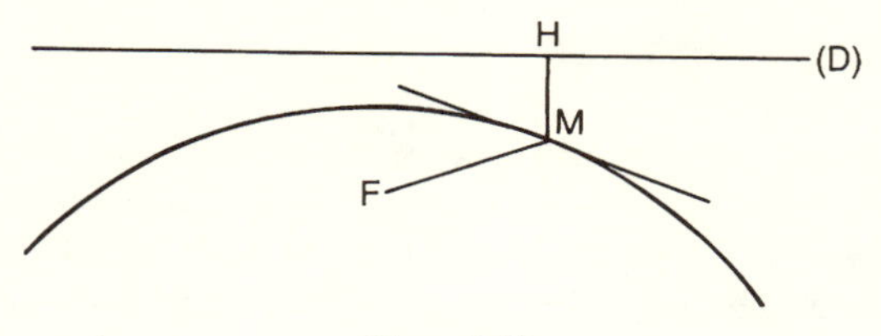

Figure 3.42

fore be considered "that the motion of the point describing the parabola is composed of two equal rectilinear motions" (ibid., 26). Consequently the bisector of the angle FMH is also the tangent in M.

A complete justification of the procedure surpasses the mathematical methods available to Roberval. With what right has he neglected the rotation of FM and the lateral displacement of HM? Roberval declares simply that "it was easier" to proceed as he had done (ibid., 26).

The tangent to the spiral is determined in an analogous way, without detour or calculation. The running point of the spiral is animated with a double motion: uniform rotation of the radius around the origin and uniform advancement along this radius. To construct the tangent direction, it is necessary to determine two segments representing the paths traversed during the same time in these two motions. For the motion of translation, one can take as representative the segment that the point would traverse on the radius during a complete turn, and that representation will be valid for any position of the point since this motion is uniform. But how to measure the circular motion, since the point will have a rotation more rapid in the degree that it is farther from the origin? It suffices to consider in each case the circle which has for a radius the distance between the point and the center of the spiral:

> In B the motion is such that if it had always had an equal circular motion from A to B, it would have described a circumference of which AB is the radius during the time of one revolution. (Ibid., 52)

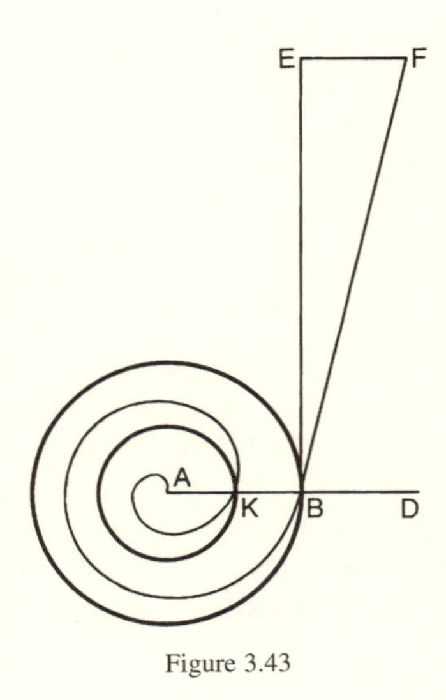

Figure 3.43

Consequently, if the segment AK or KB serves to measure the rectilinear displacement, the circumference centered in A and with radius AB will measure the circular displacement characteristic of the point proposed. To determine the tangent, it then suffices to construct the parallelogram of the velocities with these two components: draw at B a perpendicular segment of length equal to the circumference,[8] and at the end of this segment, in E, a segment EF equal to the segment AK and parallel to the radius AB. The tangent sought will be BF.

It is this property that Torricelli supposed in the reasoning whereby he compared the spiral and the parabola: BE is equal to the circumference of the circle with radius AB, if BF is the tangent and EF is equal to AK (see above, pp. 196–97). The young Italian demonstrated it in other texts exactly in the manner of Roberval, and his stroke of inspiration astonished Galileo (*TO,* 3:55, 60; cf. *TO,* vol. 1, pt. 2, 377–80).

A certain skill is needed to apply these procedures, as shown in the example of the parabola. The case of the quadratrix is especially confusing, and it would be wrong in that instance to attempt to choose generating motions. Those who undertake to employ these techniques without discernment, without a certain cleverness, risk falling into error. That even happened to a beginner who did not lack such cleverness—Isaac Newton (in a manuscript of 1665 [*NMP,* 1:378–80]).

The drawing of tangents is only one aspect of this geometry of motion. To have a more exact idea of the richness of this current of studies, it would be necessary, for instance, to examine a treatise published some years before the *Principia,* the *Geometrical Lectures* (*Lectiones geometricae*) of Barrow.[9] The work is entirely devoted to "the generation of magnitudes" by various sorts of "local motion" (*Lectiones*, 1). The method of tangents, which has already been mentioned, is introduced in Lecture 4, after Barrow has presented all sorts of rotations and translations along a straight line, along a curve, parallel to another straight line, and so forth, so as to produce curves, surfaces, and volumes. The first curves generated by motions (Lecture 3) are the parabola—in the manner of Galileo and Torricelli—then the spiral, the quadratrix, and the other conics.

Fluxions and the Role of Time

It is to this context that the Newtonian "fluxions" belong.[10] Curves and surfaces are conceived as generated by various motions, each of which has its velocity or "fluxion."

> I consider quantities as if they are generated by a continuous augmentation, in the manner of the distance that a moving body describes in its course. (*The Method of Fluxions*,[11] *NMP*, 3:72)

A geometrical figure is a sort of mechanism in which motion is transmitted according to the articulations of various elements. The schema is in general the following, with some variations: on a base, namely a horizontal straight line (here AB), a point B is displaced starting from an origin A. This moving point B carries with it, erected perpendicularly (or obliquely), an "ordinate" BD which varies according to the positions of B. The route of the point D is the

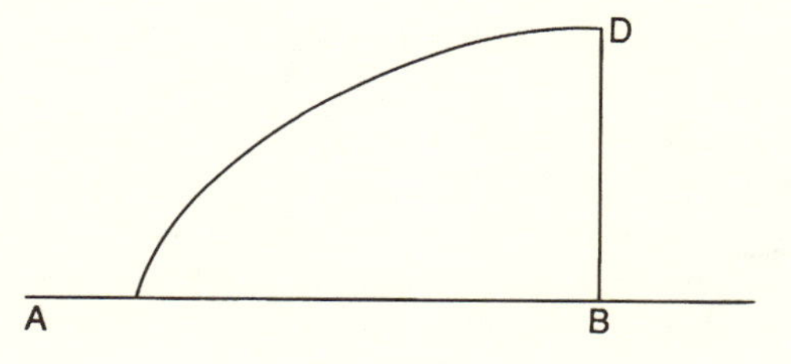

Figure 3.44

curve considered. The vertical displacement of D on BD depends on the horizontal displacement of B according to an equation, either simple or complicated, which defines the curve, or according to other relations of a nonalgebraic nature. The displacement of B, that of D, and the increase in the surface ABD, are "fluents" or "magnitudes," each of which has its "fluxion," that is to say, its rate of increase (*fluxio, velocitas, fluendi ratio*).

But this schematic presentation would mislead if it gave the impression that for Newton curves are defined by a system of orthogonal, rectilinear, "Cartesian" axes. Many other modes of reference are possible.

The questions relative to curves can, Newton declares, be reduced to two fundamental problems: to pass from fluents to fluxions and vice versa:

> All the difficulties of this sort can be reduced solely to these two problems, which can be proposed concerning the space described in a local motion accelerated or retarded in any manner:
>
> 1. The length of space being given continually (or at every moment [τὸ νῦν][12]), to find the velocity of motion at the time proposed
> 2. the velocity of motion being continually given, to find the length of space described at the time proposed (*NMP*, 3:70)

At each point of time (τὸ νῦν, the "now" of Aristotle), two quantities are associated—a distance traversed and a velocity.

Mathematically these two questions translate into Problems 1 and 2 of *The Method of Fluxions*:

> Given the relation of the fluent quantities to one another, to determine the relation between the fluxions. (*NMP*, 3:74)

> An equation involving the fluxions of the quantities having been set out, to find the relation of the quantities to each other. (*NMP*, 3:82)

A dissymmetry will be noticed in the formulation of these two problems: is it a matter of any relation whatever between two quantities, or of a relation put in the form of an equation?[13] Further on it will be seen that the terms "fluxion" and "method of fluxions" must be understood in all their generality without restriction to the study of quantities represented by analytico-algebraic expressions.

Newton gives rules for the cases in which the relation is expressed by an equation. For example, if the relation between the fluents is: $x^3 - axx + axy - y^3 = 0$, the relation of the fluxions will be: $3mxx - 2amx + amy - 3nyy + anx = 0$, with m and n designating the fluxions of x and y.[14] (In later texts, Newton will adopt his notation by "prickt letters," which is much more convenient: the fluxion of x will be $\dot{x}$, that of y will be $\dot{y}$, etc.; and this permits writing fluxions of fluxions, $\ddot{x}$, etc.)

These rules are justified some pages further on by a

> *Demonstration.* The moments of the fluent quantities (that is, their indefinitely small parts, by the addition of which they increase during each of the indefinitely small spaces of time) are as their velocities of flow. Consequently if the moment of any quantity such as x, is designated by the product of its velocity m and an infinitely small quantity o (that is to say, by mo), then the moments of the other quantities v, y, z will be designated by lo, no, ro, since lo, mo, no and ro are to each other as l, m, n, r.
>
> Now as the moments of the fluent quantities (such as x and y; these moments are mo and no) are infinitely small additions by which those quantities are augmented during each of the infinitely small intervals of time, it follows that these quantities x and y after any infinitely small interval of time will become $x + mo$ and $y + no$. And consequently the equation that designates the relation of the fluent quantities for all times indifferently, will equally designate the relation between $x + mo$ and $y + no$, and between x and y; so that $x + mo$ and $y + no$ can be substituted for these quantities x and y in the said equation.
>
> And thus let any equation $x^3 - axx + axy - y^3 = 0$ be given, and substitute $x + mo$ for x and $y + no$ for y, and there will emerge
>
> $$\left.\begin{cases} x^3 + 3moxx + 3mmoox + m^3o^3 \\ \quad - axx - 2amox - ammoo \\ \quad + axy + amoy + anox + amnoo \\ \quad - y^3 - 3noyy - 3nnooy - n^3o^3 \end{cases}\right\} = 0$$
>
> Now by hypothesis we have $x^3 - axx + axy - y^3 = 0$; when these terms are deleted and the remaining terms are divided by o, there will remain
>
> $$\begin{gathered} 3mxx + 3mmox + m^3oo - 2amx - ammo \\ + amy + anx + amno - 3nyy - 3nnoy - n^3oo = 0. \end{gathered}$$
>
> And moreover as o is supposed infinitely small, so that it can designate the momenta of the quantities, the terms that are multiplied by it will be as nothing with respect to the others [*respectu caeterorum nihil valebunt*]. I therefore reject them and there remains $3mxx - 2amx + amy + anx - 3nyy = 0$, as above in Example 1. (*NMP*, 3:78–80)

The infinitesimal element is denoted by o (a small, slanted "o," typographically distinct from zero and already employed by J. Gregory). It is an infinitely small duration, which renders possible the increments of all the magnitudes considered, for the increment of a magnitude is obtained by multiplying its own velocity by the element of time o. It is this "particle of time," as Newton calls it elsewhere, that furnishes the whole impulsion. As in the reasonings of the *Principia* or of the *De motu*, it suffices to animate the figure or the algebraic relation by the introduction of an atomic time, to put them in motion, and to determine the infinitely small "contemporary increments."

The primordial role here belongs to time. All the magnitudes are functions of time. In this sense the relation between the fluxion of a quantity and the fluxion of the base cannot be exactly identified with a simple derivative, because the displacement of the mobile point on the base is itself a function of time and itself has its fluxion with respect to time. The quantity x is not an independent variable; it is a fluent in the same way as the others, whose fluxion will be, let us say, m, and its increment $m \cdot o$. By these calculations are obtained relations between various velocities of increase. In some cases it will be useful to consider the displacement x of the point on the base in turn as a function of another displacement on another base. Newton makes use of this kinematic transformation to calculate certain recalcitrant integrals by reducing them to simpler integrals.

Although no variable is identified as independent in the figure or so designated in the writing, in practice Newton approaches the use of an independent variable. Most frequently he sets the fluxion of the base equal to 1:

> Since moreover one has the liberty of assigning any velocity to the fluxion m of the base (to this fluxion it will be convenient to relate all the others as a uniform fluxion), let us set it equal to 1. (*NMP*, 3:154–56)

The motion of one of the points, or the increase of one of the magnitudes, is by convention considered as uniform, and all the other motions are related to it. The fluxion of this magnitude is 1, and consequently its increment is $mo = 1 \cdot o = o$, that is, the infinitesimal element becomes an increment of this magnitude. From the point of view of mathematical formalism, this convention comes down to making this magnitude (often designated by x) the independent variable.

But this choice of a motion of reference has justifications that go beyond preoccupations with mathematical technique:

> Because we possess no estimation of the time except insofar as it is represented and measured by the intermediary of a uniform local motion, and furthermore because quantities, and their velocities of increase or decrease, can be compared with one another only if they are of the same kind, for this reason I shall not in what follows have any regard for the time taken formally [*ad tempus formaliter spectatum*], but among the quantities proposed which are of the same kind I shall suppose one of them to increase according to a uniform fluxion, and I shall relate all the other quantities to this as if it were the time itself, so that the name of time can rightly be attributed to it by analogy. (*NMP*, 3:72)

Because the time cannot be represented directly, one of the variations will take its place. It suffices to choose a variable to represent time; this variable will be the independent variable. This decision can be linked to the discussion in the *Principia* on absolute time and clocks: time cannot, so to speak, appear in person; it is necessary to represent it by clocks, which are all imperfect and need to be corrected (in particular the clock that is the sun needs to be corrected by the "equation" of the astronomers; see *Princ.*, 7–8). Similarly in mathematics, there must be by convention a kind of clock, that is, a variation that will take the place of time.[15]

Fluxions and "Calculus"

The theory of fluxions as thus presented has as its foundations the notion of velocity and a unique infinitesimal, which is an interval of time. Each magnitude has its velocity of being generated, which is variable from instant to instant, and its infinitesimal increase is given by the product of this velocity by the element of time. (Later, in the *De quadratura*, the element of time will itself be considered as "diminishing infinitely" [*NMP*, 7:64] in accordance with a point of view that is very close to that of the *Principia*.)

The rules and the demonstration that have been summarized possess the character of an algebraic calculus. By the letters x, y are understood any magnitudes whatever—provided, as Newton made clear, that these magnitudes are described by a continuous generation. In other texts of the seventeenth century, some of them earlier, there are analogous rules; for example, when Fermat, in order to find a maximum or minimum, "adequates" the terms A and A + E, then suppresses the expressions in which the increment E figures (Fermat, *Oeuvres*, 1:133–34 [Latin]; 3:121 [French]),[16] or when Barrow determines the tangent to a curve by considering an infinitely small arc with abscissa a and ordinate e, then rejects in the equation obtained the terms that contain ae or aa or ee as factors (*Lectiones*, 80–81).

But Newton's applications or illustrations of the method have sometimes nothing algebraic or analytic about them. The velocities of increase of all sorts of geometric magnitudes can be considered and compared without recourse to equations that would express their relations. Here, for example, is how Newton reasons to determine the area of the cycloid, in *The Method of Fluxions*:

> Thus the cycloid ADF being proposed, I refer it to the base AB, and the parallelogram ABDG being completed, I seek the surface of the complement ADG, by conceiving that it is described by the motion of the straight line DG; consequently its fluxion is equal to GD multiplied by the velocity of its motion forward, that is to say, $x \cdot l$.
>
> Yet as AL is parallel to the tangent DT, AB will be to BL as the fluxion of AB to the fluxion of the ordinate BD, that is, as 1 to l. Whence l = BL/AB, and therefore x · l = BL, and consequently the area ADG will be described by the fluxion BL. And therefore, since the circular area ALB is described by the same fluxion, they will be equal. (*NMP*, 3:204)

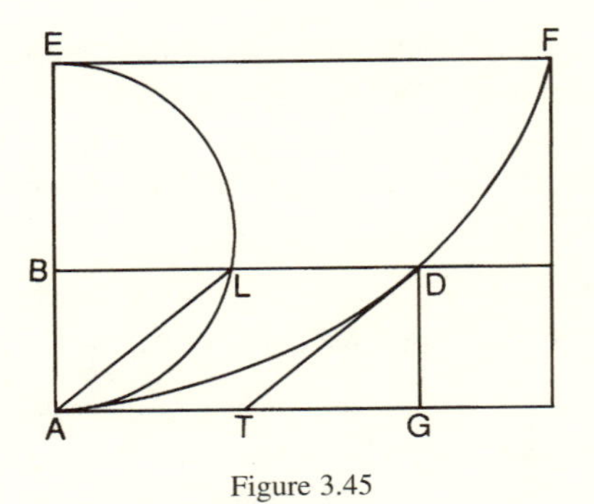

Figure 3.45

In this problem it is a vertical displacement that is chosen as the motion of reference. The point B, by rising towards E with constant velocity (the fluxion of AB is equal to 1), entrains with it all the elements of the figure and thus causes the cycloid to be generated. The objective is to evaluate the area of the complementary figure ADG under the cycloid. This surface is generated by the motion of GD, that is, that segment's parallel displacement toward the right and its elongation upward. The increase of this surface is therefore as GD × (velocity of G), that is (denoting this fluxion by the letter l, and GD by x), $x \cdot l$.

The key to the reasoning consists in the equality of the increments of the complementary surface AGD and of the portion of the circle ALB. BL generates a certain surface as it rises, while GD generates another as it advances. These two increments are always equal. Newton justifies this equality by means of the constant parallelism between AL and TD, a tangent to the cycloid. Thus, BL · (velocity of B) = GD · (velocity of G). And finally, since the two areas are described with the same fluxion starting from a common instant in which they were both zero, they are equal. From this can be deduced the result already proved by the preceding generation of scientists (see *TO*, vol. 1, pt. 1, 163–69), namely, that the cycloid has an area three times greater than the generating circle.

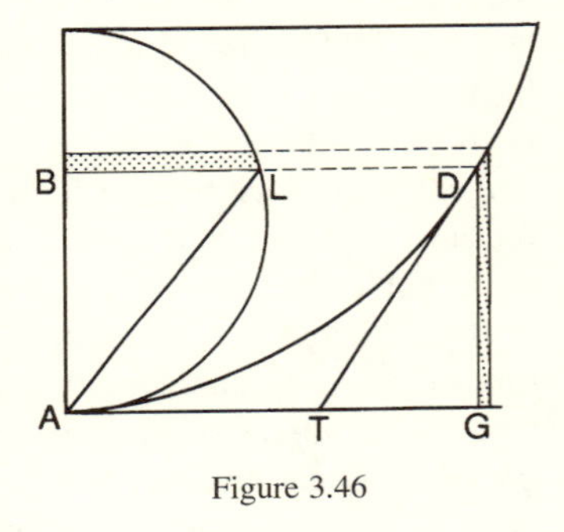

Figure 3.46

In studying this reasoning more closely it can be asked, why does the parallelism between DT and AL permit the conclusion

$$\text{fluxion of AB : fluxion of BD = AB : BL?}$$

Newton does not specify the intermediary steps, but it is very probable that he is relying on the relation that he had enunciated with respect to tangents in Problem 4 (*NMP*, 3:122): the subtangent is to the ordinate as the fluxion of the base to the fluxion of the ordinate.[17] In the present case, that relation is expressed by:

fluxion of AG : fluxion of DG = TG : DG;

because of the parallelism between DT and LA, the triangles ABL and DGT are equal, and therefore TG : DG = BL : AB.

The algebraic rules given for the solution of Problems 1 and 2 are not useful here. The question, however, is the same as that of Problem 2: passing from a relation between velocities of increase to a relation between the magnitudes themselves. But the intermediary steps in the reasoning are here essentially geometrical, while in the rules, examples, and "demonstrations" of Problems 1 and 2, they are algebraic or analytico-algebraic.

Here, a delicate point important for the purposes of the discussion should be mentioned. The modes of reasoning in the *Principia* will later be described to show how they differ from an "infinitesimal calculus." But the "method of fluxions" is not itself, strictly speaking, a calculus. This present account has been attentive to certain nuances of presentation: the curves are not necessarily related to orthogonal, rectilinear axes; the relations between fluents (or between fluxions) are not always expressed by equations; finally, the passage from fluents to fluxions or inversely is not solely based on the algebraic rules given for solving Problems 1 and 2.

The very title given here to the work—it is a traditional abbreviation of a more complete title (*De methodis serierum et fluxionum*)—risks falsifying the perspective. To call it *The Method of Fluxions* unduly unifies an extremely varied and composite work. It is not known with certainty what title Newton wished to give it (see *NMP*, 3:33 n. 3). The book begins with a long calculational preamble on the development in infinite series of quotients, radicals, and solutions of equations. These "ways of calculating" (*modi computandi*, *NMP*, 3:70) are tools of which Newton makes use in the rest of the book, when there is need, and are not organically related to the notion of fluxion. Thus, using the title *The Method of Infinite Series and Fluxions* is more faithful to the order of the work and to its diversity.

There are certainly in *The Method of Infinite Series and Fluxions* the essential elements of an "infinitesimal calculus," but one finds there many other things that fall outside this category, a very rich panoply of procedures and a large variety of examples. If the preamble, which is exclusively devoted to series (*NMP*, 3:32–71), is not considered, the unity of the work lies in the notion of the velocity of increase of a magnitude. But this notion is independent of the algorithms or algebraic manipulations that Newton proposes.

The contrast with Leibnizian presentation is striking. In the first exposition that he published of his new method,[18] Leibniz already gave to his procedures the form of an algorithm in which the new entities were introduced "blind" and defined solely by the rules with which they were to be manipulated. After some initial explanations concerning the names of the magnitudes represented in a diagram, Leibniz continues:

> Now, let a certain straight line taken arbitrarily be called dx, and let the straight line which will be to dx as . . . be called dv or the difference of the v. . . . These things being posed, the rules of the calculus will be the following:
>
> Let a be a constant quantity, da will be equal to 0, and $d(ax)$ will be equal to adx. If y is equal to v, . . . dy will be equal to dv. For *Addition* and *Subtraction*: if $z - y + w + x$ is equal to v, then $d(z - y+w+x)$, that is, dv, will be equal to $dz - dy + dw + dx$. *Multiplication*: $d(xv)$ is equal to $xdv + vdx$, that is, in positing y equal to xv, dy will become equal to $xdv + vdx$. For one is free to use a formula, such as xv, or a letter abbreviating for it, such as y. Note that x and dx are to be treated in the same manner in the calculus as y and dy, or other indeterminate letter with its differential. Note also that there is not always a regression from a differential equation, unless with certain precautions, of which we shall speak elsewhere. Then *Division*: $d(v/y)$ or (having set z equal to v/y) dz is equal to
>
> $$\frac{\pm\, vdy \mp ydv}{yy}.$$
>
> (Leibniz, "Nova methodus"; *LMS*, 6:220)

There follow other rules concerning signs, powers, and radicals, and then Leibniz proposes an application:

> The algorithm, as one may term it, of this calculus (which I call differential) once being known, it will be possible to find all the other differential equations by the ordinary calculus. (Leibniz, "Nova methodus"; *LMS*, 5:222)

Leibniz was fully conscious of his originality on this point, saying of himself:

> In truth no one before Leibniz has had the idea of constituting out of this new calculus an algorithm, whereby the imagination would be freed from the perpetual attention to figures. (*LMS*, 5:393)

For Leibniz the very foundations of the new method are algorithmic. The new entities are introduced by the rules that govern their manipulation. The reasoning on differences is not based on geometrical considerations nor on inductions concerning numbers. The new calculus is, on the contrary, intended to give access to new objects:

> Mr. Leibniz having observed that there are problems and lines which are not of any determined degree, that is, there are problems of which the very degree is unknown or required, and lines of which a single one passes continually from one degree to another, this opening made him think of a new calculus, which appeared extraordinary, but which nature had reserved for these sorts of transcendental

problems, which surpass ordinary Algebra. This is what he calls the Analysis of the Infinites, which is entirely different from the Geometry of Indivisibles of Cavalieri, and from the Arithmetic of Infinites of M. Wallis. For this Geometry of Cavalieri, which is moreover very limited, is attached to figures, in which it seeks the sums of ordinates; and M. Wallis, in order to facilitate this investigation, gives us by induction the sums of certain rows of numbers: while the new analysis of infinites concerns neither figures nor numbers, but magnitudes in general, as does the ordinary *logistice speciosa* [algebra]. It demonstrates a new algorithm, that is to say a new fashion of adding, subtracting, multiplying, dividing, extracting, appropriate to incomparable quantities, that is, to those that are infinitely great or infinitely small in comparison with others. It employs both finite and infinite equations; and in the finite it introduces the unknowns in the exponent of the powers or in place of powers or roots; and it employs a new *affectio* of variable magnitudes, which is the variation itself, distinguished by certain characters, and which consists in the differences, or in the differences of differences of several degrees, to which the sums are reciprocal, as roots are to powers.

A part of the elements of this calculus, with several illustrations, was published in the Journal of Leipzig. ("De la chainette," *LMS*, 5:259)

It is an unknown universe that opens up in these foundational texts due to the boldness of a metaphysician, a speculative mind habituated to treat systems of signs and their arrangements in complete abstraction. Leibniz, to be sure, sometimes promised more than he was in possession of. But the foundations he posed were to be the point of departure for a recasting of mathematics, in the course of which the *Principia* itself would be reformulated in a new language by Varignon, Euler, and others. The differential calculus was to be the indispensable tool of dynamics in the eighteenth century.

Newton's procedures, by contrast, retain very close links with geometrical and kinematic intuition. The algebraic calculus of fluxions is only an aspect of a much broader ensemble of methods. For illustration, excerpts are taken from two significant texts at the extremes of Newton's career: first, the earliest and most private among the numerous expositions that Newton has left (a small treatise of 1666 beginning "To resolve problems by motion . . . "), then the latest and most public (the *De quadratura curvarum* written in 1691 and published only in 1704, in modified form).

The small treatise of 1666 opens with propositions of kinematics, among which appear, in the seventh and eighth places, the Problems 1 and 2 of *The Method of Fluxions*. Here are some passages of this text:

To resolve Problems by Motion these following Propositions are sufficient: . . .

3. All the points of a Body keeping Parallel to it selfe are in equall velocity.

4. If a body move onely (angularly/circularly) about some axis, ye velocity of its points are as their distances from that axis.

5. The motions of all bodys are either parallel or angular, or mixed of y^m both. . .

7. Haveing an Equation expressing ye relation twixt two or more lines x, y, z, &c.: described in ye same time by two or more moveing bodys A, B, C, &c.: the relation of their velocitys *p*, *q*, *r*, &c may bee thus found, viz: Set all ye termes on one side of ye Equation that they become equall to nothing. And first multiply each terme . . .

8. If two Bodys A & B, by their velocitys *p* and *q* describe ye lines x and y, & an Equation bee given expressing ye relation twixt one of ye lines x, & ye ratio *q*/*p* of their motions *q* & *p*; To find ye other line y . . .

18. Could this ever bee done all problems whatever might bee resolved. But by ye following rules it may bee very often done. (*NMP*, 1:400–403; Hall, 15–19)

Here, still more clearly than in *The Method of Fluxions*, the algebraic rules for the passage from fluents to fluxions, and inversely, are incorporated in a much larger framework, which concerns the generation of magnitudes by motion according to various geometrical situations.

The text published in 1704 situated the calculus of fluxions in a framework altogether comparable but a little more explicit as to the principles of the theory. Here is the preamble of the text (which did not appear in the first version of 1691):

I don't here consider Mathematical Quantities as composed of Parts *extreamly small*, but as *generated by a continual motion*. Lines are described, and by describing are generated, not by any apposition of Parts, but by a continual motion of Points. Surfaces are generated by the motion of Lines, Solids by the motion of Surfaces, Angles by the rotation of their Legs, Time by a continual flux, and so in the rest. These *Geneses* are founded upon Nature, and are every Day seen in the motion of Bodies.

And after this manner the Ancients by carrying moveable right Lines along immoveable ones in a Normal Position or Situation, have taught us the Geneses of Rectangles.

Therefore considering that Quantities, encreasing in equal times, and generated by this encreasing, are greater or less, according as their Velocity by which they encrease, and are generated, is greater or less; I endeavoured after a Method of determining the Quantities from the Velocities of their Motions or Increments, by which they are generated; and by calling the Velocities of the Motions, or of the Augments, by the Name of *Fluxions*, and the generated Quantities *Fluents*, I (in the Years 1665 and 1666) did, by degrees, light upon the Method of *Fluxions*, which I here make use of in the *Quadrature of Curves*.

Fluxions are very nearly as the Augments of the Fluents, generated in equal, but infinitely small parts of Time; and to speak exactly, are in the *Prime Ratio* of the nascent Augments: but they may be expounded by any Lines that are proportional to 'em.

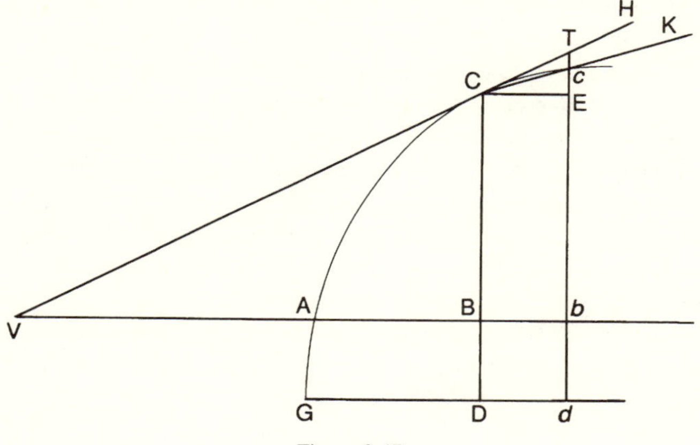

Figure 3.47

As if the *Areas* ABC, ABDG be described by the Ordinates BC, BD, moving with an uniform motion along the Base AB, the Fluxions of these *Areas* will be to one another as the describent Ordinates BC and BD, and may be expounded by those Ordinates; for those Ordinates are in the same Proportion as the Nascent Augments of the Areas.

Let the Ordinate BC move out of its place BC into any new one bcd: Compleat the Parallelogram BCEb, and let the Right Line VTH be drawn which may touch the Curve C and meet bc and BA produced in T and V; and then the just now generated Augments of the Abscissa AB, the Ordinate BC, and the Curve Line ACc, will be Bb, Ec, and Cc; and the Sides of the Triangle CET, are in the *Prime Ratio* of these Nascent Augments, and therefore the Fluxions of AB, BC and AC are as the Sides CE, ET and CT of the Triangle CET, and may be expounded by those Sides, or which is much at one, by the Sides of the Triangle VBC similar to it.

'Tis the same thing if the Fluxions be taken in the *ultimate Ratio* of the Evanescent Parts. Draw the Right Line Cc, and produce the same to K. Let the Ordinate bc return into its former place BC, and the Points C and c coming together, the Right Line CK coincides with the Tangent CH, and the Evanescent Triangle CEc in its ultimate form becomes similar to the Triangle CET, and its Evanescent Sides CE, Ec and Cc will be ultimately to one another as are CE, ET and CT the Sides of the other Triangle CET, and therefore the Fluxions of the Lines AB, BC and AC are in the same *Ratio*. If the Points C and c be at any small distance from one another, then will CK be at a small distance from the Tangent CH. As soon as the Right Line CK coincides with the Tangent CH, and the ultimate Ratio's of the Lines CE, Ec and Cd be found, the Points C and c ought to come together and exactly to coincide. For Errours, tho' never so small, are not to be neglected in Mathematicks.

By the same way of arguing, if a Circle described on the Centre B with the

Radius BC, be drawn with an uniform motion along the Abscissa AB, and at Right Angles to it, the Fluxion of the generated Solid ABC will be as the generating Circle, and the Fluxion of its Surface will be as the Perimeter of that Circle and the Fluxion of the Curve Line AC conjointly. For in what time the Solid ABC is generated by drawing the Circle along the Abscissa AB, in the same time its Surface is generated by drawing the Perimeter of that Circle along the Curve AC.

Of this Method take the following Examples.

Let the Right Line PB revolving about the given Pole P cut the Right Line AB given in Position; the Proportions of the Fluxions of the Right Line AB and PB is required.

Let the Right Line PB go out of its place PB into a new one Pb: In the Line Pb take PC equal to PB, and draw PD to AB so that the Angle bPD may be equal to the Angle bBC; and then from the Similarity of the Triangles bBC, bPD, the Augment Bb, will be to the Augment Cb as Pb is to Db.

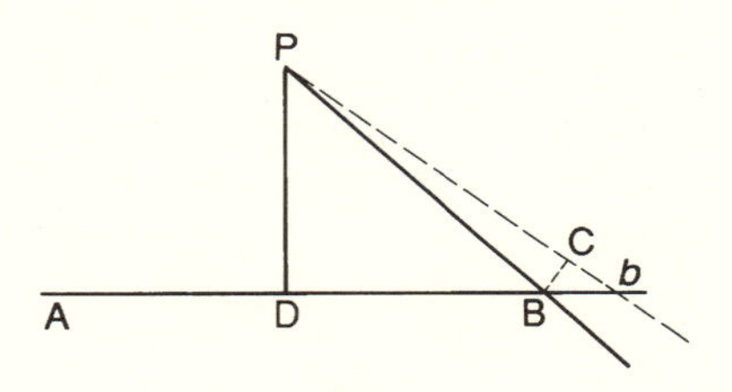

Figure 3.48

Now let Pb return into its former place PB, that those Augments may vanish, and the ultimate Ratio of the Evanescent Augments, that is, the ultimate Ratio of Pb to Db will be the same as that of PB to DB, the Angle being right; and therefore the Fluxion of AB is to the Fluxion of PB in this Ratio. (Newton, *Mathematical Works*, 1:141–42; *NMP*, 8:122–26)

After this example of rotation there follows another, analogous but more complex. And it is only then that Newton considers the case of algebraic expressions: "In the same time that the Quantity x by flowing becomes $x + o$, the quantity x^n will become $(x + o)^n$, that is, by the Method of Infinite Series's . . ." (Newton, *Mathematical Works*, 1:142).

At the beginning of the text Newton opposes his doctrine to a theory that would suppose the composition of magnitudes by the addition of parts—albeit infinitely small ones.[19] Magnitudes are produced by a continuous generation, according to all sorts of motions—the line generating the surface, the surface the solid, and finally time itself being supposed to flow continuously. The justification for taking the increase of magnitudes to be continuous is to be sought in nature itself, and among the ancients—although the traditional kinematics was not the "official" tradition of Greek mathematics. The first motive of these methods, so Newton believes and confidently tells us, was to determine the magnitudes themselves from their velocities of increase.

The examples are at first strictly geometrical: areas generated by the advance of certain segments either variable or constant, a solid generated by a disk of variable size, finally the rotation of a straight line round a pole. The evaluation of the fluxions of the straight line in rotation, in relation to the fluxion of the fixed straight line, evokes modes of reasoning that will be found again in the *Principia*: to the straight line in its new position is transferred a segment equal to the segment of the previous position, then the ultimate state of the relations between increments is considered (see below, p. 237).

To examine in more detail the rotation that Newton describes: When PB has turned into Pb, the memory of the length PB is retained in the point C. Triangle bDP, which is similar to the small triangle of the increments bCB (the similarity results from the identity of the angle at b, which is common to the two triangles, and from the equality, by construction, of the angles bBC and bPD), is constructed. Insofar as the increase BC has a certain size, that is, insofar as the rotation BPb is finite or assignable, the angle BDP, like the angle BCb to which it remains always similar, is greater than a right angle. Ultimately, when PC rejoins PB, these angles BCb and PDB are right. The triangle PDB is thus the image in the finite of what the vanishing triangle bCB has ultimately become. This form of reasoning will reappear in the *Principia*.

THE METHODS OF THE *PRINCIPIA* (1): ULTIMATE RATIOS AND FINITE WITNESSES

Against the background of the new methods of the seventeenth century, the originality of the mathematical procedures employed in the *Principia* can be clarified, and the specific traits of its reasoning brought into relief.[20] It will be shown why the *Principia* is difficult to read, and why the demonstrations are often surprising and sometimes appear too elliptical or even incomplete. A certain application of attention, of reasoning and visual imagination, is necessary to decipher the text and to follow in the figures what is said about them, but this effort does not correspond to what is required for the reading of a text of the classical tradition (to fix the ideas and give clearly defined benchmarks, the works of Euclid and of Apollonius can serve as reference or foil). Neither is this effort the well-ordered linking of operations prescribed by an algorithm; it is not a "calculus" in this sense.

Did Newton Conceal His Route?

It appears to be established today that Newton did not first write the *Principia* in a "fluxional" style—in the sense of a fluxional algorithm—before transcribing it into a purely geometrical presentation.[21] The manuscripts preparatory to the *Principia*, of which long excerpts have been studied (see the study of the *De motu*, pp. 10–57 above) carry no trace of any retranslation of this kind.

Yet a very tenacious tradition would have it that the results of the *Principia* were first "discovered" by means of differential and analytic procedures. Here is how Laplace explains the matter:

> It was by the method of Synthesis that Newton set forth his theory of the system of the world. It appears, however, that he found out most of his theorems by Analysis, of which he had pushed back the limits, and to which, by his own account, he was indebted for his general results on quadratures. But his predilection for Synthesis and his deep respect for the geometry of the Ancients led him to translate his theorems, and even his method of fluxions, into a synthetic form; and we see, by the rules and by the examples he has given of these translations, the great importance that he attached to them. (Laplace, *Exposition du système du monde*, *Oeuvres*, 6 [1884]: 464)

Whence this idea? In the mid-eighteenth century, it is expressed in particular by d'Alembert: Newton employed the geometry of the ancients "in order to conceal his route, while employing Analysis as a guide for himself" (*Essai sur les éléments*, *Oeuvres*, 2 [1805]: 320; *Mélanges*, 4:176).

Allowances should be made here for perspective: from the middle of the eighteenth century, scientists could hardly comprehend how it was possible to manage without "Analysis." But d'Alembert and Laplace based their claim on evidence of a more solid kind, and the thread of the transmission of this legend can be traced back further. The oldest evidence appears to be a text published in 1715, seemingly anonymous, but in fact written by Newton himself: the *Recensio libri*; in English, "An Account of the Book Entitled *Commercium Epistolicum*." Here is what this "anonymous" writer asserts.

> By the help of the New Analysis Mr. Newton found out most of the Propositions in his *Principia Philosophiae* but because the ancients for making things certain admitted nothing into geometry before it was demonstrated synthetically, he demonstrated the propositions synthetically that the system of the heavens might be founded on good geometry. And this makes it now difficult for unskilful men to see the Analysis by which these propositions were found out. (Hall, 11; *NMP*, 3:24; Cohen, 79)

Why did Newton insist, thirty years after the *De motu*, on propagating this version of his own history? The matter can be clarified if it is asked who the "unskilful men" were who failed to understand. A polemic was raging at this time between the partisans of Leibniz and the partisans of Newton as to which of the two was the inventor of the new calculus. In 1713 the Leibnizians had circulated an anonymous lampoon, the *Charta Volans*, which contained an extract from a letter by Johann Bernoulli, cited to provide an expert opinion on the improbability of Newton's claims:

> It seems that Newton, seizing the opportunity, very much advanced the business of series by the extraction of roots, a method he first employed, and it is very

> likely that in the beginning of all his studies he devoted himself solely to developing them nor did he then, I believe, so much as dream of his calculus of fluxions and fluents, or of its reduction to the general operations of analysis in order to serve as an algorithm or in the manner of the arithmetical and algebraic rules. The strongest evidence for this conjecture of mine is that you can find in all those letters no trace or vestige of the symbols $\dot{x}$, $\ddot{x}$, $\dddot{x}$; $\dot{y}$, $\ddot{y}$ etc. which he now employs in place of the differentials dx, d^2x, d^3x; dy, d^2y etc. Indeed, you can find no least word or single mark of this kind even in the *Principia Philosophiae Naturalis*, where he must have had so many occasions for using his calculus of fluxions, but almost everything is there done by the lines of figures without any definite analysis in the way not used by him only but by Huygens too, indeed by Torricelli, Roberval, Fermat, and Cavalieri long before. (Johann Bernoulli to Leibniz, 7 June 1713, *Corresp*., 6:4. The original Latin letter was translated into French by Leibniz in a letter to the Countess of Kilmansegg of 18 April 1716; see *Opera*, 3:459. This French version was then reproduced in the *Recueil* of des Maizeau, which had a wide diffusion; see *LMS*, 3:910 and Westfall *NR*, 760–72.)

Thus Newton had to reply to the objection: if for so long a time he had been in possession of a "calculus," why had he not used it in the *Principia*? He attempted therefore to make it believed that the new analysis had served him as a means of discovering his propositions, and that he had then translated the demonstrations into a synthetic style.

Bernoulli's description of the procedures of the *Principia* is basically accurate: everything there is done by means of "the lines of figures," as in Huygens, Roberval, or Torricelli. Fluxions play but an insignificant role; fluxion, moreover, as Johann Bernoulli expressly points out, does not imply analytic procedure. It will be necessary to await the works of the Leibnizians like Johann Bernoulli himself, Varignon, and above all, Euler, to have the results of the *Principia* translated into the language of analysis.

As for Newton's assertion that the results of the *Principia* were first discovered analytically, all the extant Newtonian manuscripts prove the contrary. The only piece which strikes a different note is a very short manuscript (*NMP*, 6:588–93), clearly subsequent to the first edition, in which Newton has sketched a fluxional calculus for the evaluation of central forces. But these few lines terminate abruptly in particular examples in which the great Newton becomes confused, erases, and fails to arrive at anything fruitful. From these attempts nothing passed into later editions of the *Principia*.

It can also be seen that the route followed in a good many of the demonstrations is far from a fluxional or differential calculus: the structure of the reasoning, in most of the cases that will be presented, is so different from such a calculus that a retranslation would require a remaking of the reasoning itself.

Since it should not be supposed that "behind" the text of the *Principia* lies another version, another demonstrative fabric more congenial to modern analytico-algebraic habits, Newton's text must quite simply be taken as it is, read on its own terms, and respected for its mode of access to mathematical objects. Historians of mathematics, even the greatest, have not been sufficiently aware that the mathematical text is denatured when it is transcribed or paraphrased in a more "modern" presentation. To be sure, present-day readers need to verify, with all necessary rigor and using their own mathematical tools, that the results enunciated in the ancient text are true; they may also welcome an easy, synthetic exposition of the purpose of the ancient demonstration and of the principal steps in its reasoning. But it would be wrong to suppose that the reasoning has simply been "abbreviated" or presented "more easily." The transcription of the demonstrations of the *Principia* in the form of the differential calculus or in analytico-algebraic notations is useful and sometimes indispensable, but it does not save readers the trouble of actually following the course of thought in the original text, and reexecuting the acts of intuition that are demanded by that text.

The mathematical procedures of the *Principia* are of a great diversity, and it is impossible to give here a truly faithful reflection of the wealth of mathematical creativity that Newton there deployed. Attention will be focused on certain more typical traits that are especially suggestive of this originality of mathematical style in the *Principia*.

Neither a Classical Geometry nor an Infinitesimal Calculus

The mathematical methods that Newton employs in the *Principia* go beyond the traditional boundaries of classical geometry. It is still a geometry, but no longer that of Euclid or Apollonius. It differs from classical geometry in two decisive respects:

1. The presence of motion
2. The admission of the infinitely small

Motion intervenes in the reasonings, even the least physical. Geometry is enriched with all that kinematics can contribute: points are displaced on lines, curves are generated as trajectories of moving bodies, circles rotate or roll, and so on.

With regard to the second point, certain elements of figures must be considered as "very small." Newton speaks of *lineola*, *linea nascens*, *linea minima*, *linea quam minima*, *distantia quam minima*; he specifies that certain segments are infinitely (*infinite*) or indefinitely (*indefinite*) small;[22] he concerns himself with nascent or evanescent arcs; he envisages two immediately neighboring positions on the same curve (*locus proximus*).[23] In a great many cases, it is a temporal rather than a geometrical infinitesimal that serves as fundamental variable: a "particle of time," in Newton's own expression, sets in motion the whole machinery, time being the independent variable in terms of which all the other magnitudes are expressed. However, because time is very often represented geometrically by virtue of certain procedures, the "particle of time" appears in the form of distance or surface.

Yet these procedures of reasoning do not constitute an "infinitesimal calculus" properly speaking. A *calculus* implies rules of writing and manipulation, a collection of algorithms. Leibniz, for example, published such rules for the treatment of infinitesimal quantities beginning in his *Nova methodus* of 1684, and what he proposed can rightly be called an "infinitesimal calculus" or a "differential and integral calculus" (see above, p. 216). But the *Principia* contains nothing of the kind, except adventitiously, in Lemma 2 of Book II, which concerns the "moments" of quantities and which Newton scarcely uses. The work does not even contain what is often called the fundamental theorem of the infinitesimal calculus: the affirmation and proof that differentiation and integration (or the investigation of tangents and that of quadratures) are operations inverse to each other.

The argumentation of the *Principia* remains resolutely geometric: the reasoning consists in interpreting the figure, in reading there certain relations of proportionality, which are then transformed according to the usual rules. The grand innovation, in relation to the ancients, consists in studying what these relations become when certain elements of the figure tend toward limiting positions or become infinitely small.

The adjective "ultimate" applies very aptly to this geometry in its contrast with the geometry of the ancients. Newton considers what the geometric relations become "ultimately" (*ultimo*); he investigates last ratios or ultimate relations (*ultimae rationes*: first or "nascent" ratios can be considered also as a sort of ultimate relation). It is in this form, in general, that the infinitely small appears in the reasonings of the *Principia*: as ultimate vestige of a finite situation.

Section 1 on Ultimate Ratios

How must the Euclidean apparatus be modified to permit such modes of reasoning? Newton is aware of leaving behind the traditional framework of ancient geometry, at least in what concerns the introduction of the infinitely small or vanishing quantities. (As for what concerns the place of motion, Newton may be less aware of innovation; in the preface to the *Principia* he categorizes geometry as a branch of mechanics based in the practice of tracing figures.) On the other hand, in Newton's view indivisibles appear "too harsh," too lacking in certainty (see above, p. 167). Thus he has given to the *Principia* a sort of mathematical preamble in section 1 devoted to "first and last ratios" or "ultimate proportions" or—more rigorously—"ultimate ratios."[24]

The eleven lemmas presented in this section are sheltered under a general justification provided in Lemma 1:

> Quantities, and the ratios of quantities, that in any finite time tend constantly to equality, and before the end of that time approach one another more closely than by any given difference, become ultimately equal. (*Princ.*, 28)

The salient point of the enunciation lies in the opposition beween "constantly . . . before the end of that time" and "ultimately." It will be permitted henceforth to jump over the infinity of steps of approximation, and to extrapolate from a certain number of successive finite situations to the final state of the relations between the magnitudes.

This recalls the procedure of the ancients, evoked by Pascal (see above, p. 202), which consists in approaching a magnitude in such a way that the difference becomes smaller than any given difference. But Euclid (for example in 12.2) does not proceed in this way. In the first place, the mode of convergence is not the same: in Newton's Lemma 1 it suffices that the quantities approach one another, while a demonstration by *double* reduction to the absurd, as in Euclid 12.2, requires establishing, first, that any difference whatever in excess can be surmounted, then, similarly, that any difference whatever in default can be surmounted (it is impossible that the magnitude A should be smaller than the magnitude B, and it is impossible that it should be greater).

The deepest rupture is in the very definition of magnitudes or quantities: the mathematics of the ancients speaks of fixed or determinate quantities, while Newton treats of quantities that can "tend . . . to," "approach," up to an ultimate situation. Variation and time are here essential to the very meaning and definition of magnitude.

What is this time of which Newton speaks? The later lemmas will show that time intervenes in two ways in the variation of geometrical magnitudes:

> It can be the discrete time of successive operations that the mathematician performs on magnitudes.
>
> It can be the continuous time of the autonomous generation of the magnitudes themselves.

To this double aspect of time corresponds the division of Lemmas 2–11 into two distinct groups according to whether the ultimate situation results from a series of operations on the geometrical object or is linked to the generation of the magnitude itself.

Lemmas 2–5

The theme of this first subdivision, or its principal thread, is not easy to discern. What are the cases of ultimate relations that Newton has retained as fundamental? Why has he chosen them?

Lemmas 2, 3, and 4 deal with a common subject: the approximation of curvilinear figures and the comparison between two areas by virtue of a dividing up into parallelograms, the breadths of which diminish infinitely. There is no question of a number that would express an area, nor of an algorithm of integration or summation.

In Lemma 2, parallelograms are inscribed in a figure, of which one of the three sides is a curve, and other parallelograms are circumscribed about the same figure on the same bases as the first parallelograms, all these bases being equal.

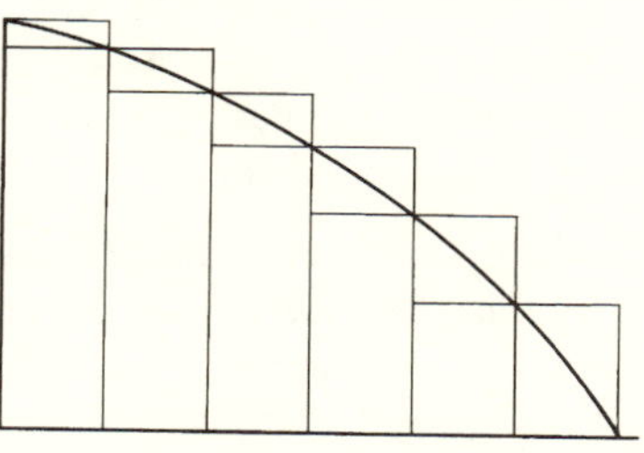

Figure 3.49

> If the breadth of those parallelograms be supposed to be diminished, and their number to be augmented *in infinitum*, I say, that the ultimate ratios which the inscribed figure, . . . the circumscribed figure, . . . and the curvilinear figure . . . have to one another are ratios of equality. (*Princ.*, 28–29)

To demonstrate this, Newton shows that the difference between the two families of parallelograms is equal to a single parallelogram having for height the sum of the heights of the small, "excess" parallelograms (this presupposes that the curve does not change from positive to negative slope or vice versa), and for base the breadth of any of the parallelograms. Since this base diminishes infinitely, the difference between the two families of parallelograms diminishes likewise and becomes smaller than any given rectangle (*fit minus quovis dato*, *Princ.*, 29).

Next, the same procedure is employed in Lemma 3 for the case of parallelograms with unequal bases: consider the parallelogram constructed on the widest of the bases and with a height equal to the sum of the heights of the excesses; it is greater than the difference between the two families, and its width diminishes ad infinitum, hence the difference does so as well.[25]

The corollaries deduce the equality between "the ultimate sum of the parallelograms" and the curvilinear figure, then between the figure made by the "vanishing chords of the arcs" and the curvilinear figure, and finally between the figure made by the tangents of the arcs and the curve. Newton concludes in the final corollary (no. 4) that the ultimate figures are no longer rectilinear, "but the curvilinear limits of rectilinear figures" (*Princ.*, 29).

Lemma 4 considers two figures in which are inscribed two "series of parallelograms" (*Princ.*, 29). If ultimately, that is to say, when the width of the parallelograms diminishes indefinitely, one obtains a well-determined ultimate ratio of the parallelograms, each to each, then the figures themselves are in this ratio. An important corollary, useful later in the work (see below, pp. 239–40), extends this procedure of ultimate comparison to all sorts of magnitudes:

> *Corollary*. Hence if two magnitudes of any kind are divided in any manner into the same number of parts; and those parts, when their number is augmented and

> their magnitude diminished *in infinitum*, come to have a given ratio to each other, the first to the first, the second to the second, and the others to the others in their order: then the wholes will be to one another in the same given ratio. For if in the figures of this lemma one takes parallelograms that are to each other as the parts of these magnitudes, the sums of the parts will always be as the sums of the parallelograms. (*Princ.*, 30)

It is no longer a matter of areas, but of any magnitudes whatever that will be represented by areas, each parallelogram representing a finite or infinitesimal part of the magnitude. Newton later applies this corollary to two linear quantities: the length of the arc of the cycloid and that of the arc of the generating circle, as we shall see. The dividing up of the two magnitudes will be interpreted in the following way: both depend on a common time (they are generated by a common mechanism regulated by the time or by a uniform rotation or translation), and their elements or "momentaneous changes" are considered during a common instant; if these contemporaneous elements of the magnitudes, that is, their nascent arcs, are in a determinate ratio, then the magnitudes themselves are also in this ratio.

This extension is unexpected: it could be believed that these enunciations had to do with the approximation of curvilinear figures by rectilinear figures. Yet one of the results of these lemmas is to render possible the determination of a finite magnitude starting from its nascent element.

As for Lemma 5, why place it in the group containing Lemmas 2–4? The text is very short, and without demonstration:

> In similar figures the sides that correspond are proportional, whether curvilinear or rectilinear; and the areas are in the duplicate ratio of the sides. (*Princ.*, 30)

The notion of similarity, and the proportions that follow from it, are thus extended from rectilinear segments and figures to curvilinear segments and figures. There is no demonstration, but it can be supposed that the basis of this enunciation is in the preceding lemmas: since the curvilinear figures are the "limits" of rectilinear figures, their relations can be treated as the relations between rectilinear figures.

The later lemmas suppose similarities between curvilinear figures, and the proportionality of similar curvilinear areas to the squares of homologous sides. In this way Lemma 5 serves as a transition from the first group of lemmas to the second.

Lemmas 6–11

The remainder of section 1, namely Lemmas 6–11, is devoted to the study of the arcs of any curves. More specifically, it is a matter of local properties of the curves in the neighborhood of a generic point. Certain magnitudes are attached to an arc; what do these magnitudes become when the arc is diminished indefinitely?

The fundamental notion is that of "continued curvature," presented in Lemma 6:

> *Lemma VI.* If any arc ABC given in position is subtended by the chord AB, and in some point A, in the middle of the continued curvature, has for tangent the straight line AD produced in either direction, and if the points A and B approach one another and come together; I say that the angle BAD, contained by the chord and the tangent, will diminish *in infinitum* and will ultimately vanish.

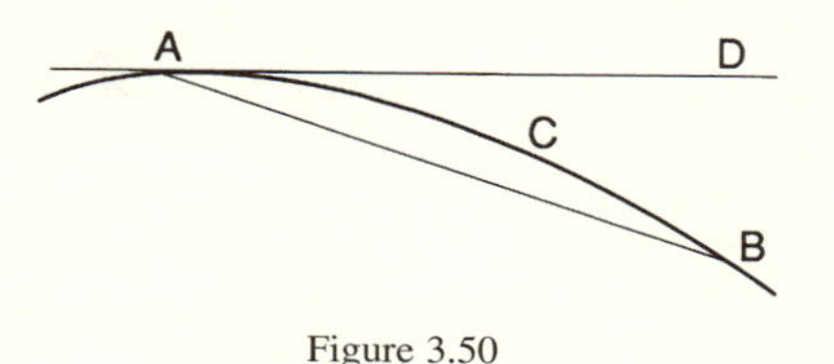

Figure 3.50

> For if this angle did not vanish, the arc ACB and the tangent AD would contain an angle equal to a rectilinear angle, and hence the curvature at the point A would not be continuous, contrary to the hypothesis. (*Princ.*, 31)[26]

It can be seen that the "curvature" would not be "continuous" if at the point A the arc and the tangent contained an angle equivalent to a rectilinear angle, that is, if A were an angular point or if the tangent did not vary continuously. "Curvature" in the technical sense of present-day mathematics is not in question (Lemma 11 will come closer to the modern use of the word).

Different aspects of this same situation are studied in the remaining lemmas of this section. When a point moving on the curve approaches a fixed point A, different segments and magnitudes come to satisfy certain ultimate relations. The focus here will be particularly on Lemmas 7 and 9, because of the very original method that Newton here applies for the first time and which he employs in other places in Book I. Lemma 9 will have as a consequence the generalization of Galileo's law of fall, which has been previously discussed; it is merely a new formulation of Lemma 2 of the *De motu* (MS D, Hall, 244–45; see above, pp. 164–65).

Lemma 7 aims to show that the arc, the chord, and the tangent are ultimately equal, that is, that their ultimate ratio is one of equality when the point B, moving on the arc, comes to coincide with the fixed point A. Here is the demonstration (according to the third edition):[27]

> For while the point B approaches the point A, let it be supposed that AB and AD are prolonged to the distant points b and d, and that bd has been drawn parallel to the secant BD; and that the arc Acb is always similar to the arc ACB.
>
> When the points A and B coincide, the angle dAb, in virtue of the preceding

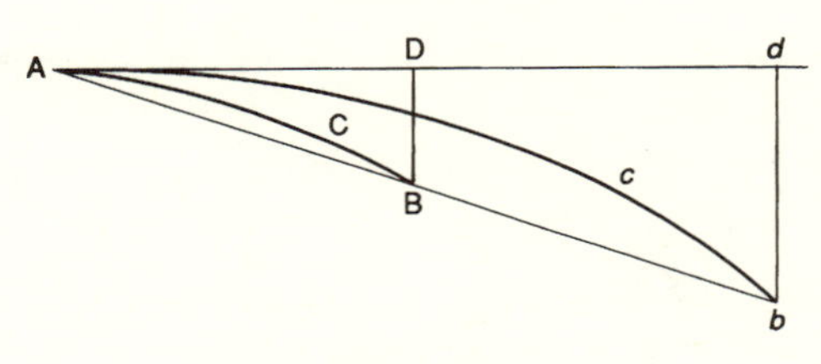

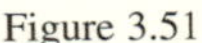
Figure 3.51

> lemma, will vanish; and therefore the straight lines Ab and Ad, which are always finite, and the intermediate arc Acb, will coincide, and will consequently be equal.
>
> It follows that the straight lines AB and AD, which are always proportional to them, and the intermediate arc ABC, will vanish and will have between them an ultimate ratio of equality. (*Princ.*, 31)

The curve is supposed drawn in advance, and the point A is fixed, as is the tangent AD. The point B traverses this curve and approaches the point A, carrying along with it the chord AB. Thus the angle BAD closes up, and in conformity with the preceding lemma (a consequence of the notion of "continued curvature") ends by vanishing. The points A, B, and D then coincide. How to study the relation between the chord AB, the arc AC, and the tangent AD in this ultimate situation?

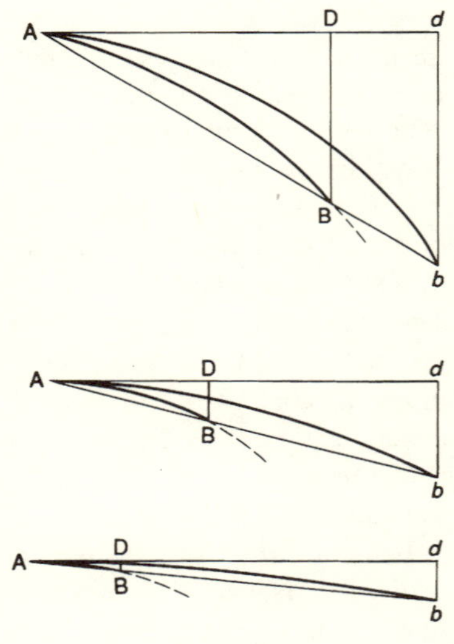

Figure 3.52

The segments AB and AD have been prolonged to the remote points b and d (at a fixed distance or in any case at a finite distance, Newton does not specify which), and constructed between these segments Ab and Ad is a curved arc similar to the arc ACB. The approach of B towards A, which diminishes and flattens the arc AB, is translated in this "copy" by a continuous deformation: the replica-arc Acb flattens as its model vanishes, but remains of finite length between A and b.

It can therefore be asserted that ultimately Ab, Ad, and the arc Acb coincide and are equal. The same thing is true of their infinitely small models. The key to the demonstration lies in the constant preservation of a similarity between the two figures, the one finite, the other vanishing: both are deformed while remaining similar to each other at each instant.

The same procedure of reasoning is employed in a somewhat more refined form in Lemma 9. This time a point of the curve is approached in two different ways. It is a matter of proving that the areas of vanishing curvilinear triangles are proportional to the squares of the sides. It is the result that was demonstrated in Lemma 2 of the *De motu* (see above, pp. 164–65). Here is the demonstration as given in the third edition:

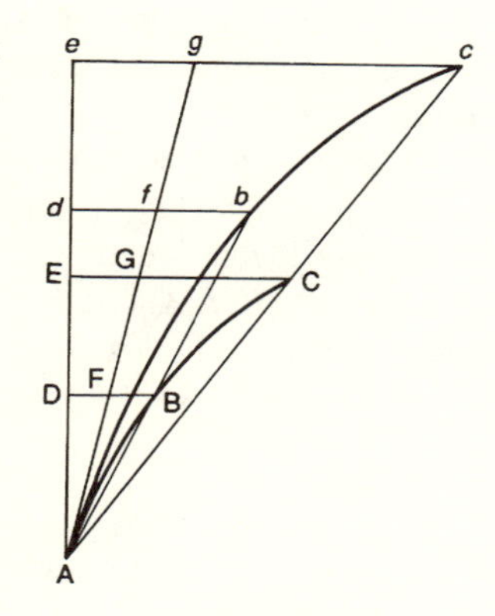

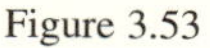
Figure 3.53

> For while the points B and C approach the point A, let it be supposed that AD is always produced to the remote points d and e, so that Ad and Ae are proportional to AD and AE, and let the ordinates db and ec be erected parallel to the ordinates DB and EC, and meeting AB and AC prolonged in b and c.
>
> Let there be drawn the curve Abc similar to ABC, and the straight line Ag, which touches either curve in A, and cuts the ordinates DB, EC, db, ec in F, G, f, g.
>
> Then, with the length Ae remaining fixed, let the points B and C rejoin the point A; the angle cAg will vanish, and the curvilinear areas Abd and Ace will coincide with the rectilinear areas Afd and Age; consequently (in virtue of Lemma V) these areas will be in the duplicate ratio of the sides Ad and Ae. But to these areas the areas ABD and ACE are always proportional, and to the sides are always proportional the sides AD and AE.
>
> Therefore the areas ABD, ACE are also ultimately in the duplicate ratio of the sides AD, AE. Q.E.D. (*Princ.*, 33)

The curve is supposed drawn in advance, and the point A is fixed. The straight line AG, tangent at the point A, and the secant straight line AE are also fixed.

The point A is approached from above in two independent—but always comparable—motions. C and B come to coincide with A each in its own manner, but in such a way that they reach A at the same instant (the enunciation of the lemma stipulates: "the points B, C simultaneously approach point A").

The two chords CA and BA come to approach the tangent and to merge with it. The curvilinear areas have then become rectilinear areas contained between the tangent and the secant. But these varying areas have disappeared (each in its own manner), engulfed in the point A. However, to know their relation at the moment in which they disappear is possible by virtue of an artifice: their image has constantly been preserved in the finite.

Throughout their displacement towards A, the points B and C, as well as the ordinates that are linked to them, have been faithfully imitated in their motion by the points b and c and the corresponding ordinates. The two essential conditions are these:

> The length Ae is fixed.
>
> Ad and Ae are constantly proportional to AD and AE.

Thus is obtained the means of following in the finite, in a sort of control-diagram, the variations of the curvilinear triangles ABD and ACE up to the moment in which they have vanished. This geometric procedure might be likened to the photographic shots taken by a so-called zoom camera, which makes it possible to follow and reproduce the motion of an object while magnifying the details at will.

The figure that accompanies this text in the *Principia* evidently gives only a very poor idea of the unfolding deformation. It is necessary to animate it following the instructions of the demonstration. Several successive figures give yet a better idea of the sorts of transformation the reader should "see" in order to follow the text.

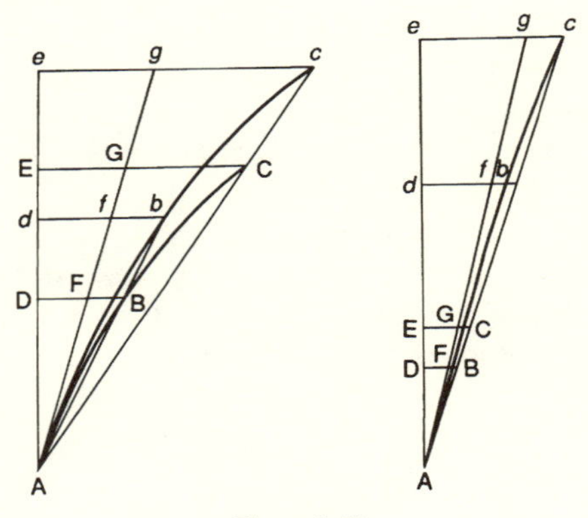

Figure 3.54

Knowing the *De motu*, the reader will have divined the physical employment of this lemma. It makes possible the extension, to forces that are variable but sufficiently regular, of Galileo's law for falling bodies. As in Galileo's diagram, the velocities are represented as ordinates and the times as abscissas. This time the acceleration is no longer represented by the constant obliquity of a straight line ABC in relation to an axis ADE but takes the aspect of a curve. Galileo's proposition is true "ultimately," "at the commencement of the time"

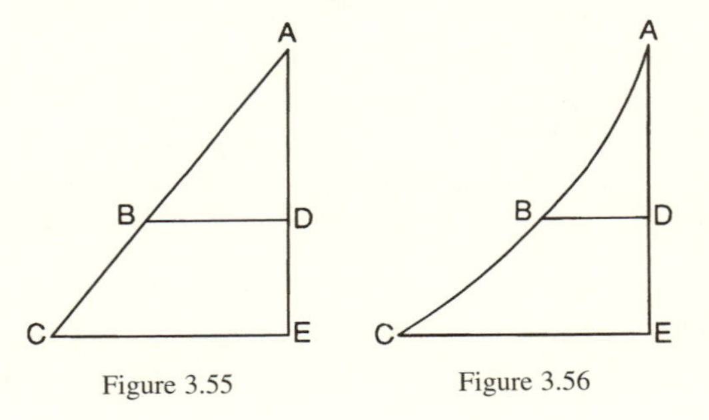

Figure 3.55

Figure 3.56

(*Princ.*, 33), because the curvilinear triangles are, at the instant of their vanishing (or of their birth) proportional to the squares of their sides. Therefore the spaces traversed under the action of variable forces are, at the very commencement, proportional to the square of the elapsed time, as Lemma 10, which follows, asserts.

These two demonstrations (Lemmas 7 and 9) illustrate a method that Newton utilized in other places. It could be called "the method of finite witnesses" because it consists in retaining a trace, a memory of the infinitesimal situation, in the form of a finite but similar configuration from which can be read the relations between the infinitesimal or vanishing elements. The triangles have disappeared, but their finite image permits describing what they resembled at the time of their disappearance.

Newton was certainly not the only one to employ such procedures. The transition from classical geometry to infinitesimal calculus presupposed considerable familiarity with and use of such modes of coping with the infinitely small. Geometers cultivated various concrete representations suggestive of the infinitesimal before subjecting it to regular procedures. The most celebrated example of these representations is without doubt the "characteristic triangle" that Leibniz "saw" in a demonstration of Pascal (*LMS*, 5:232, 399; *LMS*, 2:72, 259), and which had turned up more or less explicitly in other authors (see Hofmann, 74–75). When Leibniz imparted his discovery to Huygens, the latter recognized that he had already used the same procedure in his own works.

The reasoning and the figure are often presented as follows: The triangle constituted by a portion EE′ of the tangent to the circle and by the corresponding vertical and horizontal segments KE and KE′ is similar to (among others) the triangle constituted by the radius AD, the sine ID, and the cosine AI. Leibniz was filled with wonder to discover in this example that a figure made of "indivisibles" or "differential quantities" can remain constantly similar to a figure constituted of finite "assignable" elements. This similarity made it possible to study, by an indirect way, the relations between elements infinitely small.

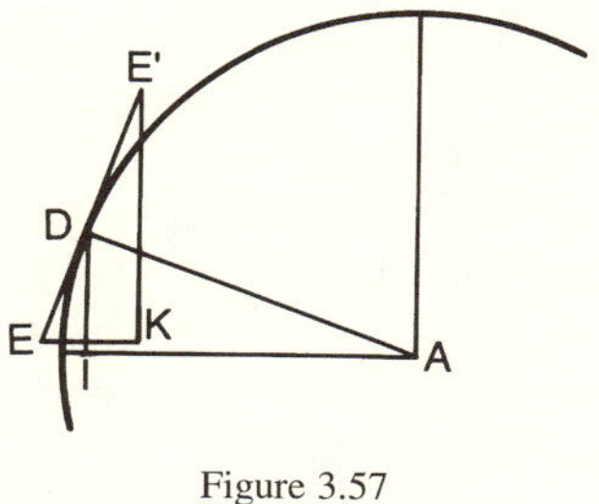

Figure 3.57

Despite the interest of the close similarity with Newton's processes, it is necessary to take note of certain disparities. The kinematic aspect is less pronounced here: the differential triangle is not an ultimate state attained at the end of a motion, and the deformations of the finite triangle with change in the position of D are less astonishing than what has been "seen" in Lemmas 7 and 9 of the *Principia*. The most decisive difference has to do with the place that such procedures occupy in the theory as a whole. Leibniz's discovery of the similarity between the differential triangle and a finite triangle is only a step in the elaboration of a work in which the algorithmic aspect finally predominates. Newton, on the contrary, cultivates similar modes of representation, particularly in the demonstrations of the *Principia*.

The Rectification of the Epicycloid (Propositions 49 and 50): An Example of the Use of Finite Witnesses

This method of "finite witnesses" is illustrated in some detail by a demonstration that is particularly refined and difficult to comprehend at first reading. The passage belongs to section 10 of Book I (*Princ.*, 145–47), which deals essentially with the isochronous pendulum.

The aim is to construct, or to describe, a pendulum whose oscillations are always carried out in the same time: the mobile body A, subject to a vertical gravity and required to move along the path ABC, must arrive in the same time at C, whether it has been released at A, at B, or at any other point of the arc.

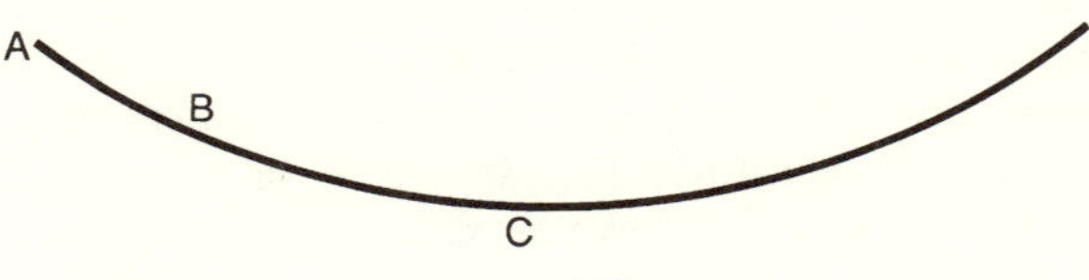

Figure 3.58

The problem had been resolved by Huygens (*Horologium*, pt. 2, Prop. 25): it suffices that the body traverse an arc of an inverted cycloid. Newton presents the same solution in a more general framework, and above all he brings into clear relief a fundamental property of the phenomenon, a property which makes understandable "why" the cycloid is appropriate: the force that draws the body from A towards C is proportional to the arc CA, or in other words the force of recall is proportional to the displacement (it is the property of the "harmonic oscillator": $\ddot{x} = -kx$).[28]

To exhibit this property, it is necessary to know the length of the arc of the cycloid. Wren had found the means of calculating this length, in other words of rectifying the cycloid, and his demonstration,[29] published by Wallis (*Tractatus duo*), is very different from the one given here by Newton; Wren's is scarcely at all kinematic and is based on a series of inequalities in which the cycloidal arc is caught between sums of portions of tangents whose difference can be diminished to zero (see Hofmann, 109–10); Newton elsewhere employs this result of Wren's without demonstrating it (*NMP*, 3:422).

The description can be based on the usual definition of the cycloid on a plane base. While the circle rolls from A to L without slipping, the trajectory is described by the point P, from its lowest position in A, rising to its highest

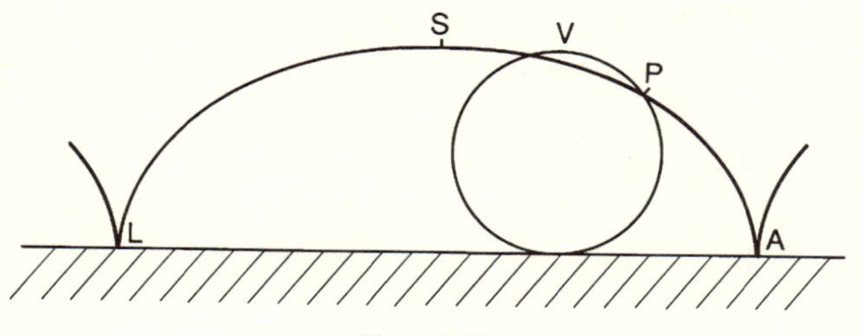

Figure 3.59

position in S, then passing onward to where it again reaches the base plane at L, and so on in repetitions of the same cycle.

Newton chooses a more complicated curve; his cycloid has a circular base, with the small circle rolling on a large circle, either on the inside (the hypocycloid) or on the outside (the epicycloid). Newton treats the two cases together, in two twinned propositions (Props. 48 and 49) that are proved by a single common demonstration. This study will be restricted to the case of the epicycloid, in which the figure is a little less difficult to make out.

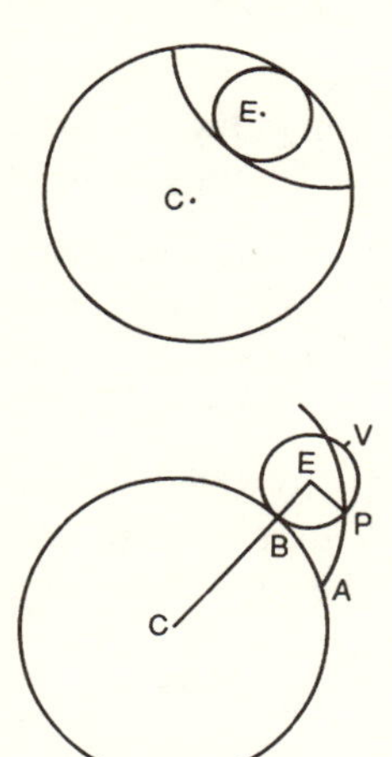

Figure 3.60

In relation to the usual case, another modification intervenes here: the force is not gravity, but a force tending towards the center C of the large circle and proportional to the distance. When the radius of this large circle is infinitely great, the simplest case, studied by Huygens, will again result, that of a cycloid on a rectilinear base with a force acting uniformly (see *Princ.*, Prop. 52, Cor. 2, 153).

The aim of the proposition is to express the length of the arc AP as a function of

> Constants, which are the radii CB of the base circle and BE of the rolling circle
>
> The angle BEP, which indicates how much the small circle has turned

The result can be expressed in the following formula:

$$\text{arc AP} : (\text{BV} - \text{VP}) :: \text{CB} : 2\ \text{CE},$$

which Newton prefers to write

$$\text{arc AP} : 2 \text{ versed sine } (\text{BEP}/2) :: \text{CB} : 2\ \text{CE}.$$

After having indicated without justification certain auxiliary constructions to be carried out on the figure (*Princ.*, 145, line 26, to 146, line 5), Newton first shows that the tangent to the cycloid is always aligned with the segment VP that joins the running point P with the summit of the rolling circle:

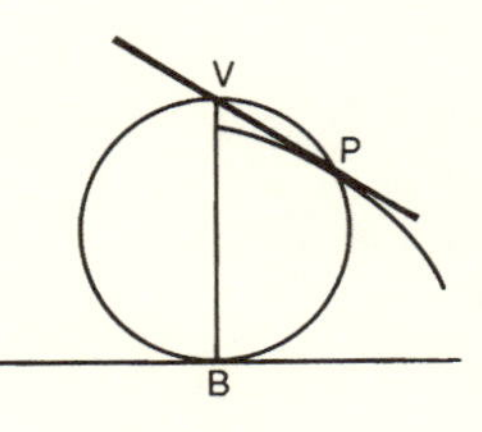

Figure 3.61

> Because the wheel in its motion always pivots about the point of contact B, it is manifest that the straight line BP is perpendicular to the curved line AP described by the point P of the wheel, and consequently that the straight line VP is tangent to this curve at the point P. (*Princ.*, 146)

Because B is the instantaneous center of rotation,[30] BP is normal to the curve, and VP which is perpendicular to it (because the angle BVP is inscribed in a semicircle) is tangent to the curve. This property and this reasoning were well known for the cycloid on a rectilinear base, notably through Descartes.

Here is now the heart of the demonstration—seven extremely cryptic lines, which will be translated further on (p. 239 below), after a preparatory explanation of the "ultimate" kinematic reasoning involved in them.

Newton requires that two auxiliary circles be constructed:

> One with center C, and radius a little less than CP, cutting the small circle in o, the cycloid in m and the radius CP in n; this circle must then inflate until it reaches the point P
>
> A second circle with center V and radius Vo, cutting VP prolonged in q

To what do these two circles correspond? For an answer it is necessary to return to the simpler case of the cycloid on a rectilinear base and to seek there the equivalent of these circles nom and oq.

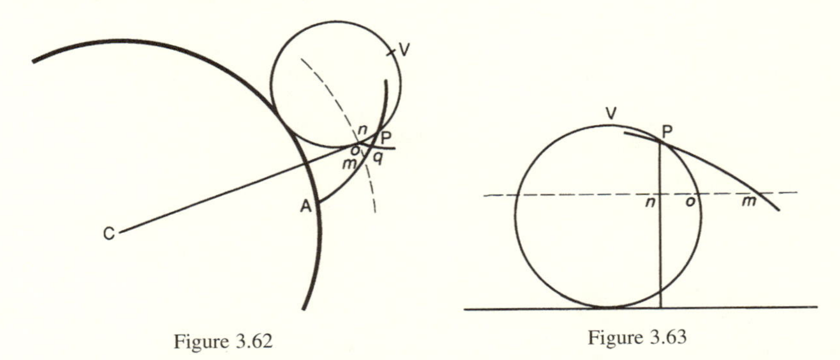

Figure 3.62

Figure 3.63

The large circle becomes a plane, and the concentric circle a parallel plane that rises until it reaches P: this plane indicates the height that the running point P had reached in the preceding instant, and the point o is the mark on the circle of this preceding position (see fig. 3.64). The function of the second circle with center V appears only if two successive states of the tangent are considered (see fig. 3.65).

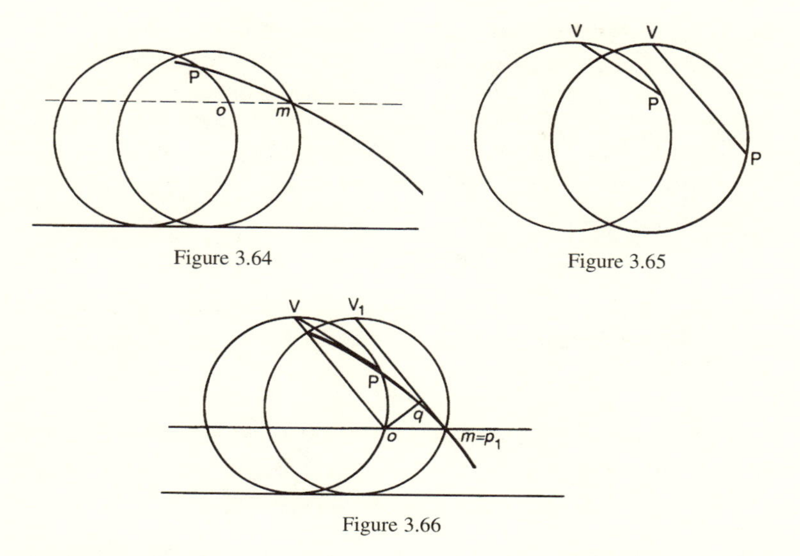

Figure 3.64

Figure 3.65

Figure 3.66

While the circle turns, the tangent VP diminishes by a certain length. This diminution will appear in the figure if the length of the first tangent is transferred to the prolongation of the second. The point o marks precisely on the circle the height of the running point in the preceding position: it suffices therefore to transfer the length Vo to VP prolonged in order to determine the length Pq, which measures the decrease of the tangent between the two positions. The segments Pq and Pm are thus "contemporaneous variations"; while the cycloid has increased by mP, the tangent has diminished by qP. Newton does not give a detailed explication of this dependence between the motions and of the significance of the auxiliary circles; he merely says that Pm is the "momentaneous increment" of the curve AP and Pq the decrement of the straight line VP.

This is the essential point of the reasoning: it is necessary to come to "see" the cycloid increasing by devouring its tangent VP, diminishing it from its maximal length when the point P is on the base, to its null length when P is at the summit S.

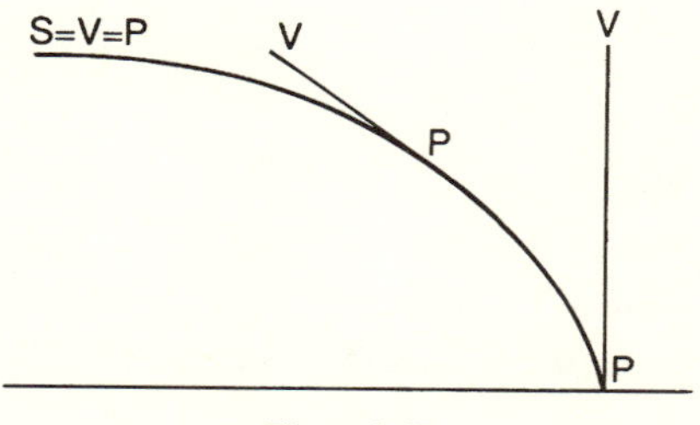

Figure 3.67

The remainder of the demonstration consists in determining the ratio between the element of the curve and that of the tangent, that is, between Pq and Pm. However, in order not to be comparing an increment with a decrement, and to introduce the angle BEP, Newton considers Pq, decrement of VP, as the increase of BV – VP (since VP decreases from BV to 0, BV – VP increases from 0 to BV). On the other hand, BV is the radius of the angle BVP, and VP is its cosine. Therefore:

$$BV - VP = BV(1 - \cos BVP),$$

or again, returning to the angle of rotation BEP which is twice as great as BVP, and to its radius BE which is twice as small:

$$BV - VP = 2\ BE[1 - \cos(BEP/2)].$$

For Newton, (1 – cos) is replaced by the notation, little used today, of the versed sine (in which the radius of reference is understood):

$$BV - VP = 2 \text{ versed sine } (BEP/2).$$

It is now necessary to find a relation of proportionality connecting Pm, the increment in the arc of the cycloid, with Pq, the increment of a certain versed sine.

It is here that Newton employs the method of "finite witnesses": the arrangement of the differential elements, connected to the running point of the cycloid, is "expressed" in the finite by a certain figure that remains constantly similar to the infinitesimal figure.

Some other auxiliary constructions were prescribed on a preceding page: VF is perpendicular to CP prolonged, PG is tangent to the circle in P, and GI is perpendicular to VP.

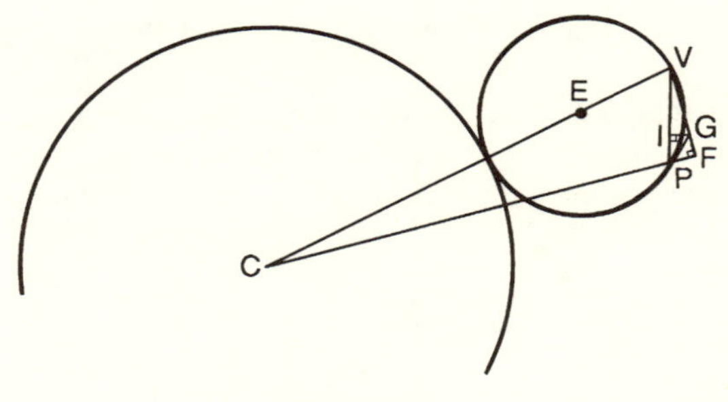

Figure 3.68

Thus is obtained a figure PFGVI, which Newton asserts to be similar to the vanishing figure constructed around the two increments Pm and Pq. Here is the decisive and extremely dense passage that was previously mentioned, in which all is said so briefly:

> Let the radius of the circle *nom* be gradually increased or diminished until it be finally equal to the distance CP; and because of the similarity between the evanescent figure Pnomq and the figure PFGVI, the ultimate ratio of the evanescent little lines Pm, Pn, Po, Pq, that is, the ratio of the momentary increments of the curve AP, the straight line CP, the circular arc BP, and the straight line VP, will be the same as that of the lines PV, PF, PG, PI respectively. (*Princ.*, 146–47)

To recognize this similarity, it is necessary to admit that the curves are prolonged by their tangents, or rather that the elements of arcs are segments of tangents: Po is a piece of PG (tangent to the circle), Pm a piece of VP (tangent to the cycloid). The angles mnP and oqP are right, because the radii CP and Vq cut the corresponding circles *nom* and *oq* at right angles. Finally, it is necessary to consider the angles vertically opposed in P.

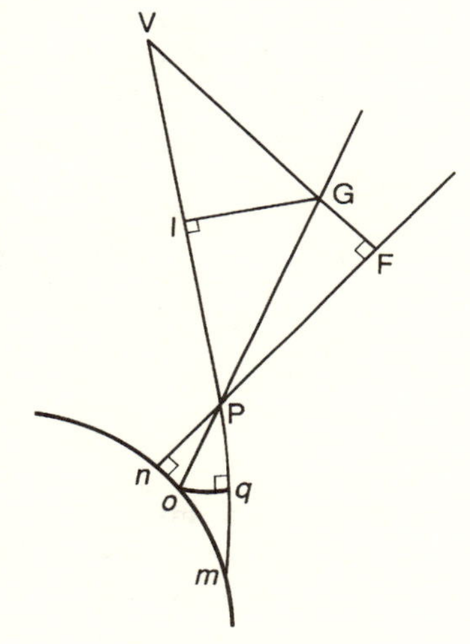

Figure 3.69

The figure PFGVI is thus the mirror image of the generating elements of the cycloid and the associated magnitudes. Whatever its position, the point P carries with it these two arrangements which are deformed symmetrically. This similarity is actual only ultimately, when the circle *nom* passes "nearly" through P (Newton speaks of the *ratio ultima lineolarum evanescentium . . . id est ratio mutationum momentanearum* [*Princ.*, 147]).

Thus the ratio between Pm and Pq comes to be determined: it is equal to the ratio between PI and PV. The latter ratio is transformed into another ratio which contains only radii of circles. This step presupposes other similarities of triangles, which need not be discussed here. Finally,

$$\text{Pq} : \text{Pm} = \text{PI} : \text{PV} = \text{CB} : 2\ \text{CE}.$$

It remains to derive, from the ratio of the infinitesimal increments, the ratio of the magnitudes themselves, that is, the ratio of AP to BV – VP (or 2 versed sine [BEP/2]). Newton justifies this step by invoking the corollary of Lemma 4 (see pp. 227–28 above).

According to this corollary, if two quantities can be subdivided in any manner into an equal number of parts such that the ratio between these parts, each to each, tends toward a determinate constant value, then the ratio between these magnitudes themselves has the same value. The conclusion is based on the comparison of areas permitted by Lemma 4; thus it is supposed that the lengths AP and BV – VP can be represented by plane areas. But how should the subdivision indicated in the lemma and its corollary be here applied? It is not a matter of dividing up a given figure into parallelograms that are made increasingly narrow, but rather of the generation of an unknown quantity by its differential element, its "momentaneous variation" [*mutatio momentanea*].

The intervention of time in the reasoning is not specified precisely. It is obvious that in a diagram agreeing with the conditions of the corollary of Lemma 4, time must figure as abscissa (the growth of arc length is a function of the time or at least of a regular motion). But what time is here in question? What independent variable should be chosen? The increase in the angle BEP by the rotation of EP is the most natural; but it is another variation, the swelling of the circle *nom*, which regulates the vanishing of the magnitudes and makes possible the ultimate similarity. Nothing would prevent reexpressing this increase of the circle *nom* as a function of angle BEP (by means of the sine of this angle). But Newton does not do it; he keeps two distinct "times" in his reasoning.

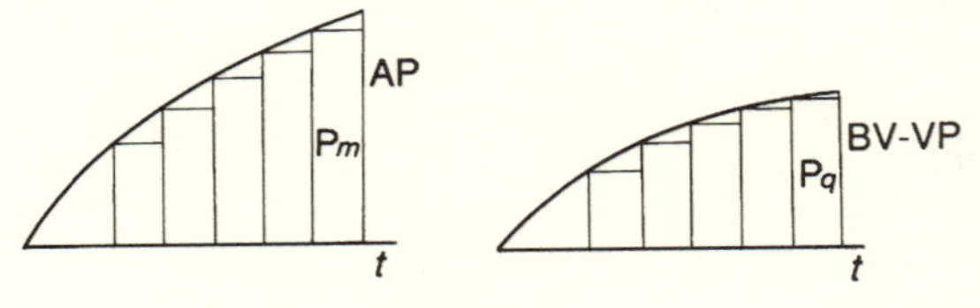

Figure 3.70

How much do these procedures differ from an infinitesimal "calculus"? What would mathematicians, trained in the differential and integral calculus have done in the present case? They would have reduced the problem to canonical procedures, first by reducing the element of arc according to some choice of coordinates, then by applying the theorem of Pythagoras to the characteristic triangle: $ds^2 = dx^2 + dy^2$. They then would have expressed the relations between x and y either directly, or through the intermediary of another variable, say t. Finally, it would have remained for them to write, for example,

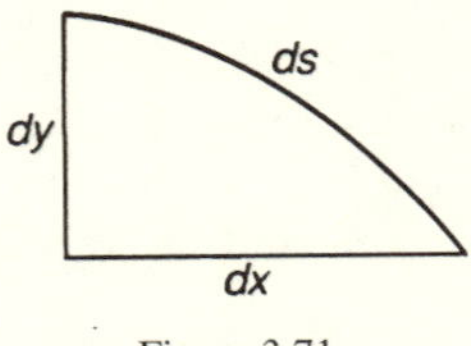

Figure 3.71

$$\int\sqrt{1 + (y')^2}\, dx,$$

and to "calculate."

The most astonishing thing is that Newton knows how to do all that, since he explained it previously in various places, in particular in *The Method of Fluxions*:

> The fluxion of the length of the curve is determined by putting it equal to the square root of the sum of the squares of the fluxions of the base and of the perpendicular ordinate.

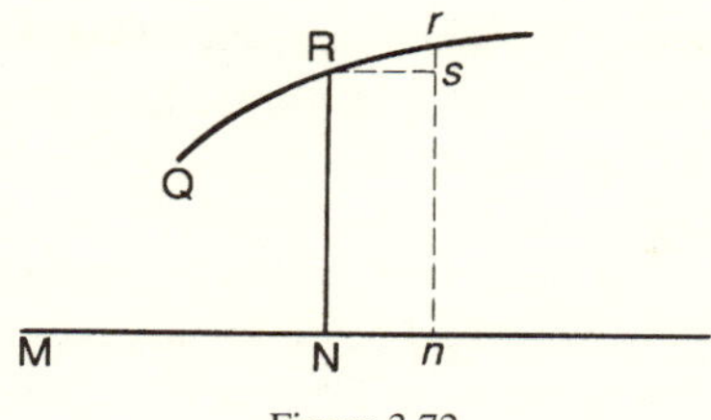

Figure 3.72

For let RN be the perpendicular ordinate on the base MN, and let QR be the proposed curve at which RN is terminated. And calling MN = s, NR = t, QR = v, and their respective fluxions p, q and l, let us conceive that the line NR advances to the next neighboring position nr [*ad locum quam proximum nr*]; if one drops on nr the perpendicular Rs, the lines Rs, sr and Rr will be the contemporaneous moments of the lines MN, NR and QR, by the addition of which they become Mn, nr and Qr. And since these moments are between them as the fluxions of these same lines, and since because angle Rsr is right therefore

$$\sqrt{Rs \times Rs + sr \times sr} = Rr,$$

we shall have

$$\sqrt{pp + qq} = l.$$

(Newton, *La méthode des fluxions*, 137; *NMP*, 3:304)

The *Principia* and Fluxions

Why did Newton not employ this procedure in the *Principia*? More generally, why is the distance apparently so great between the *Principia* and a fluxional calculus?

The example chosen is perhaps too particular: it has to do with a curve, and for curves Newton did not generally employ equations. The cycloid, like the spiral, is treated in terms more clearly kinematic or geometrical in *The Method of Fluxions* as in the *Principia*.

The distance between the two works is perhaps not so great if a double error of perspective is avoided: on the one hand the *Principia* is supposed to be a geometrical and synthetic translation of results obtained by "the new analysis"; on the other hand, *The Method of Fluxions*, and other similar writings by Newton, are taken to be only or principally expositions of a group of algorithms constituting a fluxional or "infinitesimal calculus." Newton himself did what he could to make this distorted view pass current, wishing in particular to appear equal or superior to Leibniz even in the elaboration of the new algorithm.

Scholars often contribute involuntarily to falsifying the perspective by selecting one or another example, or one or another procedure, which appears more interesting because it is more accessible to the modern reader. This is the case with the procedure of rectification that was examined on the preceding page (*NMP*, 3:304). In reality this procedure is, in *The Method of Fluxions*, mixed with others of a different character. For the spiral, Newton indicated a reasoning of a more strictly geometrical nature (*NMP*, 3:310–15). And a few pages previously, Newton had proposed a quite different means of calculating the lengths of curves with the help of evolutes (*NMP*, Prob. 10, 3:292ff.).

It remains true that Newton could have introduced into the *Principia* a considerably larger deployment of "calculatory" fluxional procedures. (At one time he thought of enriching a new edition of the *Principia* with an explanation of the theory of fluxions; see *NMP*, vol. 7, intro.) He made use of fluxions in Book II, with Lemma 2 on the "moments" of generated quantities. In Book I, the development of binomials in infinite series is invoked in places ("my method of infinite series," *Princ.*, Prop. 45, exs. 2 and 3, 137, 139). Newton even proposed to develop in series the ordinate of a curved trajectory; he asserted, as though in passing, that the force must be proportional to the term of this series in which the very small increment (*pars quam minima O*) is elevated to the second dimension (*Princ.*, Prop. 93, scholium, 221).

The need for an exposition of fluxional or infinitesimal procedures sometimes makes itself felt very strongly, and the demonstrations appear incomplete. Further on there is an example—one among others that could be cited—in which Newton presupposed tacitly a very general formula of integration, which it would be impossible to justify briefly on the basis of the lemmas of ultimate ratios (see below, p. 260); and there are also Propositions 79 and 81, in which it is a question of curved lines "of which the areas can be known by commonly known methods" [*quarum areae per methodos vulgatas innotescunt*] (*Princ.*, 204).

Why did Newton not specify such procedures in the *Principia*, and why was he so concerned to maintain the geometrical character of the work—even if the geometry there is somewhat singular? Esthetic preferences are not unconnected with this choice. During the era in which he wrote the *Principia,* Newton professed an admiration for the ancients and their style of demonstration. The *Principia* even contains a kind of manifesto for this "classical" mathematics in section 5 on the conics. Here Newton inserted a manuscript text that he had in reserve, and whose aim was to evince the supremacy of the synthetic method of the ancients over the new analysis. It was in particular the Cartesian algebra that Newton had it in mind to criticize:

> Their [the ancients'] method is much more elegant than that of Descartes. For he attains the result by means of an algebraic calculus which, if one transcribed it in words (in accordance with the practice of the Ancients in their writings) is revealed to be boring and complicated to the point of provoking nausea, and not

> to be understood. But they attained it by certain simple proportions, for they judged that what is written in a different style is not worthy to be read, and consequently they concealed the analysis by which they had found their constructions. (*NMP,* 4:277)

The touchstone of the debate was the "problem of Pappus," which Newton claimed to have resolved—for the case of four lines—without recourse to the procedures of Descartes, and in complete fidelity to the ancients:

> And thus to the four-line [locus] problem of the Ancients, begun by Euclid and continued by Apollonius, there is exhibited in this corollary, not the calculus, but the geometric composition such as the Ancients required. (*Princ.,* Lemma 18, Cor. 2, 78)

Non calculus, sed compositio geometrica: this maxim agrees with the *Principia* as a whole, at least if it is allowed that the geometry of the ancients can include quantities generated by motion and metamorphoses ending in an "ultimate" state.

To comprehend the nearly total absence of calculatory procedures, it is necessary to take account of one last element. The *Principia* is a supremely innovative book in what concerns physics, the system of the world, and the very foundations of the analysis of nature. This innovative character was sure to excite controversy. To combine with this innovative character another novelty, this time mathematical, and to make unpublished procedures in mathematics the foundation for astonishing physical assertions, was to risk gaining nothing.

The Marquis de L'Hôpital gave a proof of it *e contrario*. He, the publicizer of the new Leibnizian analysis, did not employ his own analytical techniques when demonstrating the formula for centrifugal force (L'Hôpital, "Solution d'un problème," 12–29). He established the physical result by geometrical methods; then he used his calculus of the *dx*, *ddx*, and so forth, at the moment of needing to resolve a particular problem. His *Analyse des infiniment petits pour l'intelligence des lignes courbes* had, however, appeared four years earlier.

This situation was not to last. The *Principia* was to be progressively translated into the new language, an enterprise which began precisely in 1700 with expositions by Varignon to the Academy of Sciences in Paris. A set of canonical methods inspired by Leibniz was to replace with advantage the geometrical procedures and ultimate ratios of the *Principia.* Newton's book was thus to fall into the category of outdated texts even before it had received the reception that was its rightful due. The contemporaries of Newton who could appreciate the work were very few, and the following generation, which should have furnished competent readers, became accustomed to other modes of reasoning.

There is reason to regret this outcome. Philosophers and humanists of this era and of later generations had the feeling that great marvels were contained in these pages; they were told that Newton revealed truth, and they believed it: Newtonian science became a "fact" (*ein Faktum*; see Kant, *Kritik der reinen Vernunft,* B 128). But the *Principia* still remained a sealed book. The singular character of the procedures of demonstration there is largely responsible.

There is still another reason, esthetic in nature, to regret this outcome. The style of reasoning of the *Principia* can be appreciated for itself. More than the more or less automatic methods of the "infinitesimal calculus," these forms of reasoning obliged Newton to use all his resources of invention, and they require of the reader an active intervention, an exercise in "seeing." It is necessary to learn to animate the figures and to follow the relations to the point at which the infinitesimal magnitudes vanish. This geometry, at once concrete and subtle, has its own kind of charm—if the word is not too incongruous in mathematics.

THE METHODS OF THE *PRINCIPIA* (2): THE INVERSE PROBLEM AND THE EMERGENCE OF A NEW STYLE

The *Principia* is far from being perfectly homogeneous in mathematical style. Certain passages approximate closely the methods that will have currency in the following century, with the Bernoullis and with Euler, if judged by such criteria as: the use of literal notations, the choice of reference systems with a "base" and an "ordinate," the lesser role accorded to the geometric representation of motions and forces, and finally the employment, albeit implicit, of the tools of the differential and integral calculus. The emergence of these new methods is seen in Book II, where they become necessary in treating of the resistance of mediums. The very nature of the questions sometimes necessitates going beyond the bounds of a purely geometrical investigation. A special case of this situation is presented by the inverse problem (which has been discussed in chapter 1 but set aside in order to follow Newton's exposition of the *De motu*).

The exchange of letters with Hooke had led to expectations of an early solution of the inverse problem, but then Newton chose a quite different path, appropriate only to the "direct" problem: he assumed an entirely determinate trajectory, along with a center of forces, and then evaluated the centripetal force on a point traversing this trajectory. But in so doing he replied to a different question. Newton did not always distinguish clearly between the two questions.

In the *De motu* of 1684, after Problem 3, devoted to the "direct" problem (in which it is proven that if the orbit is elliptical and the force directed towards a focus, the force is then inversely proportional to the square of the distance), Newton added a scholium that begins with:

> Therefore the major planets revolve in ellipses having a focus in the center of the Sun. (Hall, 253, 277)

But this conclusion is one of the premises of the reasoning.

It also happens that, in the *Principia*, Newton invokes Propositions 11–13 (the solutions of the direct problem leading to an inverse-square law) as if these propositions had answered the inverse problem (compare Prop. 58, Cors. 1 and 2).[31] Yet there is a sketch of a discussion, restricted to the case of conics, in Corollary 1 of Proposition 13:

> From the three last propositions it follows that, if any body P leaves the place P with any velocity along any straight line PR, and is at the same time acted upon by a centripetal force that is reciprocally proportional to the square of the distances of the places from the center, this body will be moved in some one of the conic sections having its focus in the center of forces; and conversely. For given the focus, the point of contact, and the position of the tangent, a conic section can be described, which will have at that point a given curvature. But the curvature is given from the given centripetal force and velocity of the body; and two orbits that will be tangent to each other cannot be described with the same centripetal force and same velocity. (*Princ.*, 59; the last part of the text beginning with "For given the focus . . ." is lacking in the first edition; the second edition fails to mention the dependence of the curvature on the velocity of the body.)

A somewhat more fully elaborated exposition is found in Proposition 17, which takes up nearly verbatim and develops Problem 4 of the *De motu* (see p. 46 above). The force is assumed to be as $1/r^2$, and the problem is to determine the trajectory. Here is the way Newton enunciates it:

> Supposing that the centripetal force is reciprocally proportional to the square of the distances of the places from the center, and that the absolute quantity of that force is known; the line is required that a body will describe if it leaves a given place with a given velocity in a given straight line. (*Princ.*, 63)

The method consists, as in the *De motu*, of imagining a kind of "test-particle" which traverses a determinate conic at another distance. Then in the deductive transition from the one orbit to the other, it is assumed that the other will also be a conic, but with a different parameter ("latus rectum").

One of the essential elements of this discussion is the determination of the "latus rectum" as a function of the velocity of projection and the "deflection." On a diagram inspired by those of Newton, if PR represents the velocity, or the interval that would be traversed on the tangent during the particle of time considered, and if QR is the deflection, which is proportional to the intensity of the force at P, then the "latus rectum" L is proportional to QT^2/QR (see p. 39 above).

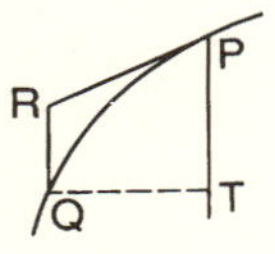

Figure 3 .73

All of this reasoning is restricted to the case of an inverse-square force, and presupposes that the solutions are conics. The only question is that of determining precisely the conic agreeing with the initial conditions. This is, therefore, an altogether partial and imperfect response to the general inverse problem, and indeed to Halley's question (which had to do only with the inverse-square case).

On the other hand, further along in Book I, in Proposition 39 and the following propositions, Newton considers this problem in a new and much more general form. The context of this new discussion contrasts sharply with that in the preceding sections of Book I. It is a sort of detour, in Book I of the *Principia*, that terminates unforeseeably in this new country. Here is the sequence of ideas and procedures: After the determination of conic orbits (sections 4 and 5), the next topic is the calculation of distance traversed in a given time on a known orbit (section 6). For this it is necessary to assign within the orbit a sector of given area centered at the focus. But this measure of area is restricted to conics and cannot therefore be valid except for inverse-square forces and for nondegenerate trajectories. In cases in which the orbit is rectilinear, that is, when the body falls directly toward the center of forces or moves away from it directly in a straight line, the same tools seem to be without leverage. Yet Newton, in a surprising way, succeeds in extending the application of the law of areas to rectilinear trajectories. This topic is dealt with in the early part of section 7 (Props. 32–37), which develops Problem 5 of the *De motu* (see above, p. 53). The result still remains restricted to the case of a force varying as r^{-2}, since the procedure consists in associating with the straight line an elliptical orbit with the center of forces at the focus.

How to resolve the general case and calculate the time of travel on a portion of a rectilinear trajectory for absolutely any force whatever? Newton has exploited the resources of the law of areas well beyond what might be expected. The new and far more general problem seems insoluble by the procedures previously employed. If the force is completely unspecified, it is impossible to associate with the trajectory a well-known curvilinear area as in Propositions 32–37.

Newton is thus engaged in an investigation of great generality and finds himself constrained to forge new tools. He comes to treat force in a more abstract and less geometrical way and to employ reasoning that is closer to the analytico-algebraic style that will flourish in the eighteenth century.

The Steps in the Construction of the Trajectory

The fundamental theory of this new approach corresponds to what will later be called the conservation of energy: the increase of the square of the velocity is proportional to the integral of the force along the path (the quantity that will be named "work"):

$$v^2 \propto \int F(x)\, dx.$$

This approach presupposes a certain manner of treating the force: with each point of the space around the center of force, a certain segment is associated which represents the intensity of the force in that point. All the segments representing the successive intensities at the different points of the line AC compose a surface proportional to the square of the velocity acquired. The velocity can thus be calculated at each point of the rectilinear path, and the evaluation of the time elapsed derived from it. The result is restricted, to be sure, to bodies that fall directly toward the center, or move away from it in a straight line.

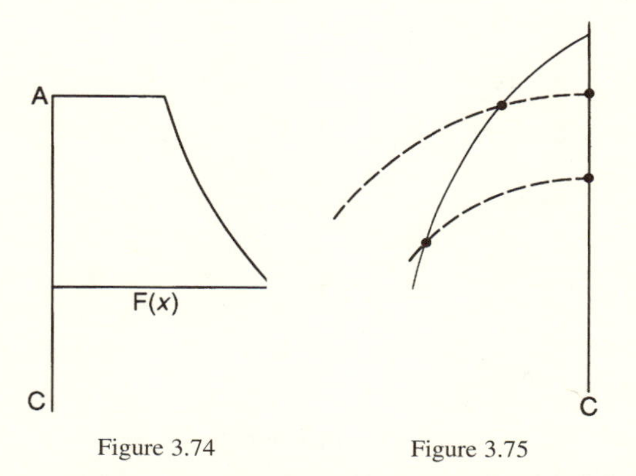

Figure 3.74 Figure 3.75

But a second remarkable theorem makes possible the extension of this integral evaluation of the velocity to oblique and curved trajectories. It is the very fruitful theorem demonstrated in Proposition 40: two bodies that have the same velocity at equal altitudes will always have the same velocity whenever their altitudes (their distances from the center of forces) are equal.

By virtue of these two theorems, new perspectives are opened up. The inverse problem is no longer beyond reach. Up to this point in the *Principia*, it has been necessary to know the trajectory in order to study the force, since the latter was manifested and measured by the divergence between the tangent and the actual trajectory.

With the new theorems, it becomes possible to study the effects of force without knowing the moving body's trajectory. One first considers a fictive body falling directly toward the center of forces, then transposes the data relative to its motion to the case of the actual body being studied. It will suffice to choose a fictive body that, at a certain altitude, has the same velocity as the actual body at the same altitude. Then the velocity of the fictive body in rectilinear fall will be known throughout its motion, by virtue of Proposition 39, and it will be the same as that of the actual body, by virtue of Proposition 40.

In this way, Newton succeeds in determining, in principle, the point at which the moving body will be located after a certain time. It is necessary to specify that this is possible in principle only, because there remain at the end of the reasoning a certain number of integrals not made explicit.

The mode of locating the trajectory is analogous to a system of polar coordinates: a first quadrature makes it possible to measure the time elapsed, and so to determine the altitude at which the body has arrived; then another quadrature makes it possible to calculate through what angle the body has turned round the center of forces.

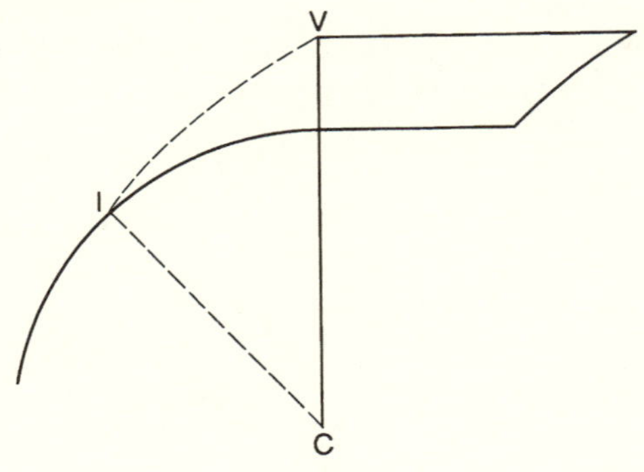

Figure 3.76

It can thus be asserted that, during the time considered, the moving body, having departed from V with a certain velocity, has been taken to I under the action of the force that attracts it towards C (Props. 41 and 42).

The Advantages of a True Integral Calculus

Is this the general solution of the inverse problem? In a sense, yes. Newton was capable, in principle, of determining point-by-point the trajectory of a body that begins to move with any initial velocity under the action of any central force depending on the distance.

But this method does not have the power of a veritable integration of the equations of motion. The latter would start from the expression of the accelerative effects of the force and would consist in first deriving—if it is possible—an integral that gives the velocity, up to a constant of integration; then from this first result deriving a second integral that makes known the position as a function of time, again up to a constant of integration.

Newton was acquainted with such procedures and used them in relatively complex cases at various places in *The Method of Fluxions*. There, Problem 2 consists in passing from fluxions to fluents, and one of the exemplar problems aims "to determine the space traversed in the total time, starting from the velocity as given for each instant of the time" (*NMP*, 3:90; see p. 210 above). Nonetheless, Newton seemed less skilled in treating problems of the second order, for which the original notation of *The Method of Fluxions* is ill-suited. In particular, in the study of curvature Newton sought to avoid fluxions of fluxions (*NMP*, 3:156ff.), and in the only known manuscript that deals with the fluxional definition of centripetal force, subsequent to the first edition of the *Principia*, the calculations are in several places erroneous and end in no conclusion (*NMP*, 6:588–98).

The advantage of a true integral calculus would be twofold:

> 1. The successive integrations of the differential equations of the force, then of the velocity, would cause to appear certain constants that correspond to the initial velocity and position and would show how these data are combined with the other terms to modify the nature of the motion.
>
> 2. The procedure of integration would lead to an analytico-algebraic expression for the trajectory as a function of a certain choice of coordinates, and the study of this expression would make it possible to know a priori the family of possible trajectories and to decide, for example, if the curves are algebraic or belong to a class of lines already known.[32] By contrast, the Newtonian method of Proposition 42, besides being very laborious, only permits determining successive points of the trajectory, in a blind, groping advance. It is impossible to trace the curve entirely, and if by chance it were algebraic or of another well-known species, nothing would permit it to be known along this route.

It is rather in the works of the disciples of Leibniz, among the adepts of the new "differential algorithm," that is found these sorts of general solution. Leibniz himself was conscious of the power of these tools; as he boasted to Huygens:

> You have remarked what is beautiful in our differential calculus, . . . namely that it gives us general solutions that lead naturally to the transcendentals, but which in certain cases cause the transcendence to disappear, and one discovers that the line is ordinary. (Leibniz to Huygens, 11 October 1693, *Mathematische Schriften*, 1:164)

One of the practitioners of these new methods, Johann Bernoulli, took up the imperfections of the *Principia* with respect to the inverse problem (in letters to Leibniz and in a communication to the Academy of Sciences in 1710). The mathematician of Basel reproached Newton for his premature or erroneous assertions concerning the determination of trajectories for an inverse-square force and proposed a quite new demonstration on this subject (approximately at the same time as another savant named Hermann).

These criticisms and new proofs, however, make evident Newton's merit and the irreplaceable contribution of the *Principia*. Indeed, if Johann Bernoulli corrected and improved on Newton, it was in supporting himself on the results and the procedures of the *Principia*. Similarly Varignon, who, starting in 1700, presented to the Academy of Sciences a number of studies on motions under the action of central forces, drew abundantly from the *Principia* in order to develop a theory more Leibnizian in style.[33] It was even exactly to the propositions on which I have just commented that Bernoulli and Varignon directed their attention: the first demonstrated Proposition 40 while simplifying a step, and the other reformulated Proposition 39.

The Theorem of Summation of Proposition 39

Here is the text of this proposition, so important and so new:

> *Proposition 39*. Given a centripetal force of any kind, and granting the quadrature of curvilinear figures, there is required, for a body ascending or descending in a straight line, on the one hand the velocity at each position, and on the other, the time it takes to arrive at any place. And conversely.
>
> Let a body E fall from any place A in the straight line ADEC; and from its place E let there always be erected the perpendicular EG, proportional to the centripetal force tending in that place to the center C. And let BFG be the curved line which the point G perpetually touches. In the beginning of the motion let EG coincide with the the perpendicular AB; and the velocity of the body in any place E will be as the straight line whose square is equal to the curvilinear area ABGE. Which it was required to find.
>
> On EG let the straight line EM be taken such as to be inversely proportional to the straight line whose square is equal to the area ABEG, and let VLM be the curved line that the point M perpetually touches, having for asymptote the straight line AB prolonged. Then the time during which the body in falling describes the line AE will be as the curvilinear area ABTVME. Which it was required to find.

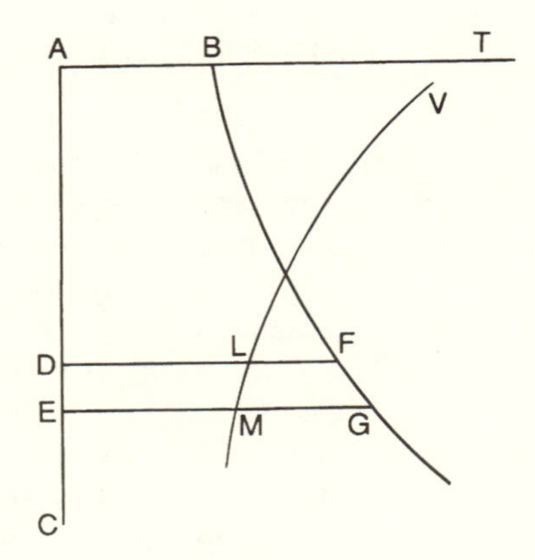

Figure 3.77

For on the straight line let there be taken the very short line [*linea quam minima*] DE of a given length, and let DLF be the place of the line DMG when the body was in D.

If the centripetal force is such that the velocity of descent is as the square root of the area ABGE, the area itself will be as the square of the velocity; that is, if for the velocities in D and E we write V and V + I, the area ABFD will be as VV, and the area ABGE will be as VV + 2VI + II; by subtraction the area DFGE will be as 2VI + II; consequently DFGE/DE will be as (2VI + II)/DE, that is, if we take the first ratios of the nascent quantities, the length DF will be as the quantity 2VI/DE, hence also as the half of this quantity IV/DE.

But the time in which the body in falling describes the very small line [*lineolam*] DE is directly as this small line and inversely as the velocity V; and the force is directly as the increase I in the velocity and inversely as the time; consequently the force is as IV/DE, that is to say as the length DF.

Therefore a force proportional to DF or EG will cause the body to descend with a velocity which is as the square root of the area ABGE. Q.E.D.

Moreover, since the time in which any very small line DE of given length is described is inversely as the velocity, and hence inversely as the square root of the area ABFD; and since the line DL—and therefore also the nascent area DLME—is inversely as this same root; therefore the time will be as the area DLME, and the sum of all the times will be as the sum of all the areas; that is to say (by the corollary to Lemma IV), the total time in which the the line AE is described will be as the total area ATVME. Q.E.D. (*Princ.*, 120–21)

Newton presents the solution of this "problem" in the manner of Euclid, carrying out the construction that leads to the required result; but the sequence of steps remains incomprehensible at this point, and it is only in a second stage that he proceeds to demonstrate that the task originally proposed has been accomplished in the construction.

The chief novelty here resides in the manner of treating force. Its action on the moving body does not appear geometrically as in Propositions 1–17; Newton restricts himself to assigning to each point of space a certain intensity of force. This intensity is represented by a horizontal segment: at the point D by DF, and at the point E by EG.

Newton specifies at the end of section 8 that the centripetal force in these propositions is supposed "to vary in its recess from the center according to any law that anyone may imagine, but that at equal distances from the center it is everywhere the same" (*Princ.*, 129). In other words, the centripetal force depends solely on distance from the center, and its variation obeys a "law," which undoubtedly implies a variation that is at least continuous, in the sense given to that word today. Therefore the successive segments that represent the force, and are erected perpendicularly on AC, have their other extremities on a certain "curve" BFG.

The demonstration of the proposition is schematically as follows:

> If the velocity is proportional to the square root of the area ABFD, then the segment DF is proportional to the product velocity × increment in velocity, divided by the element of distance traversed.
>
> The force is also proportional to this same magnitude.

Newton symbolizes the velocity by V and the increment in the velocity by I, while the element of distance traversed corresponds to DE in the figure. The reasoning thus reduces to proving that both DF and the force in D are proportional to $V \times I/\mathrm{DE}$.

It is supposed first that the velocity is proportional to the square root of the area:

$$v \propto \sqrt{\mathrm{ABFD}},$$

or

$$v^2 \propto \mathrm{ABFD}.$$

Next the letter V is chosen to designate the velocity at D, and the letter I to designate the increment in velocity between D and E. The square of the velocity at E is then proportional to the corresponding area:

$$(V+I)^2 \propto \mathrm{ABGE}.$$

The difference in the values at D and E is then:

$$(V+I)^2 - V^2 \propto \mathrm{ABGE} - \mathrm{ABFD},$$

or

$$2I \cdot V + I^2 \propto \mathrm{DFGE}.$$

Both sides of the proportionality are then divided by the element of length DE:

$$\frac{2I \cdot V + I^2}{\mathrm{DE}} \propto \frac{\mathrm{DFGE}}{\mathrm{DE}}.$$

If I and DE are very small, in the "nascent" state as Newton puts it, then I^2 can be neglected, and DFGE considered as a rectangle. Then:

$$2\frac{I \cdot V}{\mathrm{DE}} \propto \mathrm{DF},$$

or

$$\mathrm{DF} \propto \frac{I \cdot V}{\mathrm{DE}}.$$

On the other hand, the force is by definition as the increment in the velocity and inversely as the time, hence as I/time. The time itself can be expressed as a function of the length and the velocity (considered as uniform over the very small distance DE): time $\propto$ DE/V. Thus the force is as I/(DE/V) or $I \times V$/DE. It is therefore proportional to the same quantity as DF.

In the midst of this demonstration, Newton gives a specification of great value with regard to the centripetal force: it is "directly as the increment in velocity and inversely as the time" [*estque vis ut velocitatis incrementum I directe et tempus inverse*] (*Princ.*, 121; Cajori [126] in effect lacks this important point because it is made to appear as a consequence of what precedes). This property of force (centripetal or impressed) is on first view more precise and explicit than anything found in the initial enunciations of Book I concerning impressed force, and nearer, too, to the modern way of defining force.

Law II is enunciated as follows: "The change of motion [*mutatio motus*] is proportional to the motive force impressed; and is made in the direction of the straight line in which this force is impressed" (*Princ.*, 13). Apparently time does not figure in this enunciation, while the demonstration of Proposition 39 makes explicit the inverse proportionality between force and time. Moreover, "change of motion" is a vaguer notion than that of an increment in velocity for a given time.

But it is important to place these enunciations in the conceptual framework of the seventeenth century. Velocity was not conceived vectorially; hence "change of motion" was not necessarily a change in velocity. Further on, in Proposition 40, it will be seen that change of motion (the effect of force) can assume two aspects: deviation or change of velocity (*deflectere*/*accelerare*). Here, in Proposition 39, there is only a question of change in velocity, since the fall is rectilinear; in this particular case, force manifests itself solely as change in speed. Deviation was by nature more accessible to a geometric representation, while acceleration in speed required more abstract means of study.

A greater abstraction is also to be observed in the mode of reasoning and in the notation. Newton introduces magnitudes that are not at all present in the figure, such as V and I, and he imposes names on them, as in algebraic problems (see the discussion of "the imposition of names" in Newton's *Arithmetica universalis*, 89). (The reader can even be a bit confused in distinguishing between the letters designating points in the diagram and those that "name" quantities.) Also, the infinitesimal content of the reasoning is manifest not only in ultimate or nascent geometrical situations, but also in the rejection of the term I^2, the product of a nascent quantity by itself.

This passage of Book I is perhaps the one in which the emergence of a differential and analytic style is most evident. If one writes DE in the form of an element of distance dx, and V and I as v and dv, the relations that Newton announces correspond to: $F \propto dv/dt$ and $dx = v \cdot dt$. In this vocabulary the result of the demonstration could be transcribed as: $\int F(x)dx \propto \int v \cdot dv$.

This analytic transformation was proposed beginning in 1700 in the memoirs that Varignon presented to the Academy of Sciences. Varignon first proposed two "rules" for the determination of forces and velocities, in the case of rectilinear motions (with the force designated by the letter y):

> General Rules for rectilinear motions:
> 1. $v = dx/dt$
> 2. $y = dv/dt$ ($= ddx/dt$).
>
> (Varignon, "Manière générale," 32)

Some pages later, Varignon deduces from these rules the result expounded by Newton in Proposition 39:

> It is good to note that without any new hypothesis the two preceding rules give us at one blow the Proposition 39 of Book I of Newton's *Phil.Nat.Princ.Math.* Indeed, these two rules give $dt = dx/v$, $dt = dv/y$; hence $dx/v = dv/y$, or $ydx = vdv$, which gives also ($\int$ signifies sum) $\int ydx = \frac{1}{2}v^2$ or
>
> $$v = \sqrt{2\int ydx}:$$
>
> that is to say in general the velocities VH vary as the square roots of the spaces contained between AH, Fh, and the curve MF, whatever the forces y (FH); just as M. Newton demonstrated in the first part of this thirty-ninth proposition. (Ibid., 37)

Then the second part of Proposition 39 follows from

$$dt = \frac{dx}{\sqrt{2\int ydx}}.$$

The passage, in the heart of the *Principia*, from sections 1 and 2 to Proposition 39, represents a very marked evolution in the direction of increasing abstraction. With Varignon, a new step is taken in this direction toward an analytic and differential treatment of force. The geometrical intermediaries in the reasoning, which are still present in the text of the *Principia*, are no longer at all necessary in Varignon's text (the figure that Varignon proposes to accompany his text is confused, vague, and without any consequences for the reasoning).

Although he belongs to the circle of adepts of the new Leibnizian calculus, Varignon does not seem to remark that the announced result corresponds to what Leibniz, following Huygens, had called conservation of "ascensional" force or "*force vive*" (*LMS*, 6:218) and what today would be called conservation of energy: the sum of potential energy (the inverse of work: $\int F(x)dx$) and of kinetic energy ($mv^2/2$) remains constant for the system throughout its evolution in time. Force is expended along the path by producing a determinate quantity of velocity-squared multiplied by half of the mass.

Galileo had given the first point of departure for such an idea, by showing that, for a falling body, the square of the velocity is proportional to the height (since the velocity increases proportionally to the time, and the space traversed increases as the square of the time). Torricelli had employed this result to draw from it by analogy the idea that $v^2 \propto h$, for water that flows from the bottom of a receptacle filled with liquid to the height h. Huygens, starting from the same result, had conceived the principle of a conservation of "force" (*HO*, 18:477), and Leibniz constructed his dynamics on the basis of this principle.

In Newton's work (as also in Varignon's, it appears), there is only the mathematical skeleton of a reasoning about energy. Newton does not draw profit from the Galilean heritage relative to conservation and the expenditure of force (see above, p. 90). The mathematical lineaments of Proposition 39 are in fact only a skeleton; there is lacking the physical, philosophic, or conceptual flesh. A veritable energetic reasoning would treat force globally, as an entity that is transmitted, expended, and changes form. Force has none of these properties in the *Principia*, in the first place because its mode of action remains too enigmatic, and secondly because of its local character: centripetal force is calculated in the neighborhood of a point of the trajectory as the nascent difference between the inertial trajectory and the real trajectory.

The Passage from the Rectilinear to the Oblique Route in Proposition 40

Newton's enunciation—and Varignon's—is restricted to the rectilinear case. The fruitful utilization of the conclusion for a solution of the inverse problem presupposes that the results obtained can be transposed to other trajectories, oblique or curvilinear. It is this transposition that Proposition 40 permits.

> *Proposition 40.* If a body under the action of any centripetal force is moved in any manner, and another body ascends or descends in a straight line, and their velocities are equal in a case in which their altitudes are equal, then their velocities will also be equal at all equal altitudes.
>
> Let a body descend from A through D and E toward the center C, and let another body be moved from V in the curved line VIKk. From the center C, with any intervals, describe the concentric circles DI, EK intersecting the straight line AC in D and E, and the curve VIK in I and K. Join IC meeting KE in N, and on IK let fall the perpendicular NT. Let the interval DE or IN between the circumferences of the circles be very small [*quam minimum*], and let the bodies have equal velocities in D and I.
>
> Since the distances CD, CI are equal, the centripetal forces will be equal in D and I. Let those forces be represented by the very short, equal straight lines DE, IN; and if the one force IN is resolved (by virtue of Corollary II of the Laws) into two others NT and IT, the force NT, acting along the line NT perpendicular to the course ITK of the body, will not at all change the velocity of the body along this

path, but will only draw the body away from a rectilinear course, and cause it always to deflect from the tangent to the orbit, and to progress along the curvilinear path ITKk. This force is totally expended in producing this effect. But the other force, acting along the body's path, will altogether be employed in accelerating it, and in a very short given time will generate an acceleration proportional to itself.

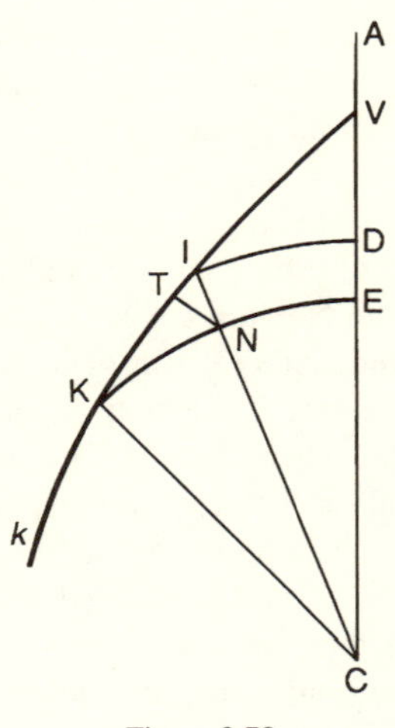

Figure 3.78

Thus the accelerations of the bodies in D and I, produced in equal times (if we take the first ratios of the nascent lines DE, IN, IK, IT, NT), are as the lines DE, IT, and therefore in unequal times they are conjointly as these lines and the times. But the times in which DE and IK are described, because of the equality of the velocities, are as the described paths DE and IK.

Consequently the accelerations, in the course of the bodies along the lines DE and IK, are conjointly as DE and IT, and DE and IK, that is, as DE squared and the rectangle IT × IK. But the rectangle IT × IK is equal to IN squared, that is, to DE squared, and it follows that equal accelerations are generated in the passages of the bodies from D and I to E and K. Therefore the velocities of the bodies are equal in E and K; and by the same argument they will always be found to be equal in any subsequent equal distances. Q.E.D.

And by the same argument, bodies of equal velocity and equally distant from the center, in their ascent will at equal distances from the center be equally retarded. Q.E.D. (*Princ.*, 123–24)

This result generalizes a proposition due to Galileo that Newton invokes in Book II of the *Principia*: "For a heavy body takes a longer time to descend to the same depth by an oblique line than by a perpendicular line, but in descending acquires the same velocity in either case, as Galileo has demonstrated" (*Princ.*, Prop. 36, 331). To Galileo, the force is uniform and the trajectories are rectilinear; here the equality of the velocities after the "descents" must be understood in a much more general sense.

Since the force depends solely on distance, it is equal at equal distances, but it is more or less oblique in relation to the direction of motion. The segment IN, which represents in I a force equal to the force in D, is decomposed into two segments: IT represents the part of the force that serves to "accelerate" the moving body, and the other segment IN represents the part of the force which "deflects" the moving body from its trajectory. This way of decomposing the force is invaluable because it indicates in what sense it is necessary to understand the expression "change of motion" in Law II: this change can be a change of "velocity" or a change of direction.[34]

Newton extends this decomposition to a somewhat different case, that of motions subject to constraints, in Corollary 1 which follows:

> *Corollary 1*. Thus if a body oscillates when suspended by a string, or is forced to move on a curved line by any impediment that is well polished and perfectly smooth, and another body ascends or descends in a straight line; then if their velocities in any altitude are equal, their velocities in any other equal altitudes will be equal. For the string of the pendulum or the impediment of the absolutely smooth receptacle do the same thing as the transverse force NT. The body by this is not retarded or accelerated, but only forced to depart from its rectilinear course. (*Princ.*, 124)

In what sense could this brief paragraph anticipate the formulation of "d'Alembert's principle" for constraint systems? That discussion would go beyond the scope of the one here and will be left to more expert "mechanists," who can accurately assess Newton's boldness: with what right does he treat these two decompositions of force equally?

The demonstration of Proposition 40 consists in showing the equality of the accelerations along DE and along IK. The acceleration is as the force and the time conjointly, that is, in the ratio of DE × (time on DE) to IT × (time on IK). On the other hand, the time is here proportional to the space traversed, since the velocities in D and I are equal, and they can be considered as constants. The accelerations are thus in the ratio of DE × DE to IT × IK. But IT × IK = $IN^2 = DE^2$. The accelerations are therefore equal over the two paths.

The same demonstration will be taken up again nearly 25 years later by Johann Bernoulli, who claims to simplify it:

> If two bodies, with masses proportional to their weights, begin to descend from the same point A with equal velocities, and with equal forces towards the same point O, the one directly following the straight line AO, and the other obliquely following the trajectory ABC that it describes: I say that at all equal distances on one side or the other from the center O of forces, as at B, E, imagining the circular arc BE described from this center O; these two bodies will always have equal velocities: so that if EG represents the velocity acquired at E by the body that descends along AO, the same EG will represent also the velocity at B of the body that describes the trajectory ABC.
>
> The demonstration of this lemma is found in M. Newton's book, *Princ. Math.Phil.Nat.*, p. 125. But there it is too muddled; I give it here more simply.
>
> Let us conceive the curve ABC and the straight line AO divided into their elements Bb, Ee, by an infinity of circles BE, be, infinitely near one another, all described from the center O. This being conceived, Mechanics shows that everywhere the force in each point along EO (which by hypothesis is the same as in each corresponding point B along BO) is to what there results from it for the moving body, following each corresponding element Bb of the curve that it traces,

as Bb is to Ee. But Mechanics also shows that the increments in velocity that result from these forces in equal bodies are between them in the ratio compounded of these same forces, and of the elementary times employed by them to produce these increments in velocity, that is, employed to cause these bodies to traverse the linear elements Ee, Bb; and that at the commencement in A, where these velocities are supposed equal on both hands, these elements of time are between them as the first of these small lines Ee, Bb: it follows thence that these increments of velocity in the first Ee, Bb, are between them in the ratio compounded of Bb to Ee, and of Ee to Bb, that is, as Bb × Ee is to Ee × Bb. Therefore at the end of these first elements of the lines AO, ABC, the increments of velocity following these first linear elements will here be equal to each other.

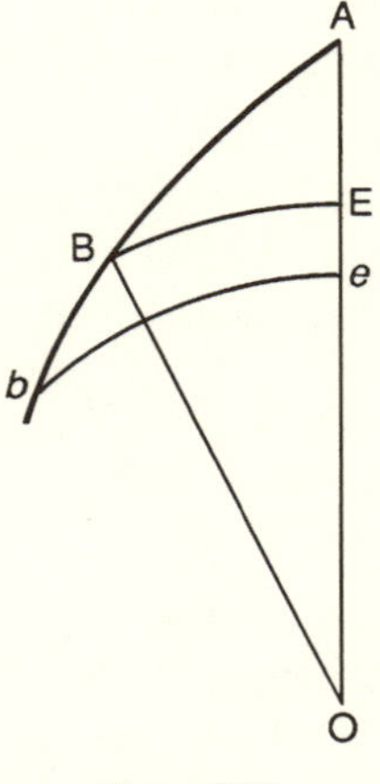

Figure 3.79

Similarly, they will be demonstrated equal to each other at the end of the second elements of these lines AO, ABC; at the end of the third; at the end of the fourth, etc., thus taken two by two at equal distances from the point O. Therefore at equal distances from this point O, the velocities along each of the lines AO, ABC, being thus made equal by hypothesis at the beginning, and receiving an equal number of increments which were two by two equal to one another, will also be equal to each other. Which it was necessary to demonstrate. (Bernoulli to Hermann, 7 October 1710, *Opera*, 1:472–74)

Wherein lies the simplification that Bernoulli contributes? To evaluate the ratio between the force along the oblique trajectory and the force along the vertical, Newton invokes the decomposition permitted by Corollary 2 of his laws and makes abstraction of the deflecting component TN. Bernoulli obtains this same ratio of the forces more immediately, by virtue of "Mechanics."

He claims thus to link himself, more directly than Newton, to the tradition of machines and weights. It is from this, he asserts, that derive the two principles that he invokes (*les Méchaniques font voir, les Méchaniques faisant voir aussi*). The first, which could be called the principle of the inclined plane, often explained and discussed in the course of the preceding century,[35] was indubitably part of Mechanics. Here time and distance play no role, or at least it is not necessary to introduce them (the ratio of the forces can be evaluated at equilibrium, as Stevin does). The second principle that Bernoulli claims to take from Mechanics is much newer; it is the relation between the increment in velocity and the time, the relation announced by Galileo after so much effort and so many detours, and which he formulated only for the case of gravity. Assigning to these two principles the same rank, Bernoulli reduces to banality, so to speak, the second law of the *Principia* and the formulations that prepared the way for it.

By virtue of these two mechanical principles, force is here found to be in the inverse ratio of the time of passage; the increment in velocity is therefore as the product of the ratio of the lengths and of the inverse ratio, hence as Bb × Ee : Ee × Bb. The velocities after a "linear element" will therefore be once again equal.

In passing from "elements" to finite distances, Bernoulli is hardly more careful than Newton. The reasoning is valid only for very small distances on which the velocity does not vary. But Newton and Bernoulli assume that it is permissible to iterate this reasoning for a succession of intervals, so as to obtain a finite distance (any attempt to translate this sort of reasoning into the terms of "non-Archimedean" infinitesimals is likely to encounter grave obstacles).

The slicing up of the trajectories is not completely the same for Newton and Bernoulli, but in either case is difficult to justify if a critic is overly rigorous, an emulator, say, of Berkeley. Bernoulli took care to divide the lines AO and ABC "into their elements" such as Ee and Bb; but why should an element of the curve be longer than an element of the straight line? Newton for his part decided that the interval between the two concentric circles was "the smallest possible"; but how, under these conditions, could two other smaller intervals IT and TK be distinguished in the interval IK? (It is worth noticing that, in contrast to most of the physical reasonings of Book I of the *Principia*, the infinitesimal that serves as fundamental variation here is not a time, but a length.)

Before turning to the solution of the inverse problem, Newton applies the results so far obtained to a more limited question: the determination of velocities. He supposes that there exists, at some place along the trajectory, a point of maximal distance at which the body has a null velocity. He supposes also that the force is proportional to a certain power of the distance. The velocity of this body at any other point can then be found by integration:

> *Corollary 2.* If a quantity P is the maximum distance from the center to which the body can ascend, whether in oscillating or in revolving in any trajectory, when it is projected upwards from any point of the trajectory with the velocity it there has; and if on the other hand the quantity A is the distance of the body from the center in any other point of its trajectory, and if the centripetal force is always as some power A^{n-1} of this quantity A, of which the index $n - 1$ is any number n diminished by unity; the velocity of the body in any altitude A will be as
>
> $$\sqrt{P^n - A^n},$$
>
> and therefore is given. For the velocity of rectilinear ascent or descent (by Proposition 39) is in this ratio. (*Princ.*, 124)

A fictive body departs with zero velocity from the same distance as the maximum P of the body investigated and is directed straight towards the center of forces. The velocity of this second, fictive body is known by virtue of Proposition 39, and it will be the same as that of the body studied whenever their distances are the same.

Since the force is proportional to the power $n - 1$ of the distance (n is a "number," which excludes at least fractional exponents), the velocity of either body will be, at the distance A, as

$$\sqrt{\int_{\mathrm{P}}^{\mathrm{A}} x^{n-1}\,dx}\,.$$

Newton asserts, without justification, that this expression can be replaced by

$$\sqrt{\mathrm{P}^n - \mathrm{A}^n}\,.$$

Nothing in the preceding results of the *Principia* permits the establishment of this formula of integration, which must be admitted without proof or discussion (the constant $1/n$ and the choice of sign are of little importance in a proportionality).

The Determination of the Trajectory in Proposition 41

Proposition 41. Supposing a centripetal force of any kind, and granting the quadrature of curvilinear figures, it is required to find, first, the trajectories in which bodies will be moved, and secondly, the times of their motions in the trajectories found.

Let any force tend to the center C, and let it be required to find the curve VIKk. Let there be given the circle VR, described about C with any radius CV, and about the same center let there be described any other circles ID, KE, cutting the trajectory in I and K and the straight line CV in D and E. Draw the straight line CNIX cutting circles KE, VR in N and X, and the straight line CKY meeting circle VR in Y.

But let the points I and K be very near [*vicinissima*] to each other; and let the body go from V through I and K to k; and let the point A be that place from which another body must fall, in order to acquire in place D a velocity equal to the velocity of the prior body in I.

With all things remaining as in Proposition 39, the very short line IK, described in the very least [*quam minimo*] given time, will be as the velocity, and therefore as the square root of the area ABFD; and the triangle ICK, proportional to the time, will be given; consequently, KN will be inversely as the height IC; that is, if there by given any quantity Q, and the height IC is named A, as Q/A. This quantity Q/A we shall name Z, and suppose the magnitude of Q to be such that in some one case

$$\sqrt{\mathrm{ABFD}} : \mathrm{Z} = \mathrm{IK} : \mathrm{KN},$$

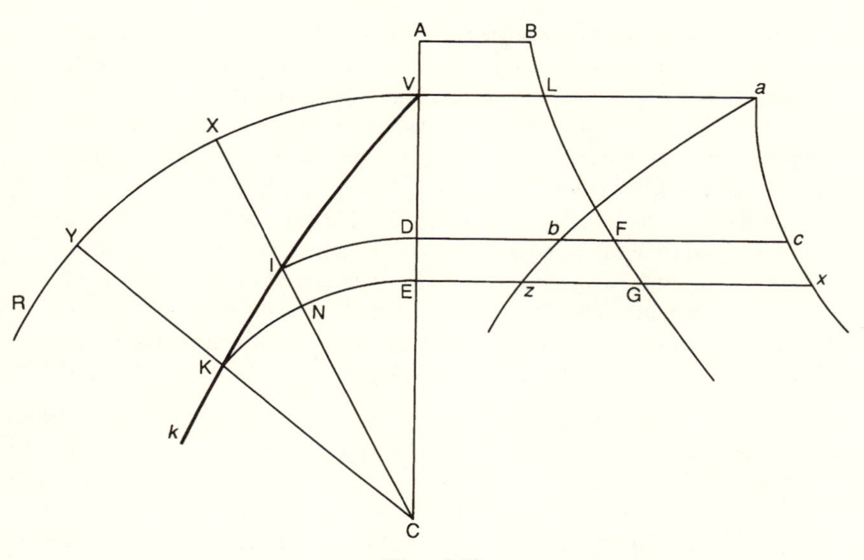

Figure 3.80

and then in all cases

$$\sqrt{\text{ABFD}} : \text{Z} = \text{IK} : \text{KN},$$

and

$$\text{ABFD} : \text{ZZ} = \text{IK}^2 : \text{KN}^2,$$

and *separando*

$$(\text{ABFD} - \text{ZZ}) : \text{ZZ} = \text{IN}^2 : \text{KN}^2,$$

and therefore

$$\sqrt{(\text{ABFD} - \text{ZZ})} : \text{Z or Q/A} = \text{IN} : \text{KN},$$

and

$$\text{A} \cdot \text{KN} = \text{Q} \cdot \text{IN}/\sqrt{(\text{ABFD} - \text{ZZ})}.$$

Since

$$\text{YX} \cdot \text{XC} : \text{A} \cdot \text{KN} = \text{CX}^2 : \text{AA},$$

the rectangle XY · XC will be equal to $\text{Q} \cdot \text{IN} \cdot \text{CX}^2/\text{AA}\sqrt{(\text{ABFD} - \text{ZZ})}$.

Therefore if in the perpendicular DF there be taken Db, Dc always equal to $\text{Q}/2\sqrt{(\text{ABFD} - \text{ZZ})}$ and to $\text{Q} \cdot \text{CX}^2/2\text{AA}\sqrt{(\text{ABFD} - \text{ZZ})}$ respectively, and there be described the curved lines ab, ac, which perpetually touch the points b, c; and at the point V there be erected Va the perpendicular to the line AC, cutting off the curvilinear areas VDba, VDca, and there also be erected the ordinates Ez, Ex: then because the rectangle Db · IN or DbzE is equal to half the rectangle A · KN, or to the triangle ICK; and the rectangle Dc · IN or DcxE is equal to half the rectangle YX · XC or triangle XCY; that is, because the nascent particles DbzE, ICK of the areas VDba, VIC are always equal, and the nascent particles DcxE, XCY of the areas VDca, VCX are always equal: therefore the generated area VDba will be equal to the generated area VIC, and hence proportional to the time; and the generated area VDca will be equal to the generated sector VCX.

Therefore if any time be given during which the the body has been moving away from V, there will be given the area VDba proportional to it, and thence will be given the altitude of the body CD or CI, and the area VDca, and the sector VCX equal to it, with its angle VCI. But if the angle VCI and the altitude CI are given, so also is given the place I, in which the body will be found at the end of that time. Which it was necessary to find. (*Princ.*, 125–27)

The essential features of the demonstration are not easy to discern at first sight. The body in motion starts from V in a certain direction. Where will it be found after a certain finite time? This point will be located by a kind of circular system of coordinates: a distance to the center C, and a rotation around C. These two magnitudes will be determined by the intermediary of two areas, limited by two curves abz and acx, which have to be constructed.

First, associated with the motion of the body starting from V is the motion of another body starting from A with zero velocity and falling towards C (as in the preceding demonstrations). The velocity of this second body will be known through the quadrature of the area ABFD ("the quadratures of curvilinear figures" are allowed in advance). The point A is so chosen that the velocity of the second body and that of the first will be equal at equal altitudes.

The given element of time is represented by the area of the nascent sector ICK. Since the area is given, the height KN of this triangle (considered as rectilinear) is inversely as its base IC. On the other hand, IK is as the velocity in I (for a given element of time), and therefore it is as the square root of the area ABFD.

The ratio IK/KN can therefore be expressed in the form (square root of the area ABFD)/(inverse of the altitude IC), that is, in Newton's notation:

$$IK/KN = \sqrt{(ABFD)}/Z, \text{ or } IK/KN = \sqrt{(ABFD)}/(Q/A),$$

where A designates the altitude IC, and Q a constant adjusted once for all.

After some transformations of this proportionality, the product A · KN comes to be expressed as

$$A \cdot KN = IN \cdot Q/\sqrt{(ABFD - ZZ)}.$$

This product A · KN is (by hypothesis) the area of the nascent sector ICK.

Similarly is obtained an expression for the product YX · XC, which represents the nascent circular area:

$$XY \cdot XC = IN \cdot Q \cdot CX^2/AA\sqrt{(ABFD - ZZ)}.$$

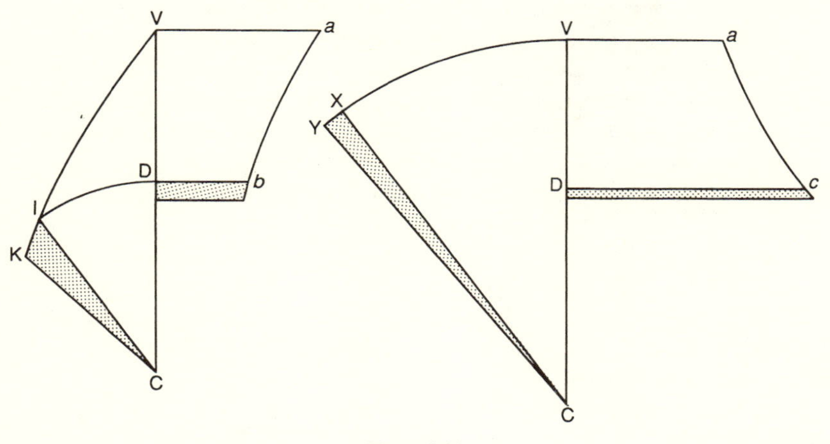

Figure 3.81

Next are constructed two curves that have precisely these two expressions for element of area:

1. If IN, or DE, is the element of the abscissa, and if Db is equal to $Q/\sqrt{(ABFD - ZZ)}$, then the elementary area $DE \cdot Db$ will be equal to the sector ICK.

2. If, with DE again as the element of the abscissa, the ordinate $Dc = Q \cdot CX^2/AA \cdot \sqrt{(ABFD - ZZ)}$, then the elementary area $DE \cdot Dc$ will be equal to the nascent sector XCY. (A is known by virtue of the preceding curve.)

The points b and c run along the curves, and the elements of area delimited by these curves and the points D and E on the abscissa, namely DbzE and DcxE, are always equal, respectively, to the areas of the corresponding sectors (ICK and XCY). As particles of the areas are equal, the total areas are equal also.

Therefore VDba represents the time elapsed, and VDca represents the circular area VCY (with constant radius). The knowledge of the first area makes it possible to know at what altitude CD the body is found after the proposed time has elapsed. The second area indicates through what angle the radius VC has turned around C. Thus the position I is determined by means of a distance CD and an angle VCX.

A detailed discussion of this demonstration and its possible use would range too far afield. It would be necessary, in particular, to clarify the significance of the ratio IK : KN (which designates, it could be said, the inclination of the trajectory with respect to the concentric circles; see Corollaries 1 and 2 of Proposition 41), and the nature of the constant Q (see Proposition 42) in order to show that the curves ab and ac are well determined.

Thus, at the cost of three quadratures can point I be located, in which the body is found after a certain interval of time. But very little is revealed about the global form of the trajectory or about its eventual analytic expression and the role that constants will play in it. Johann Bernoulli, by contrast, discusses very carefully the nature and role of the constants of integration in the solution of this "inverse problem" and reproaches his rival Hermann with having been too negligent on this point. And he also boasts of the superiority of his method for determining the class to which the solution-curves belong:

> My equation shows, moreover, whether the trajectory is algebraic or not, given any hypothesis whatever as to the forces. (Bernoulli, *Opera*, 1:475)

In strict truth, it was therefore Johann Bernoulli who answered Halley's question, and it is he who should have received the forty-shilling book promised by Sir Christopher Wren.

The cumbrousness and lack of generality in Newton's procedures serve to emphasize the merits of the new modes of "calculus." The geometry of the *Principia* has become too narrow a framework for the study of forces.[36]

CONCLUSIONS

Mathematics as an Inductive Tool

"Direct problem," "inverse problem": Newton proposed demonstrative means for passing from motions to forces and—in a more uncertain and imperfect way—from forces back to motions. Only the mathematical aspect of these operations has been considered, not the physical significance and scope they have in establishing "the constitution of the System of the World" in Book III.

The "direct" passage makes it possible to "infer" the unique force that moves the planets, the satellites, projectiles, and so forth. *To infer* is also "to gather together," according to the connotation of the Latin *colligere*—to unify the diversity. Here, briefly, are the essential steps of this deduction.

That the force retaining the planets (and the satellites) tends toward the sun (toward the principal planets) is proved by the regular sweeping out of areas around these centers (first part of Propositions 1, 2, and 3 of Book III). It is the converse of the law of areas: no longer the deduction of the regular sweeping out from the centripetal force, but rather of a centripetal force and of its center from the regular sweeping out. (This converse or inverse, absent from the *De motu*, is demonstrated as Proposition 2 in Book I of the *Principia*.)

Then the precise form of the law of force is deduced from the trajectory. It might be expected that Newton would here apply Proposition 11 of Book I (Proposition 3 of the *De motu*): if the body moves on an ellipse with the centripetal force directed toward a focus, then the force varies as $1/r^2$. In fact the first three propositions of Book III proceed otherwise. The "phenomena" indicate that the planets and satellites obey Kepler's third law (the ratio between the cube of the major axis and the square of the time of revolution is constant). It is therefore possible to apply Corollary 6 of Proposition 4 of Book I, relative to uniform circular motion (see p. 29 above: the corollaries to Theorem 2 of the *De motu*). Newton was satisfied at this stage with a circular approximation for the orbits.

He indicated as though in passing another possible route for the deduction: the immobility of the apsides of the orbits, or at least the slowness of their motion, which shows that the force varies nearly as $1/r^2$. This result follows from section 11 of Book I, concerning rotating orbits:

> If the body after each revolution returns to the same immobile apse, . . . the decrease in the force will be in the duplicate ratio of the altitude [*erit . . . decremen-*

> *tum virium in ratione duplicata altitudinis*],[1] as was demonstrated in the preceding. (*Princ.*, 140)

This route is only suggested in the case of the planets; for the moon, the argument is decisive (Proposition 3).

Once these various centripetal forces have been attested to, their unification results from an "induction" founded on the "Rules of reasoning in philosophy," especially the second which prescribes assigning the same causes to natural effects of the same kind (*Princ.*, 387). It is thus possible to assert that these several bodies "gravitate" or "are heavy" toward their respective centers, and that all these forces are the same as terrestrial gravity. The extension is begun with the moon, whose motion is compared quantitatively with that of projectiles; the extension is then propagated to the other celestial bodies (*Lunam gravitare*, Prop. 4, *Princ.*, 396; *planetas circumjoviales gravitare, . . .* etc., Prop. 5, *Princ.*, 399; *corpora omnia in planetas singulos gravitare*, Prop. 6, *Princ.*, 400).

Such is the path that leads from motions to forces, or rather to *the* force, following the procedures of what has been called the direct problem. How did Newton utilize the inverse passage within the framework of "The System of the World" presented in Book III?

The logical sequence of Book III is more complex than that often given in the short summaries of it. After the deduction of the universal force of gravitation, Newton traversed the path in the opposite direction and inferred the celestial motions from the forces, demonstrating in Proposition 13 what is sometimes called "Kepler's first law":

> The planets move in ellipses which have their foci at the center of the Sun. (*Princ.*, 409)

This inverse path is that of the deduction a priori of the motions: the "principles" of the motions, once known, make it possible "to infer a priori these motions themselves" (*Jam cognitis motuum principiis, ex his colligimus motus coelestes a priori*, *Princ.*, 409).

The journey in the opposite direction is not simply a confirmation of the phenomena; it makes possible the attainment of a precision, an exactness of determination that the observations could not provide. From the phenomena it is known only that the planets verify the uniformity of the sweeping out of areas and Kepler's harmonic law. One is restricted in this first step to a circular form for the orbits. But once the variation of the force as $1/r^2$ has been announced, it can be asserted that the planets describe ellipses and not circles (if the solidity of the "inverse" argumentation given in Cor. 1 of Prop. 13 of Book I is accepted; see p. 245 above).

Similarly, the immobility of the apsides of the orbits can now be demonstrated a priori (Prop. 14 of Book III). The departures from this conclusion—

the observed precession of the apsides—must arise from the actions of the planets on each other (*Princ.*, 411).

This double itinerary, from motions to forces, then from forces to motions, makes possible a very tight articulation of data and reasoning, an extremely refined adjustment between theory and observation, assisting and correcting each other, turn and turn about.

Newton described this to-and-fro motion in the Preface to the *Principia*:

> For all the difficulty in philosophy seems to consist in this: that from the phenomena of motions we should investigate the forces of nature, then that we should demonstrate from these forces the other phenomena. And to this end the general propositions that we have treated in Books I and II are directed. In Book III we propose an example of this [method] in the explication of the System of the World. For there, from the celestial phenomena, by means of the propositions mathematically demonstrated in the preceding books, are derived the forces of gravity, by which bodies tend toward the Sun and the several planets. Then, from these forces, by means of propositions that are also demonstrated mathematically, are deduced the motions of the planets, comets, Moon, and the seas. (*Princ.*, preface)

Books I and II in a certain sense lack physical content; they serve to develop the mathematical tools that are used in Book III. The mathematical demonstrations of these two books make possible the passage from motions to forces, and from forces to motions.

Newton described mathematics as an instrument of passage, an intermediary between the phenomena and the forces. No exaltation of a *mathesis universalis*, of a universal language. The glory of mathematics lies in its power to accomplish so much with so few principles (according to the beautiful formulation of the preface: *Ac gloriatur geometria quod tam paucis principiis aliunde petitis tam multa praestat*).

The solidity of the inductive fabric is due to its mathematical framework, which makes it possible to establish an extremely tight network in which observation and theory advance on and regulate each other. Laplace will insist on this to-and-fro movement between theory and observation: only theory, in advancing on observation, makes it possible to assemble phenomena usefully.

The image of a fabric, of a network, accords with this mode of reasoning in which different elements reinforce each other mutually, in contrast to a linear deduction, like the chains of inferences advocated by Descartes, in which the whole is as fragile as its weakest link. The fruitfulness of the operations of direct and inverse passage comes from the multiplicity of possible routes. The path in the opposite direction is not necessarily the same, and different approaches to the force turn out to converge on a single result. From Book I the attentive reader can perceive the plurality of ways leading to an inverse-square law of force:

Kepler's third law, combined with the formula of evaluation of the force for uniform circular motion

The elliptical trajectory

The immobility of the apsides

To these could be added another way, which Newton only began to open up: the consideration of the formal properties of potential, which will become "field theory." Newton proved that a force varying as $1/r^2$ has special spatial properties. The theorem that brought him to the highest point in this domain is the one that makes it possible to replace a sphere having radially symmetrical density by its center (the gravitation is the same as if the matter were concentrated at the center [Prop. 74 of Book I]). But beyond this result, important features of the modern theory of forces are foreshadowed in sections 12 and 13 on the attractive force of nonpunctiform bodies.

Newton may have hesitated to explore this direction of investigation because it required that he represent the spatial extension of force, in the manner of Kepler's speculations on the diffusion of the solar virtue. Yet the mode of subsistence of centripetal force, and its connection with matter and body, remained entirely problematic.

Force and Matter: The Analogical Way and the Refusal of Mechanism

What reality did Newton grant to centripetal force? The most constant and fundamental element to which he firmly held was the conception originating from Galileo: centripetal force is simply that which generalizes gravity, that is to say, it adds a new motion at each instant, generating equal velocities in equal times (velocities that come to be added to or subtracted from the velocity already acquired). The universality of "gravity" changes nothing: nothing more is disclosed about gravity itself; it is known only that wherever matter is, there gravity is also.

Force by itself is nothing substantial. Nothing is said about its presence and its diffusion around the center, independently of the action that it exerts on other bodies. Nor is there any mention of its deposition or conservation or transformation. The reasonings of Torricelli on the transmission and extinction of momenti and those of Kepler on the expansion of the "species" of the sun have no place in the *Principia*.

The source of the ever-renewed impulsions, the "fountain of momenti," remains a mystery. Yet, are there not central bodies that cause the deflective action, directing other bodies toward themselves? Moreover, the greater the quantity of matter in the central body, the greater the intensity of the action of the force. In sections 12 and 13 of Book I, Newton investigated in detail this increase in intensity of the central force, as a function of the addition of intensities associated with each particle of the attracting body.[2] Must not matter itself be attractive?

Newton took the greatest precautions at the point of associating force with the matter of the body. It is only an "analogy" which makes possible the linking of the intensity of the force and the quantity of matter in the attracting body:

> By these propositions [on the mutual attraction of two or several bodies, demonstrated in section 11] we are led to the analogy between the centripetal forces and the central bodies, toward which these forces are directed. For it is agreeable to reason that the forces that are directed toward bodies should depend on the nature and quantity of the latter, as is the case in magnetism. And every time that there occur cases of this sort, the attractions of the bodies are to be estimated by assigning to each of their particles their proper forces, and inferring [or collecting: *colligendo*] the sums of these forces. (*Princ.*, 188)

It is permitted to assign a force to each particle of matter, and to add together these forces, only by a simple right of analogy. The word "analogy" belongs to the same level as "induction" in Newton's vocabulary. In commenting on his third rule of reasoning in philosophy, Newton demanded that one not "depart from the analogy of Nature, since it is wont to be simple and always consonant with itself" (*Princ.*, 388). Similarly here, it is the uniformity of the phenomena that leads, if one wishes to be in accord with what is reasonable, to the evaluation of the intensity of the force directed toward a body by the quantity of the matter in the body.

But gravity is not a quality of matter; it does not belong to matter. The only force *belonging to matter* is the passive, inherent force, by which matter tends to persevere in its motion:

> I do not at all assert gravity to be essential in bodies. By inherent force I understand only the force of inertia. (*Princ.*, 389)

To discern clearly the link between force and matter, it would be necessary to have an exact idea of the "physical causes" of gravitation. Several explanations are possible, among which Newton refused to decide, at least in the *Principia*:

> I here employ the word attraction in a general sense for any endeavor whatever of bodies to approach one another, whether this endeavor arises from the action of the bodies, either in seeking to approach each other or in agitating each other by emitted spirits; or whether it arises from the action of the aether or the air, or of any medium whatever, whether corporeal or incorporeal, in impelling in any manner the bodies swimming in it towards each other. (*Princ.*, 188)

Several indications suggest that Newton had preferences and leaned toward an explanation in which force and matter are clearly separated. (A more detailed discussion would have to take account of the evolution of Newton's ideas on this point.[3]) Thus the idea of a fluid is dismissed in Query 28 of the *Opticks*:

> A dense Fluid can be of no use for explaining the Phaenomena of Nature, the Motions of the Planets and Comets being better explain'd without it. It serves only to disturb and retard the Motions of those great Bodies. . . .
>
> And for rejecting such a Medium, we have the Authority of those the oldest and most celebrated Philosophers of *Greece* and *Phoenicia*, who made a *Vacuum*, and Atoms, and the Gravity of Atoms, the first Principles of their Philosophy; tacitly attributing Gravity to some other Cause than dense Matter. Later Philosophers banish the Consideration of such a Cause out of natural Philosophy, feigning Hypotheses for explaining all things mechanically, and referring other Causes to Metaphysicks: Whereas the main Business of natural Philosophy is to argue from Phaenomena without feigning Hypotheses, and to deduce Causes from Effects, till we come to the very first Cause, which certainly is not mechanical. (Newton, *Opticks*, 368–69)

Mechanism is one hypothesis among others, or more exactly, it leads to supposing hypotheses. Those who have wished to explain everything by material principles have been carried along into the realm of fiction (Cartesians are chiefly meant). In contrast, the most respectable among the ancients admitted a nonmaterial principle. Philosophers who have proceeded otherwise have encouraged atheism. If, on the contrary, there is recognized from the first stages of physical explanation the presence of immaterial principles, then is initiated the gradual ascent towards the supreme cause, which is certainly not "mechanical." (The reader of the *Principia* is invited to this same ascent—through natural causes towards God—in the general scholium that concludes the book in the second and third editions.)

But if force is not mechanical or material, of what sort can it be? Newton rejected Descartes' strict division and otherness as between extension and thought; he admitted all degrees of "thickness" between the grossest matter and the most subtle spirit. There would be a place in this universe for various "active principles."

The Secularism of Force and the Privilege of the Law of Areas

In the *Principia*, Newton remained very discreet; all his conjectures were entrusted solely to manuscripts, except for the few that appeared in the Queries of the *Opticks*. The theory of the *Principia* is neutral to an astonishing and exceptional degree. This neutrality has become a habit over the three hundred years of mathematical physics. But was it not the first time that an author writing of natural philosophy concluded a work of this size and significance by confessing that he was unable to find "the causes" (*Princ.*, 530)?

Whence this neutrality? The explanatory physical models that Newton had thought of no doubt appeared too unsatisfactory, or too odd, to be submitted to public discussion. Consideration must also be given to traits peculiar to Newton the man—his prudence; his being moved by convictions that were, to say the least, not very orthodox; his being bruised by earlier controversies. His taste for dissimulation (or, more generously construed, his extreme modesty) has made of him perhaps the first of the scientists of the modern style: he carries on his activity as a natural philosopher without letting his private convictions show.

Not that he was satisfied with this neutrality. The *Principia* was written to discover the true causes and to advance towards the supreme cause, but the task was still incomplete ("I have not yet assigned the cause," *Princ.*, 530), and other classes of phenomena (magnetism, cohesion, etc.) yet remained that would need to be encompassed in a correct theory.

The *Principia* constitutes the columns of an unfinished edifice, but posterity will little by little become habituated to this incompleteness. According to Koyré's formulation, "the thought of the eighteenth century thus reconciled itself with the inexplicable" (Koyré *EN*, 202). Through Newton's work, a certain level of theory became autonomous. Once the mathematical physicists have determined the center of forces, once they know the law of the variation of the force and have succeeded in unifying certain families of phenomena, they can, on their own accounts, reflect on "causes" or "physical reasons" (to use Newton's term); but that will have no effect on the unfolding of their reasonings in mechanics or dynamics.

Scientists inspired by very different beliefs can accept this theory, use it, and develop it without bringing into play their private convictions as to the modes of action of this force. Around 1750–60, for example, there were three scientists continuing the Newtonian work without agreeing at all on the physical or philosophical foundations of the system. Euler, Bošković, and d'Alembert differed in their opinions on the ultimate justification of gravitation, and the very notion of force had a different meaning for each of them. Euler believed that impenetrability must be the sole source of forces (*Lettres à une princesse d'Allemagne*, 177–96). Bošković considered that action at a distance is at least as intelligible as action by contact (*Theoria philosophiae naturalis*, paras. 101–2). D'Alembert, finally, saw in force only an "abbreviated manner of expressing a fact" (*Traité de dynamique*, preface, xviii).

The term "secularism" characterizes this situation very well. Scientists can have all sorts of private opinions as to the ultimate realities (or even have none at all), yet a certain common cultural life is possible—with procedures for putting to the proof, and rules governing the confrontation of ideas at a certain level. Mathematics plays a privileged role in the "neutralization" of the study of force. It is, so to speak, an instrument of "de-reification" (one "supposes" nothing, Descartes said in the *Discours de la méthode*, *Oeuvres et Lettres*, pt. 2, 138).

The law of areas, which is the first theorem of both the *De motu* and the *Principia*, is, as it were, the emblem of a new conception of force, disembarrassed of phantasms and indifferent to physical causes. It has been noted above how significantly Newton's lack of this law affected his letters to Hooke in 1679. The regular sweeping out of areas made possible the appearance of time in the diagrams—the translation of interval of time into the sizes of segments, whatever the velocity of the body on its trajectory.

If Newton placed this theorem at the beginning of the demonstrative sequence of the *De motu* and of the *Principia*, although it was not indispensable (in the *De motu*) to the establishing of the second proposition, it is no doubt because he saw in it a theorem of the highest importance, striking and pregnant with consequences. At the beginning of Book III, the converse of the law of areas opens the deduction that leads from phenomena to forces. It permits determining the center of forces without other suppositions. Newton explained this employment of the law of areas in the scholium following Proposition 3 of Book I:

> Since the equable description of areas is the index [*index*] of a center towards which tends the force by which the body . . . is drawn back from its rectilinear motion and retained in its orbit, why may we not, in what follows, employ the equable description of areas as an indication of the center around which all curvilinear motion is performed in free spaces? (*Princ.*, 43)

What is a center of forces? It is simply a point towards which the continuous deviation of the motions is constantly directed. By what is a particular point of space recognized as a center of forces? By the uniform sweeping out of areas around this point. It is not necessary to know the nature of this central point or by what it is occupied nor to construct hypotheses as to the action exercised by a source or an attracting body. It suffices that this privileged point be detected by means of a regular effect of deviation geometrically determined.

Placed at the head of the theorems of the *Principia*, the law of areas is a presentation of force; it specifies the essential features of centripetal force and indicates at the start that this force is henceforth a mathematical entity.

NOTES

❁

Preamble

1. Statutes of the Philosophical Society of Oxford; see Scott, *Mathematical Work of Wallis,* 9. The Philosophical Society of Oxford became the Royal Society of London.

2. The distinction here drawn between "attraction" and "impulsion" is to be found in Newton's own formulations (*vires seu attractivas seu impulsivas*, *Princ*., preface, line 40; *trahuntur, impelluntur vel utcumque tendunt*, *Princ*., Def. 5, 3; *attractiones et impulsus*, *Princ*., 6; etc.). It occurs in the first line of the *De motu* and in the definition of centripetal force (*corpus attrahitur vel impellitur*). The success of the *Principia* depends crucially on not deciding between the two branches of this alternative: it is not necessary to know whether the *vis centripeta* is attractive or impulsive. This sharp opposition between attraction and impulsion is of great value both pedagogically and philosophically, but it becomes somewhat blurred in any detailed study of the history of ideas. Thus Kepler, who is undoubtedly a "magnetic philosopher," is classified by Newton as among the Continental vorticists in contrast to Englishmen and some others favoring attraction (*Philosophi recentiores, aut vortices esse volunt, ut Keplerus et Cartesius; aut aliud aliquod sive impulsus sive attractionis principium, ut Borellus, Hookius et ex nostratibus alii* [Newton, *De systemate mundi*, 2:6]). Leibniz thought Descartes borrowed his vortices from Kepler (*LMS*, 6:148). Concerning the magnetic philosophy, see the chapters of S. Pumfrey and W. Bennett in Taton and Wilson, *The General History of Astronomy*, 2A:45–53, 222–30.

3. Concerning loose formulations for that law, see Conclusions, note 1.

4. *Astronomia nova*, (*GW*, 3:250); *Epitome astronomiae Copernicanae*, (*GW*, 7:304); see above 75–76.

5. That the speeds of the planets are less the farther they are from the sun was Copernicus's primary intuition; it was fundamental for Kepler (*Mysterium cosmographicum*, especially chap. 20) and for Galileo (EN, 7:144–45, 354), who discovered a similar harmony in the satellites of Jupiter.

6. In appreciation of what Newton had accomplished by 1676, eleven years before the *Principia* appeared, Leibniz in that year wrote as follows: "Newton's discoveries are worthy of his genius, as his optical experiments and catadioptric tube [Newton's reflecting telescope] have abundantly shown. His method for obtaining the roots of equations and the areas of figures by infinite series differs completely from mine" (*Corresp*., 2:57–58; cf. Hofmann, 225–49).

7. See Galileo, Fourth Day, EN, 8:283–84. The connection is suggested by J. A. Bennett, based on Wren's personal annotations on his copy of the *Discorsi*; see Bennett, *Mathematical Science*, 61.

8. See Hall, 239, and *NMP*, 6:18 n. 52.

9. There exist at least four manuscript texts:

De motu corporum in gyrum in ULC (University Library of Cambridge) Add. 3965, fols. 55–62r; published as MS B in Hall, 237–67; also in *NMP*, 6:30–75, and in Herivel, 257–89

(without title) in ULC Add. 3965, fols. 63–70 (MS C in Hall)

De motu sphaericorum corporum in fluidis in ULC Add. 3965, fols. 40–54; MS D in Hall; excerpts in *NMP*, 6:74–80

Newton, "Isaaci Newtoni propositiones de motu," 35–51.

Photographic reproductions of these manuscripts are now available in the collection *The Preliminary Manuscripts for Isaac Newton's 1687 "Principia," 1684–1685: Facsimiles of the Original Autographs Now in Cambridge University Library, with an Introduction by D. T. Whiteside*. The manuscript cited in this book is on pages 3–11 under the subtitle: "The fundamental 'De motu corporum in gyrum' (Autumn 1684) (Add. 3965.7, fols. 55r–62*r)."

10. That is, if the text that he deposited in the Cambridge University Library is official evidence of his teaching in the autumn semesters of 1684 and 1685. This manuscript, entitled "Lucasian Lectures 'de motu corporum,'" represents an intermediary step between the manuscripts *De motu* and the final manuscript of the *Principia* as sent to the Royal Society for publication. (See *NMP*, 6:229–408.)

11. P. Costabel: "It is a matter rather of intuition than of demonstration, and it would be a mistake to believe that Newton at this epoch knew how to formulate and resolve problems by the integration of differential equations." (Costabel, "Newton and Leibniz's Dynamics," 115).

12. D. T. Whiteside: "because he subsequently, in Proposition 41 of the *Principia*, first book, gave a complete solution of the problem of determining the resulting orbit in a given central-force field" (Whiteside, *Mathematical Principles*, 16).

13. In arguing to this effect, J. Lohne is impishly provocative:

> But Halley had to wait for three months till he got—not the deduction—but a tract containing material for a course of nine lectures.
>
> Why this delay? and why did not Newton do as promised, send his solution of the inverse square problem?
>
> We believe that Newton did not possess the solution when Halley visited him, and that he was unable to work it out in the months that followed. Was it to disguise this that he sent a whole volume of theorems and into the mass of them smuggled his two small words (*Principia*, first edition, 55) "et contra," hoping that readers would not notice that this was an assertion without proof?
>
> For what do we find in the *Principia*? One of the theorems *assumes the planetary curve to be an ellipse* where the planet moves according to Kepler's second law. On that supposition, Newton derives an acceleration k/r^2.
>
> It looks as if Newton planned to derive orbits and motions on the sole supposition of central forces and the necessary axioms. From these he deduced (pp. 37–38) the law of areas. . . . Then he should have shown that the Keplerian ellipse was a necessary consequence of a central force law k/r^2. But instead he adopted the ellipse and demonstrated the acceleration there to be k/r^2. The attentive reader is not convinced: Might there not be motions on other curves also, compatible with a force law k/r^2? This problem was discussed by John Bernoulli who published the correct deduction of the ellipse in 1710. . . .
>
> We would like to ask another question: What if Newton had solved "the inverse square problem" as he apparently planned to do? We guess he would have sent his solution to Halley—and sent nothing more. We may indeed be grateful that there was a problem Newton could not solve, and so was forced to write and publish his *Principia* to overcome his rival Hooke. Newton, always procrastinating, was reluctant to finish and publish his books, unless his honor was at stake. (Lohne, "Hooke versus Newton," 35–36)

14. See Newton's letter to Flamsteed of 12 January 1684/85: "Now I am upon this subject, I would gladly know y[e] bottom of it before I publish my papers." The immediate question concerned the orbital radii of the satellites of Saturn; but the letter also sought information on the mutual perturbations of Jupiter and Saturn and on the paths of the comets of 1664 and 1680 (*Corresp.*, 2:413).

15. The search for a good method to determine the orbits of particular comets had forced him to return to Book I: "In Autumn last I spent two months in calculations to no purpose for want of a good method, wch made me afterwards return to y[e] first Book & enlarge it wth divers Propositions some relating to Comets others to other things found out last Winter" (Newton to Halley, 20 June 1686, *Corresp.*, 2:437).

16. Here I take exception to the chronological description given in Westfall *NR* (387–88, 394)—a reasoned account, to be sure, supported by detailed argument. In no known manuscript before the *De motu* did Newton utilize or even mention the law of areas. The dating of the manuscript later sent to Locke is very important in this connection (see n. 62 of chapter II).

CHAPTER I. THE *DE MOTU* OF 1684

1. See Newton's letter to Halley of 20 June 1686 (*Corresp.*, 2:338).

2. It will be demonstrated that this idea is already clearly present in Galileo's *Dialogo* and in Huygens' manuscript *De vi centrifuga*—which would not be published, however, until 1703.

3. *Discorsi*, Third Day, Theorem 2 on accelerated motion (EN, 8:209–10; see above p. 108).

4. See the beginning of Book II of the *Horologium oscillatorium* (*HO*, 18:123) and Hypothesis 3 of the *De motu corporum ex percussione* (*HO*, 16:32–33); see also pp. 114–17 above.

5. The question of the *direction* of QR will be discussed on pp. 28–29.

6. In other words, one seeks to transform the expression $QR/(SP^2 \cdot QT^2)$ into another expression consisting solely of a constant times SP raised to a certain power.

7. The text, entitled "On Motion in Ellipses," is given in Hall, 293–301, and *NMP*, 6:xxvi n. 44. See below, n. 62 of chapter II.

8. The situations dealt with in the *De motu* are much simpler than those treated in the *Principia*: all the material bodies are points, mass does not enter into the discussion, the sun attracts but is not itself attracted, there is no mutual action of the planets on one another, and the experimental and observational data are much more schematic.

9. A facsimile of the original manuscript, Add. 3965.7, fols. 55r–62*r, is given in Newton, "The Fundamental 'De motu,'" 3–11. References to the *De motu* will cite the folio numbers of this manuscript.

10. The correspondences between the *De motu* and the *Principia* are as follows:

De motu	*Principia*
Theorem 1	Proposition 1
Theorem 2	Proposition 4
Theorem 3	Proposition 6
Problem 1	Proposition 7, corollary 1
Problem 2	Proposition 10
Problem 3	Proposition 11

Theorem 4	Proposition 15
Scholium	Proposition 18
Problem 4	Proposition 17
Scholium	Proposition 31
Problem 5	Proposition 32
Problem 6	Book II, Proposition 2
Problem 7	Book II, Proposition 3

11. Wallis, *De motu* (1669–1670), chap. 1; cf. Newton in *De gravitatione* (Hall, 114, 148): "Force is the causal principle of motion and rest."

12. Descartes, *Le monde*, AT, 11:38, and *Principia philosophiae*, bk. 2, para. 40.

13. See *Princ.*, 389 (Book III, commentary on Rule 3): "Nonetheless I do not at all affirm that weight is essential to body. By inherent force I understand solely the force of inertia." See also *Opticks*, query 31, where Newton contrasts "a *Vis inertiae*, accompanied with such passive Laws of Motion as naturally result from that Force" with "certain active Principles, such as is that of Gravity" (Newton, *Opticks*, 401). See also the letters to Richard Bentley of 1692 and 1693 (*Corresp.*, 3:233ff.).

14. See the letter to des Maizeaux cited in *NMP*, 6:31.

15. "Law of gravity," "center of gravity," in place of "law of centripetal force," "center of centripetal force" (see *NMP*, 6:43 n. 31, 44 n. 33).

16. The step thus taken is not really a large one since the *De motu* represents centripetal force as the limit of a series of impacts (see above, pp. 24–25).

17. Hall, 272–73.

18. See Cohen, *The Newtonian Revolution*, 261–63.

19. See MS D of Hall, 243, and *NMP*, 6:76. This discussion assumes that the manuscript sent to Locke is subsequent to the *De motu* (cf. Hypothesis 2 in Hall, 293).

20. See Koyré *EG*, pt. 3, "Galilée et la loi d'inertie."

21. Descartes, *Principia philosophiae*, AT, 8:62–65 (*deflectere* is on p. 63, line 25).

22. Compare *Princ.*, 22, and Huygens, *Horologium oscillatorium*, pt. 2, Hyp. 3, *HO*, 18:125.

23. See Russell, "Kepler's Laws of Planetary Motion," 1–24.

24. I am assuming here, as previously, that the manuscript sent to Locke is later (in agreement with Hall and Whiteside and in opposition to Herivel and Westfall). See note 62 of chapter II.

25. At this point Newton invokes Hypothesis 1, which concerns the absence of resistance; why not also Hypothesis 2, that is, the principle of inertia? Perhaps there is here a simple error of transcription.

26. Newton designates the segments CD and cd by the expression *spatia superata*, which has been translated hypocritically as "distances gained." The Latin term is in fact ambiguous, and I have sought to maintain the ambiguity. In the seventeenth century, *spatium superatum* could designate a distance traversed; but here CD and cd are not, properly speaking, distances traversed. (Perhaps Newton considered them as component paths actually traversed; such will be the case in the next proposition.) But *superare* can also be used for the motion of one body exceeding that of another (see, for example, Kepler, *GW*, 7:428, 438). *Spatium superatum* could therefore signify a difference between two distances.

27. See D. T. Whiteside in *NMP*, 6:38–39.

28. Cor. 2 of Prop. 1 in the 2d and 3d eds., *Princ.*, 40.

29. Flamsteed to Newton of 27 December 1684 (*Corresp.*, 2:40). From the time of the *Sidereus nuncius*, Galileo had stated that, among the satellites of Jupiter, those farther from the planet appeared to move more slowly, and in the *Dialogo* he confirmed this relation by citing the approximate periods of the satellites (EN, 7:144–45). In several ways during the seventeenth century, the Galilean satellites of Jupiter played the role of a miniature model of the "system of the world" (they did, for instance, for Borelli, whom Newton willingly cited as one of his predecessors).

30. Here Newton later added that the lines TQ and PR when prolonged meet in Z and that the resulting triangles ZQR and ZTP, together with the triangle SPA, are similar. The remainder of the demonstration then follows. See ULC, Add. 3965.7, fol. 56, and *NMP*, 6:43 n. 32.

31. By QRL Newton means the rectangle QR × RL.

32. By PVG Newton means the rectangle PV × VG.

33. See manuscript ULC, Add. 3965.7, fol. 57. The text of the *Principia* (p. 52, line 30) is identical except that the letter *v* is in minuscule, and certain *q* are written *quad*: *QT quad* in place of QT^q.

34. Newton does not employ here the word *focus*, which had been introduced by Kepler, but rather *umbilicus* (navel). It is an isolated oddity, for Newton elsewhere (see, for example, *NMP*, 2:122) uses *focus* to designate this point.

35. Newton designates the compounding of ratios by the sign "+": "AC ad PC + L ad GV + CP^q ad CD^q . . ." The text of the *Principia* (*Princ.*, 55, lines 20–22) is more direct: the result of compounding the ratios is simply the ratio between all the antecedents or prior terms multiplied together and all the consequents or second terms multiplied together. Concerning the persistence of the language of proportions, see the articles of E. Sylla ("Compounding Ratios") and E. Grosholz ("Some Uses of Proportion in Newton's *Principia*").

36. But who has decided that there were "three laws" of Kepler? (The article by J. L. Russell, "Kepler's Laws of Planetary Motion," gives no information on this point.)

37. The reader is reminded that by a three-letter expression such as KPH Newton means a rectangle: KPH = KP × PH.

38. Kepler had already replaced the one problem by the other (the semi-ellipse by the semicircle) in his earliest formulation, given in chap. 60 of the *Astronomia nova*.

39. In a first draft, later deleted, Newton identified the center as the center "of the Earth," and the falling body as "a weight" (see *NMP*, 6:62 n. 85).

40. The rhetorical question imitates Galileo's formulation in an analogous case (the "bowl" of the *Discorsi*; see above, p. 173ff.).

41. The last two problems (Problems 6 and 7, Add. 3965.7, fols. 61–62), which concern motion in a resisting medium, will not be considered.

Chapter II. Aspects of Force before the *Principia*

1. Aristotle, *Physics*, 7.5.

2. Isaac Newton, *Arithmetica universalis*, "Resolutio quaestionum arithmeticarum," Probs. 6 and 7, *NMP*, vol. 4. The problems cited are not included in the small treatise by Kinckhuysen that served Newton as the nucleus of this work (see *NMP*, vol. 2).

3. The discussion of this text and the assessment of its precise meaning would range too far afield; see F. De Gandt, "Force et science des machines," 96–127. See also the strong reservations expressed by Robert Wardy against a "whig"—that is to say, progressivist and continuist—interpretation of this text (Wardy, "Le jeu des nombres aristotéliciens," 121–50). It is true, however, that Leonardo da Vinci and many others were inspired by this Aristotelian passage.

4. *Vis motrix, orbit rotundo sufficiens, ex cujus vigore et constanti fortitudine tempus revolutorium oritur* (*Astronomia nova*, Chap. 2, *GW*, 3:68). For another example, see Galileo, EN, 7:474: if the body moved remains the same, and the motive virtue also remains the same, then more time will be necessary for a larger circle.

5. It is very improbable that Newton read Kepler directly, except perhaps for the *Optics* (but see Westfall [*NR*, 83], who raises objections to the late evidence on this subject). He knew Kepler's laws through popularized presentations (see Russell, "Kepler's Laws of Planetary Motion," 1–24).

6. The deepest and most complete study of Kepler remains Koyré's *La Révolution astronomique*, but it is less detailed in its treatment of the *Epitome*. There exists a very useful English translation by C. G. Wallis of Books 4 and 5 of the *Epitome* in vol. 16 of *Great Books of the Western World*, (Chicago: Encyclopaedia Britannica, 1952).

7. The interpenetration would not occur in the Copernican arrangement. Compare the much earlier discussion of Maimonides on the incompatibilities between orbs and motions (*Guide*, II, paras. 4 and 24). Kepler gives a third list of arguments, still in the *Epitome* (*GW*, 7:294–95), in which he insists on the following difficulty: What is it that sustains the spheres? Can one imagine that they repose one on another? Kepler accepts Tycho's arguments against the orbs, but he has others that are his own (see the *Apology of Tycho*, ed. N. Jardine, in N. Jardine, *The Birth of History*, 99).

8. Descartes wrote a little later: "there is only one and the same matter for the heavens and the Earth" (*Principia philosophiae*, para. 22, AT, 8:52).

9. *imitationem manuariam*: Kepler's Latin attributes to Ptolemy the idea of planetary models actually constructed to represent the motions of the heavens by a machine, like the mechanical planetarium that Archimedes constructed (Ptolemy, *Syntaxis*, frag. 17). But the Greek text is much less clear. Toomer translates: "considering the complicated nature of our devices" (*Ptolemy's Almagest*, 600). See Lloyd, "Saving the appearances," 215: "in *Syntaxis* XIII.2, the devices (*'επιτεχνηματα*) that Ptolemy says may be found troublesome are simply the astronomical hypotheses themselves."

10. See Aiton, "Johannes Kepler and the Astronomy," 49–71; and *Apology of Tycho*, edited from the Pulkowo manuscript, translated, and commented on by N. Jardine in N. Jardine, *The Birth of History*.

11. Galileo maintained the same idea at the end of the First Day of the *Dialogo*: "the human intellect does understand some [propositions] perfectly, and thus in these it has as much absolute certainty as Nature itself has. Of such are the mathematical sciences alone; that is, geometry and arithmetic, in which the Divine intellect indeed knows infinitely more propositions, since it knows all. But with regard to those few which the human intellect does understand, I believe that its knowledge equals the Divine in objective certainty, for here it succeeds in understanding necessity, beyond which there can be no greater sureness" (*Dialogue Concerning the Two Chief World Systems*, 103; EN, 7:129; note that Koyré's translation in *EG*, 284 is very unfaithful to the original). On this subject two passages from letters by Kepler are cited in Koyré *RA*, 378–79

n. 15. The "univocity" of mathematics according to Kepler has to be qualified: if the straight line is the same for God and for man, God alone adequately knows the curve (see *Mysterium*, chap. 2; Simon, *Kepler astronome, astrologue*, 133).

12. See *Harmonice mundi*, IV.7, as commented on by G. Simon in *Kepler astronome, astrologue*, chap. 4: "les raisons du panpsychisme." (But Simon does not distinguish clearly enough between *anima* and *intellectus*: the soul is not "directive" for Kepler.) In the *Epitome*, the arguments concerning the soul of the earth are assembled at *GW*, 7:92.

13. This is the theme of chap. 20 of the *Mysterium* of 1596, to which Kepler in 1621 added the note that I have quoted. Compare a letter of 1595 to Maestlin, cited in Koyré *RA*, 386 n. 35: *motrix anima, ut dixi, in sole . . .*

14. *GW*, 7:377ff. Kepler wanted to separate what has its source in the "libration" from what has its source in the rotation.

15. The sun differs slightly from most magnets: in its case, the exterior surface corresponds to one pole, and the center to the other pole (*GW*, 7:300).

16. *GW*, 7:371. Kepler uses *sagitta* and *sinus versus* indifferently.

17. *GW*, 7:372. Kepler asserts that this property is confirmed by observation.

18. The calculation of forces leading to the elliptical orbit is very tortuous in the *Astronomia nova* (chaps. 57–59; see Koyré *RA*, 253–79, and Simon, *Kepler astronome, astrologue*, 377–86). The description in the *Epitome* is much simpler, notably in not including the different steps of the path toward the discovery of the elliptical path or the successive contributions of observation.

19. If, moreover, there were a virtue emanating from the sun, why had Kepler denied that it decreases as r^{-2}? (Boulliau, *Astronomia Philolaica*, 23–24).

20. The paths are circles properly speaking, but the planet passes continuously from one circle to another along the surface of a cone (ibid., 26–28). The law of areas is replaced by a kinematic procedure of equation in the manner of the ancients: the uniform rotation of a vertical plane around an axis passing through the empty focus of the ellipse (ibid., 34–36). The discussion of Kepler's arguments on pp. 21–24 is translated by Koyré in *RA*, 371–75.

A procedure of equation inspired by Boulliau (uniform rotation around the empty focus, which is not exactly the same thing as rotation of a plane as proposed by Boulliau) was widely used by astronomers in the middle of the century, as is shown by the title of Halley's first memoir; translated, it reads: "Direct and geometrical method for finding the aphelia, eccentricities, and proportions of the orbs of the primary planets, without supposing, as have astronomers hitherto, the equality of angular motion about the second focus" (*Philosophical Transactions*, 11:683–86). A detailed investigation of the various procedures of equation of planetary motion between Kepler and Newton would require an examination of numerous technical texts, such as those of J. Horrocks, S. Ward, V. Wing, and so forth (see Wilson, "Predictive Astronomy in the Century after Kepler," in Taton and Wilson, *The General History of Astronomy*, 162–206).

21. More detailed inquiries would be necessary: what could one find in the medieval discussions, for instance, concerning the transport of force by a *species*, or the action of fluids?

22. How to translate *tradidi*? Is it necessary to consider that the content of the *Principia*, up to this cited page, is the simple exposition and transmission (*traditio*) of already established principles? The verb *tradere*, which is very rare in the *Principia*,

does not appear to connote necessarily the idea of transmission. (I am indebted to Professor I. B. Cohen for communicating to me his computerized *Concordance* of the *Principia*, which gives for the word *tradere* the following occurrences, identified by the number of the page in the third edition followed by the number of the line: 21, 17; 73, 7; 386, 5; 517, 29. There are three occurrences in the preface by Cotes.)

23. *Experience* is in the singular (versus Cajori and du Châtelet [Newton, *Principes*]).

24. Versus Cajori, 21: the "uniform force of its gravity" for *gravitas uniformis*. An important discussion of the translations of this passage is to be found in Cohen, "Newton's Use of 'Force,' or Cajori versus Newton: A Note on the Translations of the *Principia*," *Isis* (1967): 226–30. There are three English versions of this passage by Newton himself; all of them insist upon the ever-renewed, self-identical action of weight "in each of the equal particles of time"—the conceptualization I have called "Galileo's lesson."

25. This classification is taken from P. Duhem's *Les Origines de la statique*, which remains excellent, above all if it is corrected with the aid of W. R. Knorr, "Ancient Sources," fasc. 2.

26. The preceding summarizes my analysis of the beginning of the *Mechanical Problems*: De Gandt, "Force et science des machines," 116–26. See also De Gandt, "Les Mécaniques attribuées," 3.

27. This passage may be compared with the Aristotelian text (*Physics* 7.5) cited previously: there are four things to consider in motion—the force, the weight, the distance, the time.

28. Galileo, *Mechanics*, EN, 2:159. This discussion of momento owes much to P. Galluzzi, *Momento. Studi galileiani*. See De Gandt, "Galilée et la naissance de la dynamique," 319–24.

29. In the same posthumous Sixth Day (EN, vol. 8) added to the *Discorsi*.

30. Galluzzi has stressed this divergence between Torricelli and Galileo and the hesitations in Galileo's writings that give evidence of it (*Momento. Studi galileiani*, 402–3).

31. [Aristotle?] poses this question (as Question 19) in the *Mechanical Problems*. Concerning this question, see F. De Gandt, "L'analyse de la percussion," 53–78.

32. Torricelli clarifies in other texts the link between momento and acquisition of velocity (see Torricelli to Mersenne, January 1645, *Opere*, 3:253). His thesis can be considered as equivalent to the principle that a constant force generates an acceleration.

33. See the discussion by Galluzzi, *Momento. Studi galileiani*, 322–29.

34. At least if one translates correctly Galileo's *quod si parallelae trianguli AEB usque ad IG extendantur* (see the beginning of the third paragraph of my translation of the theorem): "Now if the parallels . . . are prolonged" and not "Because if the parallels . . . are prolonged."

35. *Totidem velocitatis momenta absumpta esse . . .*

36. *Patet igitur aequalia futura esse spatia . . .*

37. *Quantum e gradibus istis constans repraesentans magnitudo spatium quoque decursum repraesentare possit* (Barrow, *Lectiones*, 9).

38. See De Gandt, "Les Indivisibles de Torricelli."

39. See Roero, "Jakob Bernoulli," 1:78 n. 6.

40. See my discussion of the composition of forces in the *De motu*, above p. 20.

41. A thorough account of the Cartesian critique should include, in particular, a discussion of the letter to Mersenne of 16 February 1643, in which Descartes, wishing to justify a law of the flow of fluids equivalent to Torricelli's law, admits that the distance traversed under the action of gravity is proportional to the square of the time, at least at the very beginning of the motion (*AT*, 3:617–31, especially 619–20).

42. The *Discours de la cause de la pesanteur* appeared as an appendix to the *Traité de la lumière* in 1690, but it reflects much earlier discussions, in particular those that occurred at the Académie des Sciences of Paris in 1669 (see in *HO*, 19:628–45, the memoirs read by Roberval, Frénicle, Buot, and Huygens, with the observations of Roberval, Mariotte, du Hamel, and Perrault; Roberval and Frénicle allowed that gravity involved attraction).

43. Yoder rightly remarks that the interest Huygens takes in centrifugal force is not sufficiently explained by his commitment to the Cartesian program (Yoder, *Unrolling Time*, 34–35). Every Copernican, including Galileo (see above, p. 143ff.), was obliged to deal with the objection that unattached objects about the earth would be extruded owing to the earth's rotation.

44. On the word *vis*, see the keen and paradoxical remark of Westfall *FNP*, 177: the choice of this term by Huygens is related to his preoccupation with statics; *vis* is the term used in the theory of simple machines.

45. The theorems are given in *Horologium oscillatorium*, 1st ed., 159–61: *De vi centrifuga ex motu circulatori theoremata*. The small treatise *De vi centrifuga* appeared in 1703 in the posthumous publication *Christiani Hugenii Opuscula posthuma*, B. de Volder and B. Fullenius, eds., (Leyden, 1703), 401–28; it is republished with a French translation in *HO*, vol. 16 (the extracts given in my text are retranslated from the original Latin). A deeper study of the *De vi centrifuga* would require a fresh examination of the different manuscripts that were combined in the edition of 1703. See Yoder, "Christiaan Huygens' Great Treasure," *Tractrix* 3 (1991): 1–13 (especially p. 2: "in its present condition, the *de Vi Centrifuga* is a treatise that Huygens never wrote").

46. In the corresponding theorem of the *Horologium* (1st ed., 160), the words *vis centrifuga* replace *conatus recedendi a centro*. On the comparison of centrifugal force and gravity, see the remarks of Westfall *FNP*, 175–76, which concern Proposition 8 of the *De vi centrifuga*.

47. See the manuscript given as an appendix to the *De vi centrifuga* (*HO*, 16:323–26). The document relative to the pierced table is found in *HO*, 16:327–28; see also *Discours sur la cause de la pesanteur*, *HO*, 21:130, and the letter to C. Huygens of 2 December 1667, (*HO*, 6:164). Huygens also proposes various mechanisms with oblique tubes or bowls of special forms (*HO*, 16:307–8; see the article by Costabel cited below, n. 28 of chapter III). These tubes are taken up again by Leibniz in his *Dynamica*, Prob. 2, Props. 28 and 29, *LMS*, 6:451–52. On Hooke's experiments relative to motion in spherical bowls, and so forth, see Lohne, "Hooke versus Newton," 15–17, and Pugliese, "Robert Hooke," 181–206.

48. It would be a mistake to regard vortex hypotheses as simply antithetical to Newtonian ideas. True, at the end of Book II of the *Principia*, Newton presented a sketch of a refutation of vortices that was largely qualitative in character (*Princ.*, 384–85). Newton was nevertheless interested, privately, in various explanations of gravity, in particu-

lar that proposed by Fatio—who rightly cites Huygens among those by whom he was inspired (see Gagnebin, "De la cause de la pesanteur," 106–60; see also Hall, 205–7, 313).

49. Curiously, it appears that Huygens did not relate the study of evolutes to that of curvature (see Yoder, *Unrolling Time*, 104).

50. From the *Discours de la cause de la pesanteur* (*HO*, 21:461), it appears that, according to Huygens, acceleration is not essential to gravity: as long as the "celerity" of the fluid matter greatly surpasses that of the cannonball, there is acceleration (the "pressure" that the latter "feels" is always approximately the same in successive instants). But in principle, nothing prevents the cannonball from going rapidly enough so that its velocity becomes comparable to the effect that it receives from the fluid matter, in which case there will no longer be any cause for it to accelerate. (The text is ambiguous: what, precisely, is it that is to be compared—velocities? efforts?)

51. Borelli, in his *Theoricae mediceorum planetarum a causis physicis deducta*, had assumed that the trajectories of the planets result from a sort of equilibrium between a "quasi-magnetic virtue" that attracts them and the thrust arising from the circular motion, which causes them to recede (see Koyré *RA*, 474–75). Newton cites Borelli among the precursors justifying a rejection of Hooke's claims (*Corresp.*, 2:437–38).

52. The question had been discussed in the *Dialogo* of Galileo (EN, 7:90–91): "It is manifest that the Moon, as drawn by a magnetic virtue, constantly regards the terrestrial globe with one of its faces." Aristotle long before, in the treatise *De caelo* (2.8.290a4), had considered the fact that the moon always turns the same face toward the earth as an argument in favor of solid orbs. On the history of this question from Aristotle to Newton, see Gabbey, "The Case of the Rotating Moon," 95–130.

53. In Italian: *Ma a deviare un mobile dal moto, dove egli ha impeto, non ci vuol egli maggior forza, o minore, secondo che la deviazione ha da esser maggiore, o minore? cioè, secondochè nella deviazione egli dovrà nell'istesso tempo passar maggiore, o minore spazio?* (EN, 7:242)

54. Did Newton read Galileo's *Dialogo*? The text translated on pp. 139–41 above (Herivel, 195–96) could be an extension of pp. 240–42 of the *Dialogo*. Yet in 1692–93, Newton seemed no longer to remember that the cosmogonic myth that he then discussed with Bentley had its origin in the *Dialogo* (Koyré *EN*, 254).

55. The representation of centrifugal and centripetal forces that equilibrate one another led to a good many absurdities and naiveties in the exposition of the Newtonian doctrine (see, among many others, the *Institutions Newtoniennes* of Sigorgne and Hegel, *Les Orbites des Planètes*. On the other hand, such an analysis of curvilinear motions is not totally without value nor irremediably vitiated, as shown by the development and application of it made by Maclaurin for a semiqualitative determination of orbits (*Sir Isaac Newton's Discoveries*, 4:305ff.).

56. "And thô his correcting my Spiral occasioned my finding ye Theorem by wch I afterward examined ye Ellipsis; yet am I not beholden to him for any light into y^{t} business but only for ye diversion he gave me from my other studies to think on these things." (Newton to Halley, 27 June 1686, *Corresp.*, 2:447).

57. On the English Magnetic Philosophy, see Bennett, *Mathematical Science*, 26–27 and 56–60 for citations of a number of significant texts by Wilkins and Wren. For information about Hooke and references to passages in his writings, see the very rich work of Aiton, *The Vortex Theory*, 94–98 and 118–19.

58. This is very strong evidence that, along with other indications, shows that Newton at the date of this letter was ignorant of or rejected the law of areas. May not "the theorem" that Newton discovered on the occasion of this correspondence have been the law of areas itself? (See above p. 154ff.)

59. Certain developments on this theme during the course of the seventeenth century are presented in Koyré, "The Problem of Fall."

60. Nevertheless, Newton up to 1681 continued to describe the celestial motions with the aid of the pair of forces, *vis centrifuga*/attraction, each alternatively overpowering the other; see *Corresp.*, 2:361.

61. In his earliest manuscripts, Newton had already attempted to evaluate sums of "motions generated." For example, he thereby deduced a measure of the total force required in reversing the direction of motion of a body when it moves through a semicircle. See Herivel, 147, and Westfall *FNP*, 350–52.

62. Here, again, it is presumed that the text sent to Locke—or its archetype—is not earlier than the *De motu*. See *On Motion in Ellipses* in Hall, 293–301, and the discussion between Herivel and the Halls (Herivel, "Newtonian Studies. III," 23–33; Herivel, "Newtonian Studies. IV," 13–22; and the Halls' response, "The Date of *On Motion in Ellipses*," 23–28). According to the conclusion of the Halls, it is most probable that this text was written in 1690 for a particular purpose. In the perspective of my study, I note that Hypothesis 2 of this text, which corresponds to Law II of the *Principia* ("The alteration of motion is ever proportional to the force by which it is altered") is absent from the first versions of the *De motu*. Along the same lines, Hypothesis 3 and the law of areas are formulated more rigorously than in the *De motu*.

63. From 1665, Hooke had proposed the idea that comets return, claiming that the comet of 1664 was the same as the comet of 1618 (see Pepys, *Diary*, 4:341; Dugas, *La mécanique au 17ème siècle*, 357; Bennett, *Mathematical Science*, 65, 69). Like Flamsteed later, Hooke combined magnetism with vortices.

Chapter III. The Mathematical Methods

1. The text given by all three editors—Hall, Herivel, and Whiteside—has "ADEG" here, but the reasoning as a whole suggests rather "ADEC."

2. The commentary on section 1 of the *Principia* will return at some length to this very original procedure of similitude between an evanescent situation and its finite replica—called "the method of finite witnesses." It is presented in a more developed manner and utilized more extensively in that section, particularly in Lemmas 7 and 9. See above, p. 225ff.

3. It would be necessary to add a hypothesis for Galileo's reasoning to be valid. Galileo appears to claim that if the resultant is finite and the terms infinitely numerous, each must be without magnitude. However, an infinite series can converge towards a finite sum although each term remains of finite, assignable size.

4. See Clavelin, *La philosophie naturelle de Galilée*, 440–52, on Galileo's discussion of cohesion and the limits that the notion of infinity imposes on the total intelligibility of the real.

5. See Andersen, "Cavalieri's Method of Indivisibles," 291–367, and De Gandt, "Cavalieri's Indivisibles," 157–82.

6. Koyré, "Bonaventura Cavalieri," 345.

7. There do exist authors more respectful of Cavalieri's restrictions. For example, van Schooten seems relatively careful in the indivisibilist demonstrations of the *De organica conicarum sectionum in plano descriptione*, 12 (on the circle and the ellipse), 41 (on hyperbolic areas).

8. The procedure assumes that one knows how to draw a rectilinear segment equal to a circumference. Archimedes had used the spiral in precisely the opposite way: the tangent to the spiral enabled him to calculate the length of the circumference (*Spirals*, Prop. 18, *Opera*). The problem is then: how to determine the tangent? Certain commentators believe that he also would have been able to make use of a kinematic procedure (cf. Heath, *A History of Greek Mathematics*, 2:557).

9. See the translation (pp. 109–11 above) of a passage from Barrow's *Lectiones geometricae* concerning force and acceleration. On the probable role of the *Lectiones geometricae* in the invention of the method of fluxions, see in particular the indications given by D. T. Whiteside in *NMP*, 1:344 n. 4.

10. On the role of motion in the mathematics of Newton, see De Gandt, "Duratio, fluxio, aequatio," 353–70, and "Newton: la justification," 147–57.

11. For this brief presentation, I rely principally on the *Tractatus de methodis serierum et fluxionum* (*NMP*, 3:32–329), written in 1671 and published much later. The very title is not established with certainty (see *NMP*, 3:33).

12. Newton first wrote the Aristotelian term, *τὸ νῦν*, then cancelled it.

13. This dissymmetry was suppressed in the translation by Buffon, who was often unfaithful to the text elsewhere (see, for example, p. 26).

14. Newton at once treats implicit equations, and his rules come down to writing the "complete differential." The example of the equation given above is the first proposed by Newton (*NMP*, 3:74).

15. The reader is referred on this subject to De Gandt, "Temps physique et temps mathématique," 87–105.

16. Newton probably learned of Fermat's method indirectly through the Latin commentaries on the *Géométrie* of Descartes (see *NMP*, 1:149).

17. Cf. Cléro and Le Rest, "La naissance du calcul infinitésimal," 160–61. See above, p. 219 (triangle VBC).

18. Leibniz, "Nova methodus pro maximis et minimis," 467–73. The translation of and commentary on this essay by Dupont and Roero, "Leibniz 84," can be consulted with profit.

19. Is this a discreet attack against Leibniz, who would here be accused of manipulating sums and differences, while Newton better respected the continuous nature of magnitudes? Leibniz would thus be put in the same rank as the partisans of indivisibles.

20. A French version of this chapter appeared in *Revue d'Histoire des Sciences* (1986): 195–222, and a Spanish version appeared in *Mathesis* (1990): 163–89.

21. See Whiteside, *The Mathematical Principles*.

22. In the *Principia* there does not appear to be any difference between "infinitely" and "indefinitely." In the context of a metaphysical discussion with Descartes, however, Newton makes a distinction between the two terms (Hall, 102, 135).

23. These departures from the Euclidean tradition are not concentrated in particular passages of the *Principia*. In Book I, all the sections, with the exception of the short section 5 (which deals only with the determination of conics by focus and directrix),

satisfy at least one of the criteria proposed: recourse to the infinitesimal, recourse to motion (beyond, of course, permission to draw a circle or even a conic). Sections 12, 13, and 14 are not of a kinematic nature, but Newton there has constant recourse to the infinitely small or the evanescent in his calculations of potential.

Section 5, which appears very classical and even archaic in its style, is not truly Euclidean because of Lemmas 21 and 22 on the transformations of figures by motions.

The highpoint of non-Euclideanism is attained in sections 2–3 and 7–10, which combine kinematics very closely with the use of infinitesimals.

24. Newton ordinarily employs *ratio* rather than *proportio* to designate a relation with respect to size. I shall avoid the word "proportion" when there is risk of confusion: is it a "ratio" of two terms, or the relation of "proportionality" that relates four terms?

25. This procedure, which Cavalieri calls "reuniting the excesses," was already used by Archimedes (*Spirals*, Prop. 21, *Opera*; *Letter to Eratosthenes*, Prop. 15). See, for example, Cavalieri, *Geometria indivisibilibus*, bk. 2, Prop. 22 and bk. 6, Prop., 6; or Torricelli, Lemma 15 of the *De dim.par.* (*TO*, 1:128).

26. Newton's figure is more complicated than the one given here because it is designed to illustrate the following lemma as well.

27. I have simplified the figure so as to retain only the features necessary for this demonstration.

28. Huygens also discovered this property, but later (*HO*, 18:489). On this investigation of the reasons of isochronism, see Costabel, "Isochronisme et accélération," 13–15.

29. See Whiteside, "Patterns," 333–35.

30. It is necessary to distinguish carefully between the center of curvature and the instantaneous center of rotation (compare the erroneous assertions of Whiteside, *NMP*, 6:387 n. 273, and of Baron, *The Infinitesimal Calculus*.

31. Johann Bernoulli remarked in a letter to Leibniz that the converse of Prop.10 was asserted without demonstration (see Leibniz, *Mathematische Schriften*, 3:853).

32. A truly complete resolution would require a discussion of the uniqueness of the solutions. But the concern for rigor on this point began only at the time of Cauchy.

33. See Aiton, "The Inverse Problem," 85: "Varignon's originality, however, should not be overestimated, for Newton supplied the foundation for such an analytical development in his Propositions 39 and 40." See the texts of Varignon reproduced and commented on in Blay, *La naissance de la mécanique analytique*.

34. A similar precision is found in a text of Torricelli: *aut inflectere . . . aut accelerare vel retardare* (*TO*, 2:159).

35. See for example Descartes to Mersenne, 13 July 1638 (AT, 2:222): for a body subject to gravity and restricted to an oblique action, "the proportion between the straight line that this motion describes and that which indicates by how much this body would during this time approach the center of the Earth, is the same as that between the absolute gravity and the relative." Similarly Galileo in the *Discorsi* (EN, 8:216): the ratio between the two momenti is the inverse of the ratio of the lengths (see above, p. 105ff.). Or again, Stevin, *Oeuvres mathématiques*, 448.

36. Newton gives only one "concrete" example—in Corollary 3 of Proposition 41. He determines in a rather "qualitative" manner the shape that the trajectory must have for an inverse-cube force—by means of an associated conic and with the choice of the initial point V of a very particular direction of propulsion (perpendicular to CV).

Conclusions

1. Newton's slightly loose formulation cannot lead to error. When he speaks of a "decrease of the force in the duplicate ratio of the altitude," one naturally understands that the force decreases when the square of the altitude increases, that is to say, that the force is *inversely* proportional to the square of the altitude.

2. Huygens sagely remarked that here was an important threshold in the argumentation of the *Principia*: he accepted the notion of the *vis centripeta*, but he rejected the principle that "all the small parts that one can imagine in two or more different bodies attract or tend to approach each other mutually" (Appendix to the *Discours de la cause de la pesanteur* in *Traité de la Lumière*, 159).

3. See McGuire, "Force, Active Principles and Newton's Invisible Realm."

BIBLIOGRAPHY

Aiton, E. "The Inverse Problem of Central Forces." *Annals of Science* 20, no. 1 (1964): 81–99.

———. *The Vortex Theory of Planetary Motions*. New York: American Elsevier, 1972.

———. "Johannes Kepler and the Astronomy without Hypotheses." *Japanese Studies in the History of Science* 14 (1975): 49–71.

Andersen, K. "Cavalieri's Method of Indivisibles." *Archive for History of Exact Sciences* 31 (1985): 291–367.

Apollonius. *Opera*. 2 vols. Edited by J. L. Heiberg. Stuttgart: Teubner, 1974.

Archimedes. *Opera*. 3 vols. Edited by J. L. Heiberg. Leipzig: Teubner, 1910–15.

Aristotle. *Physics*. Edited by W. D. Ross. Oxford: Oxford University Press, 1950.

Aristotle. *De Caelo*. Edited by W. Allen. Oxford: Oxford University Press, 1955.

[Aristotle?]. *Mechanical Problems*. In *Aristotle: Minor Works*. Translated by W. S. Hett. Cambridge: Harvard University Press, 1963.

Arnauld, A. *Nouveaux éléments de géométrie*. Paris, 1667.

Baron, M. E. *The Origins of the Infinitesimal Calculus*. Oxford: Pergamon Press, 1969.

Barrow, I. *Lectiones geometricae*. 2d ed. London, 1674. Reprint, Hildesheim: Olms, 1976.

Bennett, J. A. *The Mathematical Science of Christopher Wren*. Cambridge: Cambridge University Press, 1982.

Bernoulli, Jean. *Opera*. Lausanne, 1742.

———. *Briefwechsel*. Edited by O. Spiess. Basel: Birkhäuser, 1955.

Birch, Thomas. *History of the Royal Society*. London, 1756–57.

Blay, M. *La naissance de la mécanique analytique, la science du mouvement au tournant des XVIIe et XVIIIe siècles*. Paris: Presses Universitaires de France, 1992.

Bloch, L. *La philosophie de Newton*. Paris: Alcan, 1908.

Borelli, G. *Theorica mediceorum planetarum a causis physicis deducta.* Florence, 1666.

Bošković, R. J. *Theoria philosophiae naturalis*. Vienna, 1758.

Boulliau, I. *Astronomia Philolaica*. Paris, 1645.

Boyer, C. *History of the Calculus*. New York: Dover, 1949.

Cavalieri, B. *Geometria indivisibilibus continuorum nova quadam ratione promota.* Bologna, 1635 (1st ed.), 1653 (2d ed.).

———. *Geometria degli indivisibili.* Translated by L. Lombardo-Radice. Turin: U.T.E.T., 1966.

———. *Exercitationes geometricae Sex*. Bologna, 1647. Reprint Rome: Cremonese, a cura dell'U.M.I., 1980.

Clavelin, M. *La philosophie naturelle de Galilée*. Paris: Armand Colin, 1968.

Cléro, J. F., and E. Le Rest. "La naissance du calcul infinitésimal au XVIIè siècle." *Cahiers d'histoire et de philosophie des sciences* vol. 16. Paris: Centre National de la Recherche Scientifique–Centre de Documentation en Sciences Humaines, 1980.

Cohen, I. B. *Introduction to Newton's "Principia."* Cambridge: Harvard University Press, 1971.

Cohen, I. B. *The Newtonian Revolution.* Cambridge: Harvard University Press, 1980.

Copernicus, N. "Revolutions of Heavenly Spheres." Translated by C. G. Wallis. Vol. 16, *Great Books of the Western World.* Chicago: Encyclopaedia Britannica, 1952.

———. "Facsimile du manuscrit du *De Revolutionibus.*" *Oeuvres complètes.* Vol. 1. Varsovie: Académie polonaise des sciences. Edition du C.N.R.S., 1973.

Costabel, P. "Newton and Leibniz's Dynamics." In *The annus mirabilis of Isaac Newton.* Edited by R. Palter. Cambridge: MIT Press, 1970.

———. "Isochronisme et accélération 1638–1687." *Archives internationales d'histoire des sciences,* (June 1978): 3–20.

D'Alembert, Jean Le Rond. *Traité de dynamique.* Paris, 1743.

———. *Mélanges.* Amsterdam: Chatelain, 1770.

———. *Essai sur les éléments de philosophie, Oeuvres.* Vol. 2. Paris: Bastien, 1805. Reprint, Hildesheim: Olms, 1965.

De Gandt, F. "Duratio, fluxio, aequatio, trois aspects du temps newtonien." *Archives de Philosophie* 44 (1981): 353–70.

———. "Force et science des machines." In *Science and Speculation. Studies in Hellenistic Theory and Practice,* edited by F. Barnes et al. Cambridge: Cambridge University Press, 1982.

———. "Galilée et la naissance de la dynamique: recherches de P. Galluzzi sur le terme 'momento'." *Revue d'histoire des sciences,* (1983): 319–24.

———. "Temps physique et temps mathématique chez Newton." In *Mythes et représentations du temps.* Paris: Centre National de la Recherche Scientifique, 1985.

———. "Les Mécaniques attribuées à Aristote et le renouveau de la science des machines au XVIe siècle." In *L'aristotélisme au XVI*[e] *siècle. Etudes philosophiques* 3 (1986): 391–405.

———. "Les Indivisibles de Torricelli." In *L'oeuvre de Torricelli. Science galiléenne et nouvelle géométrie.* Edited by F. De Gandt. Nice: Presses de l'Université de Nice, 1989.

———. "L'analyse de la percussion chez Galilée et Torricelli." In *L'oeuvre de Torricelli. Science galiléenne et nouvelle géométrie.* Edited by F. De Gandt. Nice: Presses de l'Université de Nice, 1989.

———. "Cavalieri's indivisibles and Euclid's canons." In *Revolution and Continuity: Essays in the History and Philosophy of Early Modern Science,* edited by P. Barker and R. Ariew. Washington D.C.: Catholic University of America Press, 1991.

———. "Newton: la justification des infiniment petits et l'intuition du mouvement." In *Infini des mathématiciens, infini des philosophes.* Edited by F. Monnoyeur. Paris: Belin, 1992.

Descartes, R. *Geometria latine versa* (With commentaries in Latin). 2 vols. Amsterdam, 1659–61.

———. *Oeuvres.* 10 vols. Edited by C. Adam and P. Tannery. Paris, 1897–1913.

———. *Oeuvres et Lettres.* Paris: Bibliothèque de la Pléiade, Nouvelle Revue Française, 1963.

De Witt, J. "Elementa curvarum linearum." In Descartes, R., *Geometria.* 2 vols. Amsterdam, 1659–61.

Dobbs, B. J. *The Foundations of Newton's Alchemy*. Cambridge: Cambridge University Press, 1975.

Drake, S., and I. E. Drabkin. *Mechanics in Sixteenth-Century Italy*. Madison: University of Wisconsin Press, 1969.

Dugas, R. *La Mécanique au 17ème siècle*. Neufchatel: Le Griffon, 1954.

Duhem, P. *Les Origines de la statique*. 2 vols. Paris: Hermann, 1905–6.

Dupont, P. and C. S. Roero. "Leibniz 84, il decollo enigmatico del calcolo differenziale." In *Storia della matematica*. Rende (Italy): Mediterranean Press, 1991.

Euclid. *Elements*. Edited by J. L. Heiberg. Leipzig: Teubner, 1969–72.

Euler, E. *Lettres à une princesse d'Allemagne*. Paris: Charpentier, 1843.

Fermat, P. de. *Oeuvres*. 4 vols. Edited by P. Tannery and C. Henry. Paris: Gauthièr-Villors, 1891–1912.

Fierz, M. "Newton's Auffassung der Mathematik und die mathematische Form der *Principia*." *Helvetica Physica Acta* 41 (1968): 821–26.

Gabbey, A. "The Case of the Rotating Moon." In *Revolution and Continuity: Essays in the History and Philosophy of Early Modern Science*. Edited by P. Barker and R. Ariew. Washington D.C.: Catholic University of America Press, 1991.

Gagnebin, B. "De la cause de la pesanteur, Mémoire de Nicolas Fatio de Duillier présenté à la Royal Society le 26 février 1690." *Notes and Records of The Royal Society* 6 (1949): 106–60.

Galilei, Galileo. *Opere*. 20 vols. Edizione Nazionale. Florence: Barbèra, 1890–1909.

———. *Dialogue Concerning the Two Chief World Systems*. Translated by Stillman Drake. Berkeley: University of California Press, 1952.

———. *Discorsi*. Edited by E. Carugo and L. Geymonat. Torino: Boringhieri, 1958.

———. *Discours*. Translated by M. Clavelin. Paris: Armand Colin, 1970.

Galuzzi, M. *Calculus and Geometry in Newton's Mathematical Work*. Universita degli studi di Milano, Quaderno 10 (1986): 1–19.

Galluzzi, P. *Momento. Studi galileiani*. Rome: Ateneo e Bizzarri, 1979.

Gilbert, W. *De Magnete*. London, 1600. Reprint Brussels: Culture et Civilisation, 1967.

Goodman, N. *Fact, Fiction and Forecast*. Cambridge: Harvard University Press, 1955.

Gregory, J. *Geometriae pars universalis*. Padua, 1668.

Grosholz, E. "Some Uses of Proportion in Newton's *Principia*." *Studies in the History and Philosophy of Science*, (1987): 209–20.

Hall, A. R. and M. B. Hall, eds. "The Date of 'On Motion in Ellipses.'" *Archives internationales d'histoire des sciences*, (1963): 23–28.

Heath, T. L. *A History of Greek Mathematics*. Oxford: Oxford University Press, 1921.

Hegel, G. W. *Les Orbites des planètes*. Translated by and commented on by F. De Gandt. Paris: Vrin, 1979.

Herivel, J. "Newtonian Studies. III." *Archives internationales d'histoire des sciences* 14 (1962): 23–33.

———. "Newtonian Studies. IV." *Archives internationales d'histoire des sciences*, (1963): 13–22.

———. "Newton's First Solution of the Problem of Kepler's Motion." *British Journal for the History of Science*, (1964): 350–54.

———. *The Background to Newton's Principia*. Oxford: Oxford University Press, 1965.

Hobbes, T. *Leviathan*. Edited by C. B. Mayherson. Baltimore, 1968.

———. *De corpore*. Aalen: Scientia, 1966.

Hofmann, J. E. *Leibniz in Paris 1672–1676: His Growth to Mathematical Maturity*. Cambridge: Cambridge University Press, 1974.

Hooke, R. "An Attempt to Prove the Motion of the Earth" (1674). In "Lectiones Cutlerianae. Vol. 8, *Early Science in Oxford*, edited by R. T. Gunther. London, 1923–45.

Huygens, C. *Traité de la lumière*. Leyden, 1690.

———. *Oeuvres*. 22 vols. La Haye: *Société hollandaise des sciences*, 1888–1950.

———. *Horologium oscillatorium*. 1st ed. Paris, 1673. Reprint Brussels: Culture et Civilisation, 1966.

Jardine, N. *The Birth of History and Philosophy of Science: Kepler's Defence of Tycho against Ursus*. Cambridge: Cambridge University Press, 1984.

Kant, I. *Kritik der reinen Vernunft*. Riga: Hartknoch, 1787.

Kepler, J. *Gesammelte Werke*. 22 vols. Munich: Beck, 1937.

———. *The Secret of the Universe*. Translated by A. M. Duncan. New York: Abaris Books, 1981.

Kitcher, P. "Fluxions, Limits and Infinite Littleness. A Study of Newton's Presentation of the Calculus." *Isis*, (1973): 33–49.

Knorr, W. R. "Ancient Sources in the Medieval Tradition of Mechanics." In *Suppl. agli Annali dell'Istituto e Museo di Storia della Scienza*. Florence: Parenti, 1982.

Koyré, A. "A Documentary History of the Problem of Fall from Kepler to Newton." *Transactions of the American Philosophical Society*, October 1955.

———. *La Révolution astronomique*. Paris: Hermann, 1961.

———. *Etudes galiléennes*. Paris: Hermann, 1966.

———. *Etudes newtoniennes*. Paris: Gallimard, 1968.

———. "Bonaventura Cavalieri et la géométrie des continus." In *Etudes d'histoire de la pensée scientifique*. Paris: Gallimard, 1973.

Laplace, P.-S. de. *Oeuvres complètes*. Edited by Académie des sciences. Paris: Gauthier-Villars, 1878–1912.

Larmore, C. "La critique newtonienne de la méthode cartésienne." *Dix-huitième siècle*, (1986): 269–79.

Leibniz, G. W. "Nova methodus pro maximis et minimis." *Acta Eruditorum*, 1684.

———. *Opera omnia*. Edited by Louis Dutens. Geneva, 1768.

———. *Mathematische Schriften*. 7 vols. Edited by C. I. Gerhardt. Hildesheim: Olms, 1971.

Leonardo da Vinci. *Les manuscrits de Léonard de Vinci*. Edited by C. Ravaisson-Mollien. Paris, 1881.

L'Hôpital, G. F. A. de. *Analyse des infiniment petits pour l'intelligence des lignes courbes*. Paris, 1696.

———. "Solution d'un problème physico-mathématique." *Mémoire de l'Académie royale des Sciences pour l'année 1700*. In *Histoire de l'Académie royale des Sciences, année 1700*. Amsterdam: Pierre Mortier, 1734.

Lloyd, G. E. R. "Saving the Appearances." *Classical Quarterly*, (1978): 202–22.

Lohne, J. "Hooke *versus* Newton." *Centaurus* 7 (1960): 6–52.

Maclaurin, C. *An Account of Sir Isaac Newton's Philosophical Discoveries (1748)*. Reprint, Hildesheim: Olms, 1971.

Maïmonides, M. *Guide des Egarés*. Translated by S. Munk. 3 vols. Paris: Maisonneuve et Larose, 1970.

Maizeaux, Pierre des. *Recueil de diverses pièces sur la philosophie, la religion naturelle, l'histoire, les mathématiques, par Messieurs Leibniz, Clarke, Newton et autres auteurs célèbres*. 2d ed., Amsterdam, 1740.

Maxwell, J. C. *A Treatise on Electricity and Magnetism*. New York: Dover Books, 1954.

McGuire, J. E. "Force, Active Principles and Newton's Invisible Realm." *Ambix* 15 (1968): 154–208.

Newton, Isaac. *Arithmetica universalis*. Leiden: Verbek, 1732.

———. "De systemate mundi." In *Opuscula* vol. 1. Edited by Castillon. Lausanne/Geneva: Bousquet, 1744.

———. "Isaaci Newtoni propositiones de motu." *Royal Society Register* 6: 218–34. In *An Essay on Newton's Principia*. Edited by R. Ball. London, 1893.

———. *Opticks*. New York: Dover Books, 1952.

———. *The Correspondence of Isaac Newton*. Edited by H. W. Turnbull. Cambridge: Cambridge University Press, 1959.

———. *Unpublished Scientific Papers of Isaac Newton*. Edited by A. R. Hall and M. B. Hall. Cambridge: Cambridge University Press, 1962.

———. *Mathematical Works*. New York: Johnson, 1964.

———. *Sir Isaac Newton's Mathematical Principles of Natural Philosophy and His System of the World. Translated by Andrew Motte*. Edited and revised by Florian Cajori. Berkeley: University of California Press, 1966.

———. *La méthode des fluxions et des suites infinies*. Translated by G. L. de Buffon. Paris: Blanchard, 1966.

———. *Principes mathématiques de la philosophie naturelle. 2 vols.* Translated by E. du Châtelet. Paris: Blanchard, 1966.

———. *The Mathematical Papers of Isaac Newton*. 8 vols. Edited by D. T. Whiteside. Cambridge: Cambridge University Press, 1967–81.

———. *Philosophiae Naturalis Principia Mathematica*. 3d ed. (with variant readings). 2 vols. Edited by A. Koyré and I. B. Cohen. Cambridge, Mass.: Harvard University Press, 1972

———. "The Fundamental 'De motu corporum in gyrum' (Autumn 1684) (Add.3965.7, ff.55r–62*r)." In *The Preliminary Manuscripts for Isaac Newton's 1687 "Principia," 1684–1685: Facsimiles of the Original Autographs Now in Cambridge University Library, with an Introduction by D. T. Whiteside*. Cambridge: Cambridge University Press, 1989.

Pascal, B. *Oeuvres*. Edited by P. Boutroux, L. Brunschvicg, and F. Gazier. Paris: Edition des Grands Ecrivains,1904–1914.

Pepys, S. *Diary*. London, 1904.

Proclus. *In Primum euclidis librum elementorum commentarii*. Edited by G. Friedlein. Stuttgart: Teubner, 1967.

Ptolemy. *Syntaxis*. 2 vols. Edited by J. L. Heiberg. Leipzig: Teubner, 1898.

———. *Ptolemy's Almagest*. Translated by G. J. Toomer. London: Duckworth, 1984.

Pugliese, Patri J. "Robert Hooke and the Dynamics of Motion in a Curved Path." In *Robert Hooke: New Studies*. Edited by M. Hunter and S. Schaffer. Woodbridge: The Boydell Press, 1989.

Roberval, G. P. de. "Divers ouvrages." *Mémoires de l'Académie royale des sciences*. Vol. 6. Paris, 1693.

Roero, C. S. "Jakob Bernoulli attento studioso delle opere di Archimede." *Bolletino di Storia delle Scienze Matematiche dell'U.M.I.*, (1983): 77–125.

Russell, J. L. "Kepler's Laws of Planetary Motion, 1609–1666." *British Journal for the History of Science*, (1964): 1–24.

Russo, F. "Pascal et l'analyse infinitésimale." In *L'oeuvre scientifique de Pascal*. Paris: Presses Universitaires de France, 1964.

Scott, J. F. *The Mathematical Work of J. Wallis*. New York: Chelsea Publishing Co., 1981.

Sigorgne, J. *Institutions newtoniennes*. Paris, 1747.

Simon, G. *Kepler astronome, astrologue*. Paris: Nouvelle Revue Française, 1979.

Stevin, S. *Oeuvres mathématiques*. Translated by A. Girard. Leyden: Elzevir, 1634.

Sylla, E. "Compounding Ratios." In *Transformation and Tradition in the History of Science*. Edited by E. Mendelssohn. Cambridge: Cambridge University Press, 1984.

Taton, R. and C. Wilson. *The General History of Astronomy*. Vol. 2A. Cambridge: Cambridge University Press, 1989.

Torricelli, E. *Opere*. 4 vols. Edited by G. Loria and G. Vassura. Faenza, 1919–44.

———. *Opere*. Edited by L. Belloni. Turin: U.T.E.T., 1975.

Valerio, L. *De centro gravitatis solidorum*. Rome, 1604.

Van Schooten, F. *De organica conicarum sectionum in plano descriptione*. Leyden, 1646.

Varignon, P. "Manière générale de déterminer les forces, les vitesses, les espaces et les temps, une seule de ces quatre choses étant donnée, dans toutes sortes de mouvements rectilignes variés à discretion." In *Histoire et mémoires de l'Académie royale des sciences*, année 1700. Amsterdam, 1734.

Viète, F. *Opera*. Edited by F. van Schooten. Leyden, 1646.

Wallis, J. *De sectionibus conicis*. Oxford, 1655.

———. *Arithmetica infinitorum*. Oxford, 1656.

———. "De motu." In *Opera mathematica*. 3 vols. Oxford, 1669–70.

———. *Tractatus Duo, prior de cycloïde, posterior de cissoïde*. Oxford, 1659.

Wardy, Robert. "Le jeu des nombres aristotéliciens." In *La physique d'Aristote et les conditions d'une science de la nature*. Edited by F. De Gandt and P. Souffrin. Paris: Vrin, 1991.

Westfall, R. S. *Force in Newton's Physics*. London: Macdonald, 1971.

———. *Never at Rest: A Biography of Isaac Newton*. Cambridge: Cambridge University Press, 1980.

Whiteside, D. T. "Newton's Early Thoughts on Planetary Motion, a Fresh Look." *British Journal for the History of Science*, (1964): 117–37.

———. "Newton's Marvellous Year 1666 and All That." *Notes and Records of the Royal Society*, (1966): 32–41.

———. *The Mathematical Principles Underlying Newton's "Principia mathematica."* Glasgow: University of Glasgow, 1970.

———. "Patterns of Mathematical Thought in the Later Seventeenth Century." *Archive for the History of Exact Sciences* 1, no. 3 (1961): 179–388.

Wisan, W. "The New Science of Motion, a Study of Galileo's *de motu locali.*" *Archive for History of Exact Sciences* 13 (1974): 103–306.

Yoder, J. *Unrolling Time: Christiaan Huygens and the Mathematization of Nature.* Cambridge: Cambridge University Press, 1988.

INDEX

ABOUT THE AUTHOR AND TRANSLATOR

François De Gandt, a senior research fellow at the Centre National de la Recherche Scientifique in Paris, is the author of studies of Aristotle, Torricelli, Huygens, and d'Alembert. Curtis Wilson, Tutor Emeritus of St. John's College in Annapolis, Maryland, has written on the history of seventeenth- and eighteenth-century astronomy.